Horst Frie

Basic One- and Two-Dimensional NMR Spectroscopy

Horst Friebolin

Basic One- and Two-Dimensional NMR Spectroscopy

Fourth Completely Revised
and Updated Edition

Translated by
Jack K. Becconsall

WILEY-VCH

WILEY-VCH Verlag GmbH & Co. KGaA

Prof. Dr. Horst Friebolin
Organisch-Chemisches Institut der Universität
Im Neuenheimer Feld 270
69120 Heidelberg
Germany

Library of Congress Card No.: applied for

British Library Cataloguing-in-Publication Data:
A catalogue record for this book is available from the British Library

Bibliographic information published by Die Deutsche Bibliothek
Die Deutsche Bibliothek lists this publication in the Deutsche Nationalbibliografie; detailed bibliographic data is available in the Internet at <http://dnb.ddb.de>.

© 2005 WILEY-VCH Verlag GmbH & Co. KGaA, Weinheim

Printed on acid-free paper

Printed in the Federal Republic of Germany

Composition: hagedorn kommunikation, Viernheim
Printing: Strauss GmbH, Mörlenbach
Bookbinding: Litges & Dopf Buchbinderei GmbH, Heppenheim

ISBN 3-527-31233-1

Foreword

Over the past decade, many, if not most, graduate students and postdoctoral fellows in organic chemistry seem to have come to regard nuclear magnetic resonance spectroscopy as a "black box". Something in which you insert an unknown and out comes the data to establish structural formula. Perhaps this trend in the way NMR is perceived is really not surprising, because how manufacturers have tried to package their instruments to be "user friendly" and because of the enormous growth in the sophistication of what NMR instrumentation can do.

Those of us who were fortunate to be in on the beginning of the applications of NMR to organic chemistry in the mid-1950's were able to more or less "grow" with the field and come into the modern arena of multipulse and multidimensional NMR with substantial experience with NMR fundamentals. Those who now wish to start using NMR in their research surely must feel, at least somewhat, overwhelmed by the enormity and sophistication of the currently available knowledge of NMR and must have further concerns at being told by the specialists, that fantastic further, even more sophisticated, developments are in the offing. A "black box" that would function with no need to worry about its inner workings must then be an attractive proposition. The problem is that to properly use modern NMR requires a lot of rather specialized knowledge. The effects of couplings, of exchange, of relaxation times, of low sensitivity, of solvent and so on, make selection of instrumental parameters for taking spectra far from routine. Serious errors or inefficient use of very expensive instrumentation come with ease.

The best way to learn NMR spectroscopy is by doing it, but textbooks, guidebooks and reference books are vitally necessary. Having written two books about NMR basics, with a third in progress, I know something about the difficulties of making available "what every NMR user should know". And it is to this objective that Professor Friebolin has made a wonderfully broad contribution. This book will be of interest and help to both those needing to learn and those needing a reference book to refresh their memories, or extend their capabilities in NMR spectroscopy. It is not a book intended to replace the treatises of Abragam or Bodenhausen, Ernst and Wokaun. However, even though it does start at a useful elementary level, it goes rather deeply into the difficult basics of multipulse and multidimensional spectroscopy. The result is material that almost every reader will find of value.

The beginner can start with the elements of chemical shifts and couplings and later proceed to more difficult matters. The expert can find ways of explaining what he is doing, without necessarily resorting to density matrices; or else, in impatience with an eager, but dull, learner, can say "Go read about it in Friebolin, then we can talk".

Much is covered in this book in meaningful detail. There is a plethora of structural parameters for proton and carbon NMR and many examples of how they can be used. Best of all, though, are very clear, meticulously written descriptions of INEPT, DEPT, INADEQUATE, COSY, NOESY, and the like, in one- and two-dimensional NMR spectroscopy. Experts may prefer mathematical equations for compactness and "it is easy to see". I prefer descriptions such as those used by Professor Friebolin, which will indeed require careful reading, rereading and drawing and redrawing ones own vector diagrams, but can lead to a real level of understanding. Such understanding in its turn can only result in improved ability to take and interpret NMR spectra.

November 28, 1990　　　　　　　　　John D. Roberts
　　　　　　　　　　　　　　　　　Pasadena, California

The Nightingale and the Lark

What can one say to those writers who blithely
fly far above the heads of most of their readers?
The same, surely, as the nightingale said to the
lark: My friend, are you soaring so high in order
that your song cannot be heard?

Gotthold Ephraim Lessing

Preface to the Fourth Edition

Only minor additions were needed for this new edition, at
least as far as the fundamentals and methods used in NMR
spectroscopy are concerned. Worthy of mention here is the
new definition of the δ value in Chapter 1, in accordance with
the recommendations of the IUPAC Committee in 2001.
Among the many new 2D experiments, I have only incorporated the ROESY Experiment in Chapter 9, since this has now
become a routine operation.

Much progress has been made, on the other hand, in explaining the structure of large molecules. I have taken this trend into
consideration and expanded Chapter 13 "Macromolecules" by
way of a comprehensive section on biopolymers. Since solid
state NMR spectroscopy has gained importance in investigating
synthetic polymers, I have included a new section providing a
rough overview of this method.

As a valuable addition to magnetic resonance tomography
(Section 14.4), which in the meantime has become a permanent
part of medical diagnostics, a combined use of tomography
and spectroscopy has developed in recent years – *Magnetic
Resonance Spectroscopy*. Thus, I have described the basics of
this method in a new section (14.4.2.2) and illustrated the
potential of such investigations using an example from clinical
practice.

I would like to express my gratitude to all those who helped
me in preparing this new edition. In particular I would like to
thank Professor J. Blümel, who critically reviewed the section
on solid state NMR (13.2.4), Dr. R. Rensch for his untiring
help in solving daily computer problems, as well as Dr. Raimund
Kleiser, who supported me in writing the section on magnetic
resonance spectroscopy (14.4.2.2). It is to him, as well as
Dr. Hans-Jörg Wittsack of the Institute of Diagnostic Radiology, Heinrich Heine University of Düsseldorf, that I am
indebted for the example shown in Figure 14-20.

In addition, I wish to thank J. Becconsall for his expert translation as well as the staff at Wiley-VCH in Weinheim, above all Dr. E. Wille, Dr. B. Bems and H. J. Schmitt, who once again was responsible for the production of this book.

Heidelberg, August 2004   Horst Friebolin

Preface
to the Third Edition

Even now, more than 50 years after the discovery of nuclear magnetic resonance, the development of NMR spectroscopy is still continuing unabated, and consequently the second edition was seen to be in need of a further thorough updating. The revised edition covers new advances in techniques, such as the use of pulsed field gradients, and methods such as HMQC, HMBC, and TOCSY (one- and two-dimensional) which have now progressed to the status of routine procedures. To avoid an undue increase in the size of the book, some procedures that have become obsolete or unimportant have fallen victim to the red pen; these include the continuous wave (CW) method of recording spectra, off-resonance decoupling, and the "relayed" experiments. Thus the changes from the second edition have mainly affected Chapters 1, 5, 6, 8, 9, and (in view of the need for a thorough revision of magnetic resonance tomography) Chapter 14.

Once more I have to thank many of my students and colleagues for their help and suggestions. My thanks are due especially to the following. First, to Dr. G. Schilling (Heidelberg), whose expertise and commitment in performing all the experiments exemplifying the procedures newly introduced into the book have been invaluable. Next I thank Prof. Dr. R. Brossmer (Heidelberg), who once again provided samples of the neuraminic acid derivative used as a model compound (by personally carrying out the laboratory preparation!). I thank Dr. R. Rensch (Heidelberg) for his constant help with all the computer aspects. Once more, as previously, I am indebted to Dr. T. Keller of Bruker for help of many kinds. I thank Dr. H. Post and Dr. D. Phillips of Bruker Medical GmbH for their help in updating and rewriting the section on magnetic resonance tomography, and for supplying new illustrations. I thank Dr. P. Ullrich (Chemical Concepts), Dr. H. Thiele (Bruker-Franzen Analytik), and Dr. H. Hofmann (ScienceServe, ACD) for their advice during the rewriting of Section 6.4 on computer-aided assignment in NMR spectroscopy.

Special thanks are again due to the translator, Dr. J. Becconsall, who, as in the previous two editions, has viewed the text with an expert and critical eye, thereby making a significant contribution to the successful outcome seen in this book.

Finally I thank the editorial and production staff of Wiley-VCH, Weinheim, who have been involved, namely Dr. E. Wille, Dr. C. Dyllick, and especially H.-J. Schmitt, who was again responsible for the careful production of the book.

Heidelberg, August 1998   Horst Friebolin

Acknowledgements

In writing this book I have relied on the active help of many people. To begin with, I must mention three names in particular: Dr. Wolfgang Baumann, Dr. Wolfgang Bermel (Bruker), and Doris Lang. As well as helping in other ways, Wolfgang Baumann recorded and prepared in suitable form all the 250 MHz and 300 MHz spectra that are reproduced; Wolfgang Bermel applied his considerable skill to recording the one- and two-dimensional 400 MHz NMR spectra in Chapters 8 and 9. In addition I thank both for their critical reading of parts of the manuscript. Doris Lang worked tirelessly on the drafting, proof-correcting, labeling and arranging of the many figures, sketches and structural formulas – a task whose size can only be fully appreciated by someone who has experienced it.

I thank Dr. Gerhard Schilling for recording the spectra reproduced in Figures 9, 13 and 17 of Chapter 8 and for many stimulating discussions. I thank Dieter Ratzel (Bruker) for recording the NMR tomograms (Figs. 11, 13 and 14 of Chapter 14) and for much additional information. I am indebted to the firm of Bruker, and especially to Dr. Tony Keller, for support in many different respects, including much time-consuming work on experimental measurements, providing material for figures, and processing of the text.

I also thank the following: Dr. Wolfgang Bremser (BASF) for the predicted spectrum of a model compound and for critically reading the section on computer-aided assignment of spectra; Dr. Hans-J. Opferkuch (DKFZ HD) for recording the 2D NMR spectras of glutamic acid (Figs. 19 and 24 of Chapter 9); Brigitte Faul and Wilfried Haseloff for recording 90 MHz ^{1}H NMR spectra at low temperatures; Dr. Peter Bischof for the picture of the model compound on the front cover; Prof. Reinhard Brossmer for providing the neuraminic acid derivative used as a test compound; Prof. Klaus Weinges for correcting Section 2.4; Prof. Dieter Hellwinkel for settling some awkward problems of chemical nomenclature. I am greatly indebted to my departmental colleagues for their constructive criticisms and stimulation. In additiona I thank Dr. Gerhard Weißhaar and Doris Lang for their critical reading of the proofs. Brigitte Rüger and Irmgard Pichler provided much appreciated help in the preparation of the first manuscript.

I received many valuable suggestion from attentive readers of the German edition. In this respect I should especially mention the comments and suggestions of Dr. Erhard T. K. Haupt, Hamburg, and of Prof. Bernd Wrackmeyer, Bayreuth. Although large portions of the German and English editions are identical in content, the expert and committed involvement of Dr. Jack Becconsall has resulted in a book which clearly bears the influence of the "translator". It is therefore my pleasant duty to thank him for his translation, corrections, suggestions and discussions.

Of those in VCH Publishers I must especially mention Dr. Eva Wille, who not only prepared my manuscript for the printing of the German edition but also, as specialist editor, worked critically through the text. I am greatly indebted to her, to Karin von der Saal as publishing editor of the English edition, and to Myriam Nothacker, who converted the manuscript into a book.

It gives me special pleasure to thank Father Franz Alferi of the Catholic parish of St. Nikolaus in Mannheim, who provided me with a room where I was able to work in total seclusion.

Finally I thank my wife and three sons, who have all endured much in last few years because of this book, and have had to make considerable sacrifices.

Heidelberg, December 1990   H. Friebolin

Abbreviations and Acronyms

ADP	Adenosine diphosphate
APT	Attached Proton Test
ATP	Adenosine triphosphate
CLA	Complete line-shape analysis
COSY	Correlated spectroscopy
CSA	Chiral shift agent
	or Chemical shift anisotropy
CW	Continuous wave
2D	Two-dimensional
DD	Dipole-dipole
DEPT	Distortionless enhancement by polarization transfer
DMSO	Dimethylsulfoxide
DNMR	Dynamic NMR
DPM	Dipivaloylmethane, (2,2,6,6-tetramethyl-heptandione)
DTPA	Diethylene-triamine-pentaacetic-acid
EXSY	Exchange Spectroscopy
FID	Free induction decay
FOD	Heptafluoro-7,7-dimethyl-4,6-octandione
FT	Fourier transform/transformation
gs	Gradient selected
HETCOR	Heteronuclear correlation
HMBC	Heteronuclear multiple bond correlation
HMQC	Heteronuclear multiple quantum coherence
HSQC	Heteronuclear single quantum coherence
INADEQUATE	Incredible natural abundance double quantum transfer
INEPT	Insensitive nuclei enhanced by polarization transfer
LSR	Lanthanide shift reagent
MAS	Magic Angle Spinning
MO	Molecular orbital
MRI	Magnetic resonance imaging
NOE	Nuclear Overhauser effect/enhancement
NOESY	Nuclear Overhauser enhancement spectroscopy
PCr	Creatine phosphate
Pfg	Pulsed field gradient
P_i	Inorganic phosphate

PMMA	Polymethylmethacrylate
ppm	Parts per million
PRESS	Point Resolved Spectroscopy
ROESY	Rotating frame Overhauser Enhancement Spectroscopy
S:N	Signal-to-noise ratio
SPI	Selective population inversion
TMS	Tetramethylsilane
TOCSY	Total correlation spectroscopy
TROSY	Transverse Relaxation Optimized Spectroscopy

Symbols used

B_0	Static magnetic field (flux density)
B_1, B_2	Radiofrequency field with frequencies v_1 and v_2
B_{eff}	Effective field at position of nucleus
C_2	Two-fold axis of symmetry
χ	Magnetic susceptibility
$^{13}C\{^1H\}$	Observation of ^{13}C resonance while ^1H is decoupled
δ	Chemical shift relative to a standard (e. g. TMS)
E	Energy
ΔE	Energy difference between two states
δE	Uncertainty in the energy of a state
E_A	Arrhenius activation energy
E_X	Electronegativity of substituent X
η	Fractional enhancement in NOE
φ	Phase difference between two vectors
Φ	Bond angle or dihedral angle
F_1, F_2	Frequency axes in a 2D NMR spectrum
G	Field gradient
$\Delta G^{\ddagger}$	Free (Gibbs) enthalpy of activation
γ	Gyromagnetic ratio; also $\gamma = \gamma/2\pi$
h	Planck constant; also $\hbar = h/2\pi$
$\Delta H^{\ddagger}$	Enthalpy of activation
I	Nuclear angular momentum quantum number (spin)
$\boldsymbol{I}$	Nuclear spin operator
^nJ	Coupling constant through n bonds
K	Equilibrium constant
k	Rate constant
k_C	Rate constant at the coalescence temperature T_C
k_B	Boltzmann constant
k_0	Frequency factor
$\boldsymbol{\mu}$	Magnetic moment (of nucleus)
μ_z	Component of $\boldsymbol{\mu}$ along the static field direction (z-axis)
m	Magnetic quantum number
M_0	Macroscopic magnetization of the sample in the static field B_0
$M_{x'}, M_{y'}$	Transverse magnetization components in the x'- and y'-directions
M_z	Longitudinal magnetization in the z-direction (static field direction)

M_X	Magnetization vector for the X nuclei
$M_H^{C\alpha}, M_H^{C\beta}$	^{1}H magnetization vectors in a two-spin system with the ^{13}C nuclei in the α and β states
$M_C^{H\alpha}, M_C^{H\beta}$	^{13}C magnetization vectors in a two-spin system with the protons in the α and β states
N	Total number of nuclei in sample
N_α, N_β	Numbers of nuclei in the α and β states
N_i	Number of nuclei in level i
ν	Frequency
ν_L	Larmor frequency
ν_i	Resonance frequency of nucleus i
ν_1	Frequency of r. f. generator (observing frequency)
ν_2	Decoupling frequency
$\nu_{1/2}$	Half-height width
P	Angular momentum of nucleus
P_z	Component of P in the z-direction
Q	Electric quadrupole moment
R	Universal gas constant
r	Interatomic (or internuclear) distance
σ	Shielding constant
$S(t), S(f)$	Signal as a function of time or of frequency
S_i	Substituent increment for predicting chemical shifts
$\Delta S^{\ddagger}$	Entropy of activation
τ	Time interval between pulses
τ_C	Correlation time
τ_l	Lifetime of a nucleus in a particular spin state or magnetic environment
τ_p	Pulse duration
τ_{zero}	Zero-crossing point (time at which $M_z = 0$ after a 180° pulse)
t_1	Variable time in a 2D experiment; usually increased in regular steps
t_2	Acquisition time
Δ	Fixed time interval in a 2D pulse sequence
T	Tesla (unit of magnetic flux density)
T	Absolute temperature (in K)
T_1	Spin-lattice or longitudinal relaxation time
T_2	Spin-spin or transverse relaxation time
T_2^*	Experimentally observed transverse relaxation time (including effect of field inhomogeneity)
Θ	Pulse angel
W_0, W_1, W_2	Transition probabilities for zero-quantum, single quantum and double quantum transitions by relaxation processes
x	Mole fraction

Contents

1 The Physical Basis of NMR Spectroscopy

1.1 Introduction

In 1946 two research groups, that of F. Bloch, W.W. Hansen and M. E. Packard and that of E. M. Purcell, H. C. Torrey and R.V. Pound, independently observed nuclear magnetic resonance signals for the first time. Bloch and Purcell were jointly awarded the Nobel Prize for Physics in 1952 for their discovery. Since then nuclear magnetic resonance (NMR) spectroscopy has developed into an indispensable tool for chemists, biochemists, physicists, and more recently medical scientists. During the first three decades of NMR spectroscopy all measurements relied on *one-dimensional* modes of observation; this gives spectra having just *one* frequency axis, the second axis being used to display the signal intensities. The development of *two-dimensional* NMR experiments during the 1970s heralded the start of a new era in NMR spectroscopy. Spectra recorded by these methods have *two* frequency axes, the intensities being displayed in the third dimension. More recently it has even become possible to perform experiments with three or more dimensions, although these are still far from being routine techniques. The importance of the position that NMR spectroscopy now occupies is illustrated by the awards of the Nobel Prize for Chemistry in 1991 to R. R. Ernst and in 2002 to K. Wüthrich, and of the Nobel Prize for Medicine in 2003 to P. Lauterbur and P. Mansfield for their pioneering research on NMR methods in chemistry, biochemistry and medicine. The new techniques that have emerged during the last few years show that developments in NMR spectroscopy are still far from coming to an end.

This book aims to explain why it is that, for chemists especially, NMR spectroscopy has become (possibly) the most important of all spectroscopic methods.

The main field of application of NMR spectroscopy is that of determining the structures of molecules. The necessary information for this is obtained by measuring, analyzing and interpreting high-resolution NMR spectra recorded on liquids of low viscosity (or in some cases on solids by using special techniques and instruments that have been developed in the last few years). In this book we will confine our attention almost exclusively to *high-resolution* NMR spectroscopy on liquids, since solid-state measurements involve quite different experimental techniques and the interpretation often brings in extra complications.

The nuclides that mainly interest us are protons (^{1}H) and carbon-13 (^{13}C), as their resonances are the most important ones for determining the structures of organic molecules. However, in the following chapters we shall meet also examples of NMR spectroscopy of other nuclides whose NMR signals can now be observed without difficulty.

In order to understand NMR spectroscopy we first need to learn how nuclei which have a nuclear angular momentum P and a magnetic moment μ behave in a static magnetic field. Following this we shall discuss the basic NMR experiment, the different methods of observation, and the spectral parameters.

1.2 Nuclear Angular Momentum and Magnetic Moment

Most nuclei possess a nuclear or intrinsic angular momentum P. According to the classical picture the atomic nucleus, assumed to be spherical, rotates about an axis. Quantum mechanical considerations show that, like many other atomic quantities, this angular momentum is quantized:

$$P = \sqrt{I(I + 1)}\, \hbar \qquad (1\text{-}1)$$

Here $\hbar = h/2\pi$, where h is Planck's constant ($= 6.6256 \times 10^{-34}$ J s), and I is the angular momentum quantum number, usually called simply the nuclear spin. The nuclear spin can have the values $I = 0, 1/2, 1, 3/2, 2 \ldots$ up to 6 (see also Table 1-1). Neither the values of I nor those of P can yet be predicted from theory.

The angular momentum P has associated with it a magnetic moment μ. Both are vector quantities, and they are proportional to each other:

$$\mu = \gamma\, P \qquad (1\text{-}2)$$

The proportionality factor γ is a constant for each nuclide (i. e. each isotope of each element), and is called the magnetogyric ratio. The detection sensitivity of a nuclide in the NMR experiment depends on γ; nuclides with a large value of γ are said to be sensitive (i. e. easy to observe), while those with a small γ are said to be insensitive.

By combining Equations (1-1) and (1-2) we obtain for the magnetic moment μ:

$$\mu = \gamma\, \sqrt{I(I + 1)}\, \hbar \qquad (1\text{-}3)$$

Table 1-1.
Properties of some nuclides of importance in NMR spectroscopy.

Nuclide	Spin I	Natural abundance[a] [%]	Magnetic Moment[b] μ_z/μ_N	Electric quadrupole moment[a] Q [10^{-30} m^2]	Magnetogyric ratio[a] γ [10^7 rad T^{-1} s^{-1}]	NMR frequency[a] [MHz] ($B_0 = 2.3488$ T)	Relative receptivity[c]
^{1}H	1/2	99.9885	2.7928	–	26.7522	100.000	1.00
^{2}H	1	0.0115	0.8574	0.2860	4.1066	15.3506	9.65 x 10^{-3}
^{3}H[d]	1/2	–	2.9790	–	28.5350	106.6640	1.21
^{6}Li	1	7.59	0.8220	–0.0808	3.9372	14.7161	8.50 x 10^{-3}
^{10}B	3	19.9	1.8006	8.459	2.8747	10.7437	1.99 x 10^{-2}
^{11}B	3/2	80.1	2.6887	4.059	8.5847	32.0840	1.65 x 10^{-1}
^{12}C	0	98.9	–	–	–	–	–
^{13}C	1/2	1.07	0.7024	–	6.7283	25.1450	1.59 x 10^{-2}
^{14}N	1	99.63	0.4038	2.044	1.9338	7.2263	1.01 x 10^{-3}
^{15}N	1/2	0.368	–0.2832	–	–2.7126	10.1368	1.04 x 10^{-3}
^{16}O	0	99.96	–	–	–	–	–
^{17}O	5/2	0.038	–1.8938	–2.558	–3.6281	13.5565	2.91 x 10^{-2}
^{19}F	1/2	100	2.6269	–	25.1815	94.0940	8.32 x 10^{-1}
^{23}Na	3/2	100	2.2177	10.4	7.0809	26.4519	9.27 x 10^{-2}
^{25}Mg	5/2	10.00	–0.8555	19.94	–1.6389	6.1216	2.68 x 10^{-3}
^{29}Si	1/2	4.68	–0.5553	–	–5.3190	19.8672	7.86 x 10^{-3}
^{31}P	1/2	100	1.1316	–	10.8394	40.4807	6.65 x 10^{-2}
^{39}K	3/2	93.258	0.3915	5.85	1.2501	4.6664	5.10 x 10^{-4}
^{43}Ca	7/2	0.135	–1.3176	–4.08	–1.8031	6.7301	6.43 x 10^{-3}
^{57}Fe	1/2	2.119	0.0906	–	0.8681	3.2378	3.42 x 10^{-5}
^{59}Co	7/2	100	4.627	42.0	6.332	23.7271	2.78 x 10^{-1}
^{119}Sn	1/2	8.59	–1.0473	–	–10.0317	37.2906	5.27 x 10^{-2}
^{133}Cs	7/2	100	2.5820	–0.343	3.5333	13.1161	4.84 x 10^{-2}
^{195}Pt	1/2	33.832	0.6095	–	5.8385	21.4968	1.04 x 10^{-2}

[a] Values from [1].
[b] z-component of nuclear magnetic moment in units of the nuclear magneton μ_N.
  Values from the Bruker Almanac 2004. $\mu_N = eh/4\pi m_p$, m_p: mass of the proton = 5.05095 x 10^{-27} JT^{-1}.
[c] Receptivity is expressed relative to ^{1}H ($= 1$) for constant field and equal numbers of nuclei.
  Values from the Bruker Almanac 2004.
[d] ^{3}H is radioactive.

Nuclides with spin $I = 0$ therefore have no nuclear magnetic moment. Two very important facts for our purposes are that the ^{12}C isotope of carbon and the ^{16}O isotope of oxygen belong to this class of nuclides – this means that the main building blocks of organic compounds cannot be observed by NMR spectroscopy.

For most nuclides the nuclear angular momentum vector P and the magnetic moment vector μ point in the same direction, i.e. they are parallel. However, in a few cases, for example ^{15}N and ^{29}Si (and also the electron!), they are antiparallel. The consequences of this will be considered in Chapter 10.

1.3 Nuclei in a Static Magnetic Field

1.3.1 Directional Quantization

If a nucleus with angular momentum $\boldsymbol{P}$ and magnetic moment $\boldsymbol{\mu}$ is placed in a static magnetic field $\boldsymbol{B_0}$, the angular momentum takes up an orientation such that its component P_z along the direction of the field is an integral or half-integral multiple of $\hbar$:

$$P_z = m\,\hbar \qquad (1\text{-}4)$$

Here m is the magnetic or directional quantum number, and can take any of the values $m = I, I-1, \ldots \ldots -I$.

It can easily be deduced that there are $(2I + 1)$ different values of m, and consequently an equal number of possible orientations of the angular momentum and magnetic moment in the magnetic field. This behavior of the nuclei in the magnetic field is called *directional quantization*. For protons and ^{13}C nuclei, which have $I = 1/2$, this results in two m-values ($+ 1/2$ and $-1/2$); however, for nuclei with $I = 1$, such as ^{2}H and ^{14}N, there are three values ($m = + 1$, 0 and -1; see Fig. 1-1).

From Equations (1-2) and (1-4) we obtain the components of the magnetic moment along the field direction z:

$$\mu_z = m\,\gamma\,\hbar \qquad (1\text{-}5)$$

In the classical representation the nuclear dipoles precess around the z-axis, which is the direction of the magnetic field – their behavior resembles that of a spinning top (Fig. 1-2). The precession frequency or Larmor frequency ν_L is proportional to the magnetic flux density B_0:

$$\nu_L = \left|\frac{\gamma}{2\pi}\right| B_0 \qquad (1\text{-}6)$$

However, in contrast to the classical spinning top, for a precessing nuclear dipole only certain angles are allowed, because of the directional quantization. For the proton with $I = 1/2$, for example, this angle is 54° 44′.

1.3.2 Energy of the Nuclei in the Magnetic Field

The energy of a magnetic dipole in a magnetic field with a flux density B_0 is:

$$E = -\mu_z\,B_0 \qquad (1\text{-}7)$$

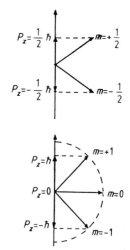

Figure 1-1.
Directional quantization of the angular momentum $\boldsymbol{P}$ in the magnetic field for nuclei with $I = 1/2$ and 1.

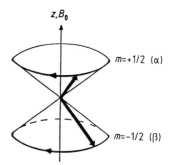

Figure 1-2.
Precession of nuclear dipoles with spin $I = 1/2$ around a double cone; the half-angle of the cone is 54°44′.

Thus, for a nucleus with $(2I + 1)$ possible orientations there are also $(2I + 1)$ energy states, which are called the nuclear Zeeman levels. From Equation (1-5) we have:

$$E = - m \gamma h B_0 \qquad (1-8)$$

For the proton and the ^{13}C nucleus, both of which have $I = 1/2$, there are two energy values in the magnetic field corresponding to the two m-values $+ 1/2$ and $- 1/2$. If $m = + 1/2$, μ_z is parallel to the field direction, which is the energetically preferred orientation; conversely, for $m = - 1/2$, μ_z is antiparallel. In quantum mechanics the $m = + 1/2$ state is described by the spin function α, while the $m = - 1/2$ state is described by the spin function β; the exact form of these functions need not concern us here.

For nuclei with $I = 1$, such as ^2H and ^{14}N, there are three m-values $(+ 1, 0$ and $- 1)$ and therefore three energy levels (Fig. 1-3).

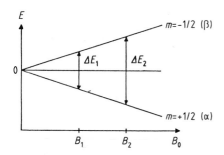

$I=1/2$		$I=1$
$^1H, ^{13}C$		$^2H, ^{14}N$

$m=-1/2$ (β) —— $E_\beta = +1/2\,\gamma\hbar B_0$ $m=-1$ —— $E_{-1}=\gamma\hbar B_0$

0 $m=0$ —— $E_0=0$

$m=+1/2$ (α) —— $E_\alpha = -1/2\,\gamma\hbar B_0$ $m=+1$ —— $E_{+1}=-\gamma\hbar B_0$

Figure 1-3.
Energy level schemes for nuclei with $I = 1/2$ (left) and $I = 1$ (right).

The energy difference between two adjacent energy levels is:

$$\Delta E = \gamma h B_0 \qquad (1-9)$$

Figure 1-4 illustrates, taking the example of nuclei with $I = 1/2$, that ΔE is proportional to B_0.

$m=-1/2$ (β)

ΔE_1 ΔE_2

$m=+1/2$ (α)

B_1 B_2 B_0

Figure 1-4.
The energy difference ΔE between two adjacent energy levels as a function of the magnetic flux density B_0.

5

1.3.3 Populations of the Energy Levels

How do the nuclei in a macroscopic sample, such as that in an NMR sample tube, distribute themselves between the different energy states in thermal equilibrium? The answer to this question is provided by Boltzmann statistics. For nuclei with $I = 1/2$, if we represent the number of nuclei in the upper energy level by N_β and the number in the lower energy level by N_α, then:

$$\frac{N_\beta}{N_\alpha} = e^{-\Delta E/k_B T} \approx 1 - \frac{\Delta E}{k_B T} = 1 - \frac{\gamma \hbar B_0}{k_B T} \qquad (1\text{-}10)$$

where k_B is the Boltzmann constant ($= 1.3805 \times 10^{-23}$ J K^{-1}) and T is the absolute temperature in K.

For protons – and also for all other nuclides – the energy difference ΔE is very small compared with the average energy $k_B T$ of the thermal motions, and consequently the populations of the energy levels are nearly equal. The excess in the lower energy level is only in the region of parts per million (ppm).

Numerical example for protons:
○ With a magnetic flux density $B_0 = 1.41$ T (resonance frequency 60 MHz) the energy difference is given by Equation (1-9) as:
$\Delta E \approx 2.4 \times 10^{-2}$ J mol^{-1} ($\approx 0.6 \times 10^{-2}$ cal mol^{-1}).
The γ-value needed for the calculation can be obtained from Table 1-1; alternatively we can anticipate Section 1.4.1 and calculate it from Equation (1-12).
For a temperature of 300 K these values give:
$N_\beta \approx 0.9999904\ N_\alpha$.
For $B_0 = 7.05$ T (resonance frequency 300 MHz) the difference is greater:
$N_\beta \approx 0.99995\ N_\alpha$.

1.3.4 Macroscopic Magnetization

According to the classical picture, a nucleus with $I = 1/2$ (e.g. ^{1}H or ^{13}C) precesses around the field axis z on the surface of a double cone as shown in Figure 1-5. If we add the z-components of all the nuclear magnetic moments in a sample we obtain a macroscopic magnetization M_0 along the field direction, since N_α is greater than N_β. The vector M_0 plays an important role in the description of all types of pulsed NMR experiments.

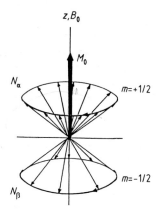

Figure 1-5.
Distribution of the precessing nuclear dipoles (total number $N = N_\alpha + N_\beta$) around the double cone. As $N_\alpha > N_\beta$ there is a resultant macroscopic magnetization M_0.

1.4 Basic Principles of the NMR Experiment

1.4.1 The Resonance Condition

In the nuclear magnetic resonance experiment transitions are induced between different energy levels by irradiating the nuclei with a superimposed field B_1 of the correct quantum energy, i. e. with electromagnetic waves of the appropriate frequency v_1. This condition enables the magnetic component of the radiation to interact with the nuclear dipoles.

Let us consider the protons in a solution of chloroform ($CHCl_3$). The energy level scheme on the left-hand side of Figure 1-3 is appropriate for this case, and transitions between the energy levels can occur when the frequency v_1 is chosen so that Equation (1-11) is satisfied:

$$h\,v_1 = \Delta E \qquad (1\text{-}11)$$

Transitions from the lower to the upper energy level correspond to an absorption of energy, and those in the reverse direction to an emission of energy (arrows labeled 'a' and 'e' respectively in Figure 1-6). Both transitions are possible, and they are equally probable. Each transition is associated with a reversal of the spin orientation. Due to the population excess in the lower level, the absorption of energy from the irradiating field is the dominant process. This is observed as a signal, whose intensity is proportional to the population difference $N_\alpha\text{-}N_\beta$, and is therefore also proportional to the total number of spins in the sample, and thus to the concentration. However, if the populations are equal ($N_\alpha = N_\beta$) the absorption and emission processes cancel each other and no signal is observed. This condition is called *saturation*.

From Equations (1-9) and (1-11) we obtain the *resonance condition:*

$$v_L = v_1 = \left|\frac{\gamma}{2\pi}\right| B_0 \qquad (1\text{-}12)$$

The term "resonance" relates to the classical interpretation of the phenomenon, since transitions only occur when the frequency v_1 of the electromagnetic radiation matches the Larmor frequency v_L.

So far we have considered nuclei with $I = 1/2$, for which there are only two energy levels. But what transitions are allowed when there are more than two such levels, as is the case for nuclei with $I \geq 1$ (Fig. 1-3, right-hand level scheme), and for the coupled systems with several nuclei which will be discussed

E

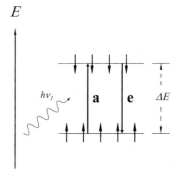

Figure 1-6.
Energy level scheme for a system of nuclei with spin $I = 1/2$. Irradiation with the frequency v_1 such that $hv_1 = \Delta E$ induces absorption (a) and emission (e) transitions.

later? The answer provided by quantum mechanics is that only those transitions in which the magnetic quantum number m changes by 1, i. e. single quantum transitions, are allowed:

$$\Delta m = \pm 1 \qquad (1\text{-}13)$$

Therefore the only transitions that can take place are those between adjacent energy levels. According to this selection rule the transition from $m = +1$ to $m = -1$ for a ^{14}N nucleus is forbidden.

Although coupled spin systems will not be treated until later, we shall point out here that a simultaneous flipping of two or more spins is forbidden. For example, in a two-spin system one could imagine, in addition to the single-quantum transitions, two other possible transitions, whereby both nuclei are initially in the α-state ($m = +1/2$) and simultaneously change to the β-state ($m = -1/2$). This would be a double quantum transition, since the sum of the magnetic quantum numbers changes by two ($\Delta m = 2$). Or again we could have one nucleus switching from the α- to the β-state while simultaneously the other switches from the β- to the α-state. In this case $\Delta m = 0$, i. e. we would have a zero-quantum transition. According to Equation (1-13) such double-quantum and zero-quantum transitions are forbidden.

1.4.2 Basic Principle of the NMR Measurement

NMR transitions in an analytical sample, producing a signal in the receiver channel of the spectrometer, occur when the resonance condition (Eq. 1-12) is satisfied. The most obvious and straightforward way to arrive at this condition and to record a spectrum is either to vary the magnetic flux density B_0 with a constant transmitter frequency ν_1 (field sweep method), or to vary the transmitter frequency ν_1 with constant flux density B_0 (frequency sweep method). In either case the recorder drive is linked directly to the field or frequency sweep, so that the recorder pen progressively traces out the spectrum. This is called the continuous wave (CW) method, as it uses uninterrupted radiofrequency power, unlike the pulsed method described in the next section. The CW method was the basis of all NMR spectrometers constructed up to about the end of the 1960s, but it has been entirely superseded by the pulsed method, for reasons that we shall see later.

The CW method was usually adequate for recording the spectra of sensitive nuclides such as ^1H, ^{19}F and ^{31}P; these have $I = 1/2$ with large magnetic moments, and also high natural abundance. Routine measurements on insensitive nuclides and those with low natural abundance, ^{13}C for example, and also on very dilute solutions, were at first out of the question. To make this possible it was necessary to develop new NMR spectrometers and new measurement techniques.

With regard to spectrometers an important innovation was the introduction of cryomagnets, which make possible considerably higher magnetic field strengths ($B_0 > 2.35$ T) than can be reached with permanent magnets or electromagnets (see Table 1-2), and therefore higher sensitivities. However, the decisive step forward was achieved through *pulsed NMR spectroscopy*, a technique whose development has been closely linked to the rapid advances made in computer technology. During the 1960s R. R. Ernst, together with W. A. Anderson [2], pioneered the application of such techniques to NMR spectroscopy (pulsed Fourier transform or PFT spectroscopy), and thereby initiated the development of a new generation of spectrometers and experiments (see Chapters 8 and 9). The text of Ernst's Nobel Lecture [3] gives a fascinating insight into this phase of the development of NMR spectroscopy. The pulsed Fourier transform method will be described in detail in the next section, as it forms the basis of modern NMR spectroscopy.

The treatment given here will be based on the classical description using vector diagrams, as already introduced in Section 1.3. Then, in Section 1.5.6, we will come to the basic principles and main features of an NMR spectrometer.

Table 1-2.
^1H and ^{13}C resonance frequencies at different magnetic flux densities B_0.[a]

B_0 [T]	Resonance frequencies [MHz]	
	^1H	^{13}C
2.35	100	25.15
4.70	200	50.32
5.87	250	62.90
7.05	300	75.47
9.40	400	100.61
11.75	500	125.76
14.10	600	150.90
16.44	700	176.05
17.62	750	188.62
18.79	800	201.19
21.14	900	226.34

[a] Values from the Bruker Almanac 2004.

1.5 The Pulsed NMR Method

1.5.1 The Pulse

In the pulse method all the nuclei of one species in the sample, e. g. all the protons or all the ^{13}C nuclei, are excited *simultaneously* by a radiofrequency pulse. What does such a pulse consist of, and how is it generated?

A radiofrequency generator usually operates at a fixed frequency ν_1. However, if it is switched on only for a short time τ_P, one obtains a pulse which contains not just the frequency ν_1 but a continuous band of frequencies symmetrical about the center frequency ν_1. However, only a part of the frequency band is effective in exciting transitions, and this part is approximately proportional to τ_P^{-1}. In NMR experiments the *pulse duration* τ_P is of the order of a few µs (Fig. 1-7).

The choice of the generator frequency ν_1 is determined by B_0 and the nuclide to be observed. For example, to observe proton

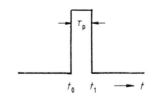

Figure 1-7.
Schematic representation of a pulse. The r. f. generator (frequency ν_1) is switched on at time t_0 and off at t_1. The pulse duration τ_P is typically several µs.

transitions at $B_0 = 4.70$ T, the required generator frequency is 200 MHz, whereas for observing ^{13}C resonances it must be 50.32 MHz. The pulse duration needed for the experiment depends on the width of the spectrum. For example, if $\tau_P = 10^{-5}$ s the frequency band is about 10^5 Hz wide. Provided that ν_1 is correctly chosen, all the frequencies in the spectrum to be recorded are contained in this band (Fig. 1-8).

The amplitudes of the frequency components of a pulse decrease with increasing separation from ν_1. Since, however, it is desirable that all the nuclei should be irradiated equally in the experiment so far as possible (see Section 1.6.3), "hard pulses" are used, i.e. short pulses of high power. The pulse duration is so chosen that the frequency band width exceeds the width of the spectrum by one to two powers of ten; the power level is typically several watts.

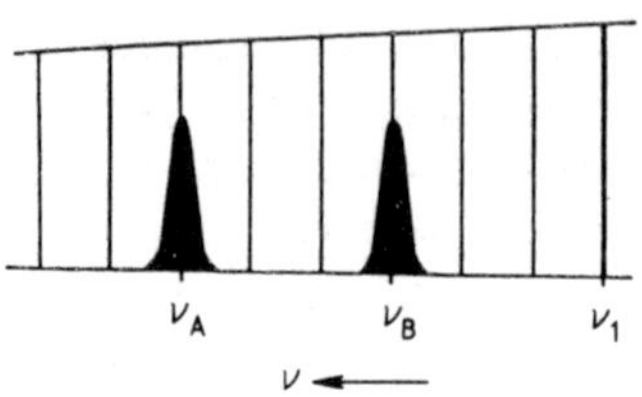

Figure 1-8.
Frequency components of a pulse. The band extends approximately from $\nu_1 - \tau_P^{-1}$ to $\nu_1 + \tau_P^{-1}$; ν_1 is the generator frequency and ν_A and ν_B are the resonance frequencies of nuclei A and B.

1.5.2 The Pulse Angle

We consider the simplest case, that of a sample containing only one nuclear species i, for example the protons in a chloroform solution ($CHCl_3$). As shown in Figure 1-5, the nuclear moments precess with the Larmor frequency ν_L on the surface of a double cone, and as a result of the difference in populations there is a macroscopic magnetization M_0 along the field direction. To induce NMR transitions the radiofrequency pulse is applied to the sample along the direction of the x-axis. The magnetic vector of this electromagnetic radiation is then able to interact with the nuclear dipoles, and therefore with M_0. So that we can more easily visualize this quantum mechanical process, the linear alternating magnetic field along the x-direction will be represented by two vectors with the same magnitude B_1, rotating in the x, y-plane with the same frequency ν_L, one of which, B_1 (r), rotates clockwise, while the other, B_1 (l), rotates anticlockwise (Fig. 1-9). The sum of these two vectors gives the alternating magnetic field along the x-direction, its maximum value being $2B_1$. Of the two rotating magnetic field components only the one which has the same direction of rotation as the precessing nuclear dipoles can interact with them (and thus with M_0); from now on this component will be called simply B_1. Under its influence M_0 is tipped away from the z-axis (the direction of the static field B_0); this tilting occurs in a plane perpendicular to the direction of B_1. However, as this plane rotates with the Larmor frequency ν_L, the complicated motion of M_0 is difficult to show in a diagram. If instead of the fixed coordinate system x, y, z we use a *rotating coordinate system* x', y', z, which rotates with the same frequency as B_1, the

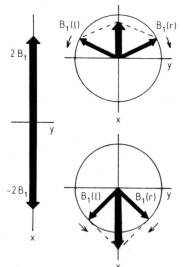

Figure 1-9.
Representation of a linear alternating field (max. $2B_1$) as the sum of two rotating fields, B_1 (r) (clockwise) and B_1 (l) (anticlockwise).

10

orientation and magnitude of B_1 in this system are fixed. As the x'-axis of the rotating coordinate system is normally defined as being along the direction of B_1, the effect of B_1 is to turn the vector M_0 about the x'-axis, i. e. in the y', z-plane. Equation (1-14) shows that the angle Θ through which it is tipped increases with the amplitude B_{1i} of the component of the r. f. pulse at the frequency ν_i of the nuclear resonance transition, and with the length of time for which the pulse is applied.

$$\Theta = \gamma \, B_{1i} \, \tau_\mathrm{P} \qquad (1\text{-}14)$$

The angle Θ is called the *pulse angle*. The majority of pulse techniques can be understood in terms of two special cases, namely experiments with pulse angles of 90° and 180°. If, as in the case just described, the direction of B_1 coincides with that of the x'-axis, these are called $90^\circ_{x'}$ and $180^\circ_{x'}$ pulses. In Figure 1-10 the position of the magnetization vector following a $90^\circ_{x'}$ pulse, a $180^\circ_{x'}$ pulse, and a pulse of arbitrary angle $\Theta_{x'}$ is shown. In the vector diagrams the direction of B_1 is indicated by a wavy line.

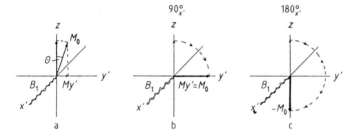

Figure 1-10.
Direction of the macroscopic magnetization vector M_0 in the rotating coordinate system: a) after a pulse of arbitrary angle $\Theta_{x'}$; b) after a $90^\circ_{x'}$ pulse; c) after a $180^\circ_{x'}$ pulse. The wavy line along the x'-axis indicates the direction of the effective B_1 field.

If B_1 is along the direction of the y'-axis, as in experiments that we shall meet in Chapters 8 and 9, we speak of a $90^\circ_{y'}$ or a $180^\circ_{y'}$ pulse; in the vector diagrams the wavy line is then along the y'-axis of the rotating coordinate system.

From these vector diagrams it is seen that the transverse magnetization $M_{y'}$ is greatest immediately after a $90^\circ_{x'}$ pulse, and is zero for $\Theta_{x'} = 0°$ or 180°. The transverse component $M_{y'}$ is crucial for the observation of an NMR signal, since the receiver coil is orientated with its axis along the y-direction, and as a consequence of the detection method used a signal proportional to $M_{y'}$ is induced in it. This signal is a maximum for a $90^\circ_{x'}$ pulse, whereas for a $180^\circ_{x'}$ pulse no signal can be observed.

Without going into the details, this can be illustrated by the following experiment. Equation (1-14) shows that the pulse angle can be increased either by increasing the amplitude of the pulse component B_{1i} or by increasing the pulse duration τ_P. To obtain the results reproduced in Figure 1-11 we kept B_{1i} constant and increased the pulse duration τ_P in steps of 1 μs; the sig-

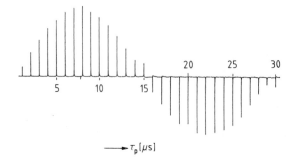

$\longrightarrow \tau_p [\mu s]$

Figure 1-11.
Dependence of the NMR signal from a water sample on the pulse angle Θ. In the experiment the pulse duration τ_P was increased in 1 µs steps. The maximum signal amplitude is obtained with a 90° pulse, which in this case corresponds to a pulse duration of about 8 µs. For $\tau_P = 15$ to 16 µs the signal amplitude is zero, as the pulse angle is then 180°. For still longer pulses the signal amplitude is negative.

nal recorded in each case is shown. The signal amplitude passes through a maximum at a τ_P value of about 7 to 9 µs, then again falls and is approximately zero at $\tau_P \approx 15$ to 16 µs. The maximum corresponds to a pulse angle of 90°, and where the signal passes through zero the pulse is 180°. The experiment also shows that the time interval τ_P needed for a 90° pulse must be doubled to give a 180° pulse.

For still greater values of τ_P a signal is again observed, but with negative amplitude, i. e. it appears in the spectrum as an inverted peak. From the vector diagram we can understand why this is so: for a pulse angle greater than 180° a transverse component $- M_{y'}$ in the direction of the negative y'-axis appears, inducing a negative signal in the receiver coil.

Up to now we have described, with the help of vector diagrams (Fig. 1-10), the effects of the pulses on the macroscopic magnetization vector M_0. But what has become of the $N = N_\alpha + N_\beta$ individual spins which together make up M_0? The condition of the spin system after the $180^\circ_{x'}$ pulse is easy to visualize: the populations N_α and N_β have been exactly reversed by the experiment, so that there are now more nuclei in the upper than in the lower energy level. However, to describe the situation after the $90^\circ_{x'}$ pulse is more complicated. In this case $M_z = 0$, and the two Zeeman levels are equally populated. However, this situation differs from that of saturation which was mentioned earlier (see Section 1.4.1), because after the $90^\circ_{x'}$ pulse a magnetization in the y'-direction is present, whereas in saturation it is not. The way in which this transverse magnetization comes about can be understood by the following picture: under the influence of the B_1 field the individual nuclear dipoles no longer precess with their axes in a uniform random distribution over the surface of the double cone, but instead a (small) fraction precess in phase, bunched together. This condition is referred to as *phase coherence* (Fig. 1-12; see also Section 7.3).

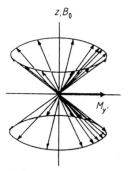

Figure 1-12.
Pictorial representation of phase coherence: after a $90^\circ_{x'}$ pulse a fraction of the nuclear spins (not all!) are bunched together in phase as they precess about the field direction z.

1.5.3 Relaxation

At the instant when the pulse is switched off the magnetization vector M_0 is deflected from its equilibrium position through an angle Θ. M_0, like the individual spins, precesses about the z-axis with the Larmor frequency ν_L; its orientation at any instant is specified in the stationary coordinate system by the three components M_x, M_y and M_z, which vary with the time t (Fig. 1-13).

The spin system now reverts to its equilibrium state by relaxation, with M_z growing back to its original value M_0, while M_x and M_y approach zero. The rather complicated motion of the magnetization vector, both during the application of the radiofrequency field and in the subsequent relaxation, was mathematically analyzed by F. Bloch. He assumed that the relaxation processes were first-order and could be described by two different relaxation times T_1 and T_2. The analysis leads to a set of equations (or a single vector equation) which describe the variation with time of M_x, M_y and M_z. If, instead of the stationary coordinate system x, y, z, we use a rotating coordinate system x', y', z which rotates at the Larmor frequency (see previous section) the equations become much simpler, as they no longer include the precession about the z-axis. For the situation after the pulse has been switched off ($B_1 = 0$), this gives the Bloch equations for relaxation in the rotating frame:

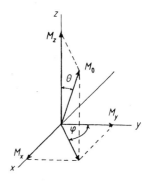

Figure 1-13.
The macroscopic magnetization vector M_0, after being turned from its equilibrium orientation through an angle Θ by applying a pulse, now precesses with the Larmor frequency ν_L. At the instant t it has the coordinates M_x, M_y and M_z, in the stationary coordinate system.

$$\frac{dM_z}{dt} = -\frac{M_z - M_0}{T_1} \tag{1-15}$$

$$\frac{dM_{x'}}{dt} = -\frac{M_{x'}}{T_2} \quad \text{and} \quad \frac{dM_{y'}}{dt} = -\frac{M_{y'}}{T_2} \tag{1-16}$$

T_1 is the *spin-lattice* or *longitudinal relaxation time*, while T_2 is the *spin-spin* or *transverse relaxation time*.

The reciprocals T_1^{-1} and T_2^{-1} of the relaxation times correspond to the rate constants for the two relaxation processes.

To illustrate how simple the description of relaxation becomes in the rotating coordinate system, we will use the above equations to consider what happens to the transverse magnetization after a $90_{x'}^{\circ}$ pulse.

After the pulse M_0 lies along the y'-axis at the instant $t = 0$ (Fig. 1-14a). Consequently $M_0 = M_{y'}$. Since the y'-axis rotates at the Larmor frequency of the nuclei, the transverse magnetization in the y'-direction remains constant, or more exactly, its magnitude only decreases with time at a rate determined by the loss through relaxation. As a result of this relaxation the precessing dipoles, which are at first bunched together (see

13

Figure 1-12), gradually fan out, so that $M_{y'}$ becomes smaller and eventually zero. This is shown schematically by the vector diagrams in Figure 1-14 b and c. Here the magnetic moments of the individual spins are represented by showing only their components in the x', y' plane (thin arrows). According to Equation (1-16) this decrease is exponential (Figure 1-14 d), and its rate is determined by the transverse relaxation time T_2.

In the discussion of one- and two-dimensional pulse techniques in Chapters 8 and 9 we shall return in detail to the question of the motion of the vectors in the stationary and rotating coordinate systems. The phenomenon of relaxation will be treated in Chapter 7.

1.5.4 The Time and Frequency Domains; the Fourier Transformation

The signal detected by the NMR spectrometer following a pulse is determined by $M_{y'}$, but it does not normally look like the decay curve shown in Figure 14 d. Because of the detection method used, such a curve would only be obtained if the generator frequency v_1 and the resonance frequency of the observed nuclei accidentally coincided. Instead the receiver produces a curve like that shown for CH_3I (**1**) in Figure 1-15. The envelope of this curve corresponds to the curve of Figure 1-14 d. In this example, where there is only a single resonance frequency for the three equivalent protons of the methyl group, the time interval between successive maxima is $1/\Delta v$, the reciprocal of the difference in frequency between v_1 and the resonance frequency v_i of the nuclei. This decay of the transverse magnetization as detected in the receiver is called the free induction decay (FID).

If the sample contains nuclei with different resonance frequencies, or if the spectrum consists of a multiplet as a result of spin-spin coupling (see Section 1.6), the decay curves of the different transverse magnetization components are superimposed and interference between the FIDs occurs. Figure 1-16 A shows such an interferogram for the ^{13}C resonance of $^{13}CH_3OH$ (**2**). This interferogram contains the resonance frequencies and intensities which are of interest to us. However, we cannot analyze the interferogram directly, as we are accustomed to interpreting spectra in the frequency domain rather than in the time domain. Nevertheless, the two spectra can each be derived from the other by a mathematical operation called the *Fourier transformation* (FT):

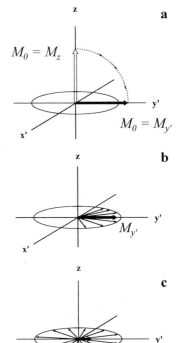

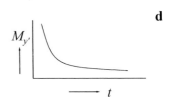

Figure 1-14.
Transverse or spin-spin relaxation. A $90°_{x'}$ pulse turns M_0 into the y'-direction (a), then the bunched precessing nuclear dipoles gradually fan out due to spin-spin relaxation (b and c). Diagram d shows the exponential decay of the transverse magnetization component $M_{y'}$.

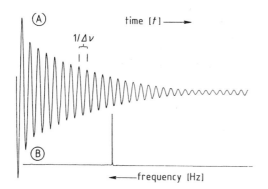

Figure 1-15.
90 MHz ^{1}H NMR spectrum of methyl iodide CH$_3$I (**1**); one pulse, spectral width 1200 Hz, 8 K data points, acquisition time 0.8 s. A: time domain spectrum (FID); the generator frequency is almost exactly equal to the resonance frequency of the sample; B: frequency domain spectrum obtained by Fourier transformation of A.

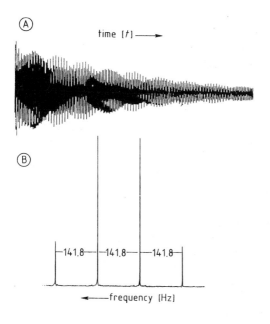

Figure 1-16.
22.63 MHz ^{13}C NMR spectrum of methanol ^{13}CH$_3$OH (**2**); solvent: D$_2$O, 17 pulses, spectral width 1000 Hz, 8 K data points. A: Time domain spectrum (FID); B: frequency domain spectrum obtained by Fourier transformation of A. This consists of a quartet, as the ^{13}C nucleus is coupled to the three protons of the methyl group.

$$g(\omega) = \int_{-\infty}^{+\infty} f(t)e^{-i\omega t}dt \qquad (1\text{-}17)$$

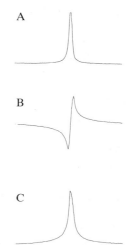

$f(t)$ corresponds to the spectrum in the time domain, and $g(\omega)$ to that in the frequency domain. $g(\omega)$ is a complex function, i. e. it consists of a real part Re and an imaginary part Im. In principle it is equally valid to display the spectrum by using either the real part or the imaginary part, since both represent the frequency spectrum. However, the phases of the signals differ by 90°, and thus in the first case an *absorption mode* signal is obtained, whereas in the second case one has a *dispersion mode* signal. In one-dimensional NMR spectroscopy it is usual to represent the spectrum by the real part and thus to display absorption mode signals (Fig. 1-17).

Figure 1-17.
A: Absorption signal
B: Dispersion signal
C: "Absolute value" signal

15

Because of the method of detection used one usually obtains after the Fourier transformation signals containing both absorption and dispersion components. However, by means of a mathematical operation called phase correction the dispersion component can be removed, so that all the signals in the spectrum are of the usual absorption form.

Figures 1-15 B and 1-16 B show the phase-corrected frequency spectra obtained by Fourier transformation of the interferograms 1-15 A and 1-16 A. The ^{1}H NMR spectrum of methyl iodide consists of a singlet, whereas the ^{13}C spectrum of $^{13}CH_3OH$ is a quartet.

In two-dimensional NMR spectroscopy, for example in displaying COSY spectra, one often uses the "absolute values" of the signals, thus giving the "magnitudes spectrum" M. For this purpose the spectrum is computed by using the expression $M = \sqrt{Re^2 + Im^2}$. This too gives a frequency spectrum with absorption-type signals, but the peaks computed in this way have a broader base than those computed from the real component. However, the great advantage of this form of presentation is that no problems arise from any phase differences between the signals that may occur. We shall return to this problem in Section 9.4.2.

The theory of the Fourier transformation, descriptions of computer programs for performing FT calculations, and other details of the technique can be found in the specialist literature cited in Section 1.8 under "Additional and More Advanced Reading".

1.5.5 Spectrum Accumulation

Usually the intensity of an individual FID is so weak that after Fourier transformation the signals are very small compared with the noise. This is especially so for nuclides with low sensitivity and low natural abundance (e. g. ^{13}C or ^{15}N), or for very dilute samples. Therefore the FIDs of many pulses are added together (accumulated) in the computer, and only then transformed. In this accumulation the random electronic noise becomes partly averaged out, whereas the contribution from the signals is always positive and therefore builds up by addition. The signal-to-noise ratio S : N increases in proportion to the square root of the number of scans NS:

$$S : N \sim \sqrt{NS} \qquad (1-18)$$

The accumulation of many FIDs – sometimes many hundreds of thousands over a period of several days – calls for very precise field-frequency stability and requires that the data from each

FID are stored in the exactly corresponding memory addresses in the computer. Any variation, including also that of temperature, causes a line broadening and a consequent loss of sensitivity (S : N). The device whose task is to ensure the necessary field-frequency stability is called the lock unit. This uses a separate radiofrequency channel to measure a nuclear resonance other than that of the actual NMR experiment; most commonly it is the ^{2}H resonance of the deuterated solvent. For this one needs a transmitter which excites the deuterium resonance (using pulses), a receiver and amplifier, and a detector (signal processor). If the magnetic field strength or the frequency changes, the resonance condition is no longer exactly satisfied, causing a reduction of the signal intensity. The lock unit responds to this automatically, by applying a field correction until the resonance condition is restored. Having thus achieved field-frequency stability for the solvent, we can assume that the same stability condition also applies for the nuclei in the dissolved molecules.

The lock signal can also be displayed on the monitor screen when needed, and it is the normal practice to use it for optimizing the magnetic field homogeneity, either manually or automatically. This is achieved by means of a "shim" unit as described in Section 1.5.6, and the procedure of making the necessary adjustments is called "shimming" in everyday laboratory jargon.

Recording an FID and storing it in digital form requires a certain length of time, which is called the *acquisition time*, and is proportional to the number of memory locations used. The number of locations chosen depends on the width of the spectrum, and it is therefore not possible to specify values which are universally valid. For a spectrum with a width of 5000 Hz and 8 K memory locations, for example, about 1 second is needed to store the data (1 K = 2^{10} = 1024). This is also the shortest possible time interval between two successive pulses (the pulse interval). The system is already starting to undergo relaxation during the storage of the data, as can be observed directly on the oscilloscope (Figs. 1-15 A and 1-16 A).

Because the magnetization decreases with time owing to relaxation, the interferogram contains a higher proportion of noise at the end of the acquisition than at the beginning. The rate at which the FID decays is determined by the relaxation time T_2 and by field inhomogeneities (ΔB_0). This fact is particularly important when choosing the pulse interval for the experiment, since in order to make precise intensity measurements the system must be fully relaxed, i. e. must return to the equilibrium condition, before each new pulse. In practice, however, in order to use the time most efficiently, the next pulse usually follows soon after the collection and storage of the FID data, before equilibrium has been established. This means, firstly, that M_z has not yet reached its equilibrium value M_0, and secondly that there may still be some residual transverse magnetization components $M_{x'}$ and $M_{y'}$ in the x', y' plane. Usually one is prepared to accept this disadvantage, but in some cases, especially in two-dimensional experiments, the residual transverse magnetization can cause artefacts.

An elegant solution which eliminates these unwanted transverse magnetization components is to use pulsed field gradients. Field gradients have long been used in magnetic resonance tomography (see Chapter 14), but their use in high-resolution NMR spectroscopy only became possible when spectrometers began to be equipped with the appropriate hardware, which includes special gradient coils within the probe-head. The method of pulsed field gradients (PFG) has now developed into a very important tool of modern NMR spectroscopy, and therefore it will be described in detail later, in Section 8.3.

1.5.6 The Pulsed NMR Spectrometer

NMR spectrometers are expensive instruments, as the experiment imposes exacting requirements on both the homogeneity and the stability of the magnet and on the performance of the electronics. In the context of this book it would be neither practicable nor appropriate to describe in detail the construction and mechanism of operation of a pulsed NMR spectrometer. The discussion which follows will therefore only briefly outline some of the basic points.

Figure 1-18 shows in schematic form the construction of a pulsed NMR spectrometer. Its components are the cryomagnet, the probe-head, the console, which contains all the associated electronics, and the computer.

The Magnet: A very important component of every NMR spectrometer is the magnet (**1**), shown here in the form of a cut-away vertical section through the cylindrical-shaped assembly. The quality of the magnet plays a key role in determining the quality of the experimental measurement, and therefore of the final spectrum. Until the early 1960s permanent magnets or electromagnets were used, giving flux densities up to 1.41 T (corresponding to an observing frequency of 60 MHz for protons); nowadays, however, by using cryomagnets it is possible to reach flux densities as high as 21.14 T, giving a frequency of 900 MHz for protons. Table 1-2 lists typical magnetic flux densities B_0 and the corresponding ^{1}H and ^{13}C resonance frequencies for NMR spectrometers that have been used in the past and for modern instruments. Very few spectrometers operating below 4.7 T (200 MHz for ^{1}H) remain in use, and all modern instru-

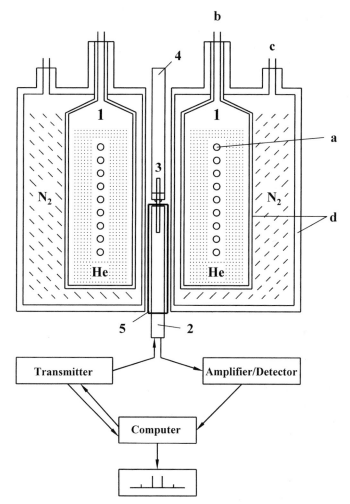

Figure 1-18.
Schematic diagram showing the
components of an NMR spectrometer with a cryomagnet. 1 cryomagnet assembly; a magnet coils;
b, c filling columns for liquid helium
and liquid nitrogen respectively;
d vacuum jackets of the inner and
outer Dewar vessels; 2 probe-head;
3 sample; 4 sample changer; 5 shim
coils.

ments are equipped with cryomagnets. With these magnets the
magnetic field B_0 is directed along the axis of the sample tube.

The Probe-Head: The heart of an NMR spectrometer is the
probe-head (**2**), into which the sample is inserted. It contains
the transmitter and receiver coils, together with additional
coils for decoupling, for the lock, and for generating field gradients, and also a preamplifier. The probe-head is introduced into
the bore of the magnet from below. The diameter of the bore is
usually 5 cm.

The sample (**3**) is usually contained in a glass sample tube
with an external diameter of 5 mm and a length of about 20
cm, and is introduced into the probe-head from above by
means of a sample changer (**4**). When necessary, one can ensure
the most efficient utilization of the spectrometer by using an
automatic sample changer which can accommodate 50 samples,
thus allowing unattended operation overnight or during the

weekend, for example. The turbine for spinning the sample (see below) and the shim coils for optimizing the field homogeneity can also be considered, in a broad sense, as belonging to the probe-head. These are housed in a separate tube which is concentric with the actual probe-head, but remains in the magnet when the probe-head is changed. Usually the sample tube is rotated rapidly about its longitudinal axis, which improves the effective field homogeneity. The sample spinning device (turbine) is at the upper end of the shim tube. In the latest generation of instruments the field homogeneity is so good that sample spinning is not always necessary, and this has made it possible to perform experiments with pulsed field gradients.

The Transmitter: The transmitter consists of a radiofrequency generator and a frequency synthesizer, which provides the various frequencies needed for the NMR experiment (the observing frequency v_1, and the decoupling and lock frequencies), by deriving all these from a fixed frequency generated by a quartz crystal oscillator. The transmitter also produces pulses of the required duration and power. All these functions are controlled by the computer.

The Receiver: As explained earlier in Section 1.5.2, a radiofrequency voltage proportional to the transverse magnetization $M_{y'}$ is induced in the receiver coil. Its frequency corresponds to the energy of the NMR transition [see Eq. (1-11)], typically several hundred MHz. In general, because the transmitter pulse can excite nuclear spins with slightly different precession frequencies in the sample, the induced voltage does not have a pure single frequency, but a spread of frequencies determined by the chemical shifts (see next section), typically within a narrow range of a few kHz above or below the transmitter reference frequency v_1. After receiving an initial boost from the pre-amplifier within the probe-head, the signal must be further amplified and processed. For technical reasons it is preferable and easier to handle frequencies that are considerably lower than v_1, and therefore the method of detection employs a useful trick which is standard practice in radio reception: using the frequency v_1 as a reference base, a separate generator (in radio parlance the "local oscillator") provides a new frequency $v_1 + v_{i.f.}$, where $v_{i.f.}$, called the intermediate frequency, is typically only a few MHz (i.e., much lower than v_1). The local oscillator frequency is "mixed" with the radiofrequency NMR signal to form a new signal whose frequency is the difference between the two. Most of the required amplification can then be carried out conveniently and efficiently by an amplifier designed to handle this intermediate frequency (i.f.). A phase-sensitive detector then compares the amplified i.f. signal with an i.f. reference voltage derived from the local oscillator, resulting in a signal made up of relatively low frequencies which are the differences between v_1 and the frequencies picked up by the receiver coil. After further amplification by a low-frequency broad-band amplifier, this interferogram or FID (like the examples seen earlier in Figures 1-15 A and 1-16 A) is sent to the computer for processing.

In this method of detection the FID and the frequency-domain spectrum derived from it are determined only by the absolute values of the differences between the signal and reference frequencies, and consequently the system cannot distinguish between a signal at a frequency higher than v_1 and one at a frequency lower than v_1 by the same amount. Thus, if two signals fortuitously had frequencies equidistant from v_1, one higher and one lower, we would obtain only one signal in the spectrum. To avoid this it is necessary to position the reference frequency v_1 off one end of the spectrum rather than in the middle of it. However, this has two disadvantages: firstly that one half of the frequency band determined by the pulse duration τ_P (see Section 1.5.1) remains unused, and secondly that the electronic noise from this unused part becomes "folded back" into the spectrum. Both these effects result in a considerable sacrifice of sensitivity. To avoid this, most modern pulse spectrometers use *quadrature detection*. This method employs two phase-sensitive detectors, one to measure $M_{y'}$, the y'-component of the magnetization vector, and the second to simultaneously measure $M_{x'}$, the x'-component, which is shifted 90° relative to the first. When both these components are used in the Fourier transformation, their different phases make it possible to differentiate between signals whose frequencies lie above and below the reference frequency. It then becomes possible to position v_1 in the middle of the spectral region to be observed, thus increasing the sensitivity by a factor $\sqrt{2}$.

The Computer: The interferogram emerging from the final amplifier contains the information of the NMR spectrum in analog form. Before one can obtain the spectrum in the usual frequency-domain form, the information must first be digitized. Measurements of the amplitude of the interferogram (voltage values) are taken at regular time intervals, converted into digital form, and stored in the computer (see Section 1.5.5, Spectrum Accumulation). The Fourier transformation is then carried out in a matter of seconds. In order to ensure that the NMR signals appear at their correct frequencies in the Fourier-transformed spectrum, the sampling interval must be short enough to give at least two data points within each period of the highest frequency component present in the interferogram. If this condition is not satisfied, the computer will treat that component as if it were one of lower frequency, so that the signal becomes back-folded in the spectrum, appearing in a position where there should not actually be a signal. The highest frequency that can be correctly handled by the system is called the *Nyquist frequency*.

Until comparatively recently the main tasks that the computer had to perform were limited to data storage and Fourier transformation. In modern instruments, however, the computer is used to control nearly all the functions of the spectrometer. These include loading programs and setting parameters for the particular experiments that one wishes to perform, automatic shimming, the analysis, simulation, prediction and interpretation of spectra, and very much more. Nowadays it is also possible for the spectrometer's computer to be linked into a network, so that FIDs and other data can be called up remotely from another workstation, where they can be processed independently [4].

1.6 Spectral Parameters: a Brief Survey

1.6.1 The Chemical Shift

1.6.1.1 Nuclear Shielding

The discussion up to this point would lead one to expect only one nuclear resonance signal for each nuclear species. If this were really so the technique would be of little interest to the chemist.

Fortunately, however, the resonances are influenced in characteristic ways by the environments of the observed nuclei. Just to make things simpler we have up to now assumed that we are dealing with isolated nuclei. However, the chemist is concerned with molecules, in which the nuclei are always surrounded by electrons and other atoms. The result of this is that in diamagnetic molecules the effective magnetic field B_{eff} at the nucleus is always less than the applied field B_0, i.e. the nuclei are shielded. The effect, although small, is measurable. This observation is expressed by Equation (1-19):

$$B_{\text{eff}} = B_0 - \sigma B_0 = (1 - \sigma) B_0 \qquad (1\text{-}19)$$

Here σ is the *shielding constant*, a dimensionless quantity which is of the order of 10^{-5} for protons, but reaches larger values for heavy atoms, since the shielding increases with the number of electrons. It should be noted that σ-values are molecular constants which do not depend on the magnetic field. They are determined solely by the electronic and magnetic environment of the nuclei being observed.

The resonance condition, Equation (1-12), thus becomes:

$$v_1 = \frac{\gamma}{2\pi} (1 - \sigma) B_0 \qquad (1\text{-}20)$$

The resonance frequency v_1 is proportional to the magnetic flux density B_0 and – more importantly for our purpose – to the shielding factor $(1 - \sigma)$. From this statement we arrive at the following important conclusion: nuclei that are chemically non-equivalent are shielded to different extents and give separate resonance signals in the spectrum.

The 90 MHz ^{1}H NMR spectrum $(B_0 = 2.11$ T$)$ of a mixture of bromoform (CHBr$_3$, **3**), methylene bromide (CH$_2$Br$_2$, **4**), methyl bromide (CH$_3$Br, **5**) and tetramethylsilane (TMS, Si(CH$_3$)$_4$, **6**), recorded by the frequency sweep method (see Section 1.4.2), is reproduced in Figure 1-19. The signal of TMS appears at exactly $90\,000\,000$ Hz $= 90$ MHz; the signals of the other compounds are found at $90\,000\,237$ Hz (CH$_3$Br), $90\,000\,441$ Hz (CH$_2$Br$_2$) and $90\,000\,614$ Hz (CHBr$_3$). Thus the proton in bromoform has the highest measured resonance frequency and the protons in TMS the lowest. It follows from the resonance condition, Equation (1-20), that the protons are least shielded in bromoform and most shielded in TMS, i. e.:

$$\sigma\,(\text{CHBr}_3) < \sigma\,(\text{CH}_2\text{Br}_2) < \sigma\,(\text{CH}_3\text{Br}) < \sigma\,(\text{TMS})$$

In accordance with a universal convention, resonance signals in NMR spectroscopy are recorded in such a way that the shielding constant σ increases from left to right.

(In a spectrum recorded by the field sweep method – constant frequency v_1 and variable flux density B_0 – to keep the signals in the same sequence on the abscissa the magnetic flux density must increase from left to right. Because of this fact it is customary for historical reasons to use expressions such as "a signal appears at higher field", or "a signal is at lower field").

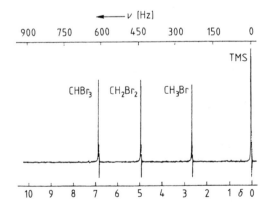

Figure 1-19.
90 MHz ^{1}H NMR spectrum of a mixture of CHBr$_3$ (**3**), CH$_2$Br$_2$, (**4**), CH$_3$Br (**5**) and TMS (**6**).

23

1.6.1.2 Reference Compounds
and the δ-Scale

In Figure 1-19 no absolute values are given for the magnetic flux density B_0 or for the resonance frequencies v_1, since the measured frequencies are only valid for the experiment carried out with $B_0 = 2.11$ T. In NMR spectroscopy there is, unfortunately, no absolute spectral scale, since the resonance frequency and the magnetic flux density are interdependent as a consequence of the resonance condition (Equation 1-20). A relative scale is therefore used, whereby one measures the frequency difference Δv between the resonance signals of the sample and that of a reference compound. Modern spectrometers are calibrated in such a way that the separation between the signals can be read off directly in Hz, or printed out directly by the computer.

Before each measurement the reference compound is added to the sample to be examined, and in such cases it is therefore called an *internal standard*. In practice one uses a solvent into which one has previously added a suitable amount of a reference compound. In ^{1}H and ^{13}C NMR spectroscopy tetramethylsilane (TMS) is usually employed for this purpose. TMS is particularly suitable from both the spectroscopic and the chemical points of view. The fact that it contains twelve equivalent, highly shielded protons means firstly that only a small amount needs to be added, and secondly that it gives only a single sharp peak, which is at the right-hand end of the spectrum clearly separated from most other resonances (Fig. 1-19). Moreover, TMS is chemically inert, magnetically isotropic, and non-associating. In addition, because of its low boiling point (26.5 °C), it can be easily removed in cases where the sample needs to be recovered.

If we measure the peak positions in the spectrum shown in Figure 1-19, we obtain for the frequency differences Δv from TMS as internal standard the values 614, 441 and 237 Hz respectively for bromoform, methylene bromide and methyl bromide. With modern spectrometers all the line positions are calculated by Fourier transformation and the frequencies are printed out directly.

But the Δv-values also depend on B_0! One therefore defines a quantity δ, the *chemical shift*, as follows:

$$\delta_{sample} = \frac{v_{sample} - v_{reference}}{v_{reference}} \tag{1-21}$$

As the numerator in this expression is typically no more than a few hundred Hz, whereas the denominator is usually several hundred MHz, the δ-value defined in this way is in general a very small number, and therefore δ-values are instead given in parts per million (ppm). Thus, Equation (1-21) is replaced by Equation (1-22), where Δv is given in Hz and $v_{reference}$ in MHz:

$$\delta_{sample} \; [\text{ppm}] \; = \; \frac{\Delta v \quad [\text{Hz}]}{v_{reference} \quad [\text{MHz}]} \qquad (1\text{-}22)$$

The convention of introducing a factor of 10^6 to give more convenient numbers and quoting δ-values in parts per million was included in a recommendation issued by IUPAC in 1972. But as ppm is not a unit of measurement in the sense of a dimension, IUPAC also recommended then that δ-values in this form should not be followed by "ppm" (since the factor of 10^6 was already contained in the definition of δ). However, that recommendation did not find general acceptance among NMR spectroscopists. Therefore, in 2001 the IUPAC Committee issued a new recommendation, namely that the chemical shift should be defined as in Equation (1-22), and that quoted δ-values should also be described as "ppm". Accordingly, that procedure is followed throughout this book.

The δ-value for the reference compound TMS is, by definition, zero, since in this case $\Delta v = 0$; thus:

$$\delta \, (\text{TMS}) \; = \; 0 \; \text{ppm} \qquad (1\text{-}23)$$

The δ-values, starting from the TMS signal, are positive to the left and negative to the right of it.

Examples:
○ For the three compounds of Figure 1-19 we calculate the following chemical shifts:

$$\delta \, (\text{CHBr}_3) \; = \; \frac{614}{90 \times 10^6} \; = \; 6.82 \; \text{ppm}$$

$$\delta \, (\text{CH}_2\text{Br}_2) \; = \; \frac{441}{90 \times 10^6} \; = \; 4.90 \; \text{ppm}$$

$$\delta \, (\text{CH}_3\text{Br}) \; = \; \frac{237}{90 \times 10^6} \; = \; 2.63 \; \text{ppm}$$

From the δ-values we can, by using Equation (1-22), calculate the interval in Hz between the resonance and the TMS signal for any arbitrary observing frequency. Thus, assuming in our example an observing frequency of 300 MHz these intervals are:

$$\text{CHBr}_3 : \Delta v = 2046 \; \text{Hz}$$
$$\text{CH}_2\text{Br}_2 : \Delta v = 1470 \; \text{Hz}$$
$$\text{CH}_3\text{Br} : \Delta v = 789 \; \text{Hz}$$

It is not always possible to add TMS to the sample under examination. For example, TMS is insoluble in water. In such cases a different reference compound is added to the sample and the results are converted to the TMS scale. Occasionally too an *external standard* is used. This is a reference compound which is sealed into a capillary and its resonance recorded along with that of the sample. For this purpose it is best to use special coaxial sample tubes which are commercially available. However, in calculations using such data and for comparing with literature values one must take into account the fact that the nuclei at the position of the reference standard experience a different shielding from those in the sample. For reference compounds of "other" nuclides see [1].

The relationships between chemical shifts and molecular structure will be described in detail in Chapter 2, and the chemical shifts of "other" nuclides apart from ^{1}H and ^{13}C will also be discussed.

1.6.2 Spin-Spin Coupling

1.6.2.1 The Indirect Spin-Spin Coupling

In the mixture of $CHBr_3$, CH_2Br_2, CH_3Br and TMS (**3–6**) we observe a singlet for each compound (Fig. 1-19), since each compound contains only one group of chemically equivalent protons. This is an unusual example; in general most of the signals exhibit a fine structure. Figure 1-20 shows a simple example, the spectrum of ethyl acetate (**7**). From left to right we see a quartet, a singlet, a triplet, and the TMS signal. The protons within each group are certainly chemically equivalent, and therefore the splitting of the signals cannot be caused by any differences of chemical type between individual protons. The cause of this fine structure will be explained below by taking examples from ^{1}H NMR spectroscopy, but the same considerations can be extended to ^{13}C and other nuclides with $I = 1/2$.

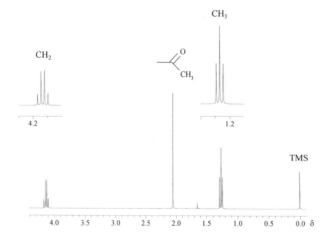

Figure 1-20.
300 MHz ^{1}H NMR spectrum of ethyl acetate (**7**) in CDCl₃.

Up to now we have taken no account of the fact that neighboring magnetic dipoles in a molecule interact with each other. In the ethyl group of ethyl acetate the two methylene protons are coupled to the three protons of the methyl group. This *spin-spin coupling* affects the magnetic field at the positions of the nuclei being observed. The effective field is stronger or weaker than it would be in the absence of the coupling, and in accordance with Equation (1-20) for the resonance condition this alters the resonance frequencies.

The fine structure seen in Figure 1-20 is caused by the socalled *indirect spin-spin coupling*, indirect because it occurs through the chemical bonds (see Section 3.6).

Nuclear dipoles can also be coupled to each other directly through space. For example, this *direct spin-spin coupling* has an important role in solid-state NMR spectroscopy. In high-resolution NMR spectroscopy, when making measurements on low-viscosity liquids, this coupling is averaged to zero by molecular motions. In the following discussion we will be concerned only with the indirect spin-spin coupling.

Before considering in detail the multiplet structure of the resonances of ethyl group protons, we must first try to understand how the spectrum is altered by the effects of indirect spin-spin coupling, taking the example of two coupled nuclei A and X. Examples of two-spin systems of this type are the molecules **H-F**, $^{13}CHCl_3$, and $Ph-CH^A = CH^XCOOH$. This model will then be extended to multi-spin systems.

1.6.2.2 Coupling to One Neighboring Nucleus (AX Spin System)

In a two-spin system AX, if we consider only the chemical shifts the spectrum consists of two resonance signals with frequencies v_A and v_X. If now A and X are coupled to each other, we find two signals for the A-nuclei and two for the X-nuclei (Fig. 1-21).

Let us first confine our attention to the two resonances for A. To understand the splitting into a doublet we must distinguish between two situations. In the first the nucleus X, coupled to A, is in the α-state, and its magnetic moment thus has a compo-

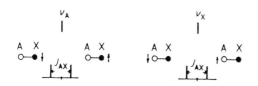

Figure 1-21.
Sketch to explain the fine structure in the spectrum of an AX spin system with a coupling constant J_{AX}. v_A and v_X are the resonance frequencies in the absence of coupling.

27

nent μ_z along the field direction; we represent this by an arrow directed upwards (A-X ↑). If on the other hand X is in the β-state, μ_z points in the negative field direction (A-X ↓; see Section 1.3). The interaction between A and X produces an additional field contribution at the position of A, and for the two states of the X-nucleus the contributions, although equal in magnitude, have opposite signs. In the first situation, therefore, ν_A is shifted to higher frequency by a fixed amount, whereas in the second situation it is shifted to lower frequency by the same amount. We are unable to predict whether an X-nucleus in the α-state will shift the A-resonance to higher or lower frequency. The assignment indicated in Figure 1-21 is arbitrarily chosen. We shall return to this problem in Section 4.3 in connection with the signs of coupling constants.

Since in a macroscopic sample the numbers of molecules with the X-nucleus in the α-state (A-X ↑) and in the β-state (A-X ↓) are nearly equal, two signals of equal intensity appear in the spectrum, i. e. the single peak in the spectrum without coupling is split into a doublet.

An analogous situation applies for the X-nucleus, since the coupling to A similarly causes two X resonances, i. e. a doublet.

The interval between the two lines of each doublet is the same for the A and X parts of the spectrum; it is called the *indirect* or *scalar coupling constant* and is denoted by J_{AX}. Since the splitting is due solely to the nuclear magnetic moments, the value of the coupling constant J_{AX} – unlike the chemical shift – does not depend on the magnetic flux density B_0. It is therefore always given in Hz.

It should be noted that the chemical shift always corresponds to the middle of a doublet, which would be the position of the signal if there were no coupling.

The spectrum of the two olefinic protons of cinnamic acid (**8**, Fig. 1-22) is of the AX type. However, the intensities within the doublet deviate to some extent from the ideal 1 : 1 ratio ("roof effect"). There are two reasons for this: firstly the signals of the proton in the α-position are slightly broadened by coupling to the ring protons, and secondly the spectrum of our two-spin system is in this case not quite of the first-order type. These complications will be discussed in more detail in Sections 1.6.2.8 and 4.3.2.

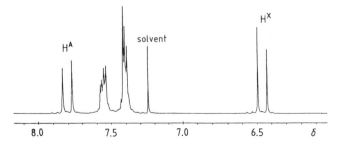

Figure 1-22.
Part of the 250 MHz ^{1}H NMR spectrum of cinnamic acid (**8**) in CDCl$_3$; δ (OH) $\approx$ 11.8.

1.6.2.3 Coupling to Two Equivalent Neighboring Nuclei (AX₂ Spin System)

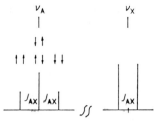

As an example of coupling between three nuclei we will consider the three-spin system CH^A-CH_2^X. Nucleus A now has two equivalent neighbor nuclei X, for which there are three possible spin orientations in the magnetic field. Considering only the z-components μ_z of the magnetic moments, the two spins of the X-nuclei can be parallel to each other, either along the field direction ($\uparrow\uparrow:\alpha\alpha$) or opposite to the field direction ($\downarrow\downarrow:\beta\beta$), or they can be antiparallel to each other ($\uparrow\downarrow:\alpha\beta$ or $\downarrow\uparrow:\beta\alpha$) (Fig. 1-23). If the X-spins are antiparallel the additional field contributions at the position of nucleus A cancel to zero, and the resonance signal is in the position which it would occupy if there were no coupling. The two parallel spin arrangements cause additional field contributions at the position of A which are equal in magnitude but opposite in sign. This results in two further resonance signals, and a triplet is therefore observed for the protons H^A. The interval between any two adjacent lines is J_{AX}. The intensities are in the ratio 1:2:1. The middle signal is thus of double intensity, since in a macroscopic sample molecules with an antiparallel arrangement of the X-spins are twice as numerous as those with a parallel arrangement in either one or the other direction.

For the two protons H^X of the CH_2^X group we obtain a doublet, as they are coupled to only one neighbor nucleus H^A. The total intensities of the triplet and of the doublet are in the ratio 1:2.

The chemical shifts δ_A and δ_X are calculated from the signal positions ν_A and ν_X in the absence of coupling, as was done in the case of the two-spin system AX; ν_A corresponds to the middle signal of the triplet and ν_x to the mid-point of the doublet. As an example Figure 1-24 shows the spectrum of benzyl alcohol (**9**), in which the triplet at $\delta \approx 5.3$ is assigned to the proton

Figure 1-23.
Sketch to explain the splitting pattern observed for a three-spin AX₂ system; the arrows indicate the orientations of the two X spins.

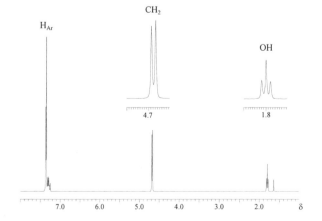

9

Figure 1-24.
300 MHz ^1H NMR spectrum of benzyl alcohol (**9**) in CDCl₃.

attached to oxygen (**H**^{A}), and the doublet at $\delta \approx 4.4$ to the two protons of the methylene group (**CH**$_2^X$).

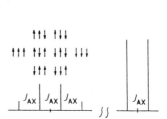

1.6.2.4 Coupling to Three or More Equivalent Neighboring Nuclei (AX$_n$ Spin System)

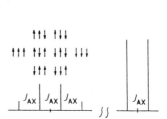

Figure 1-25.
Sketch to explain the splitting pattern observed for a four-spin AX$_3$ system; the arrows indicate the orientations of the three X spins.

The splitting patterns for couplings to three or more equivalent neighboring nuclei can be constructed in the same way as was shown in the previous section for two equivalent neighboring nuclei. If a proton is coupled to three neighbors X, e. g. to the three protons of a methyl group, as in **CH**^{A}-**CH**$_3^X$, one expects to find for **H**^{A} a quartet with the intensity distribution $1:3:3:1$ (Fig. 1-25). For the methyl protons **H**^{X} one again finds a doublet, since these are coupled to only a single neighbor, **H**^{A}. A spectrum of this type is obtained for paraldehyde (**10**), which is the trimer of acetaldehyde (Fig. 1-26).

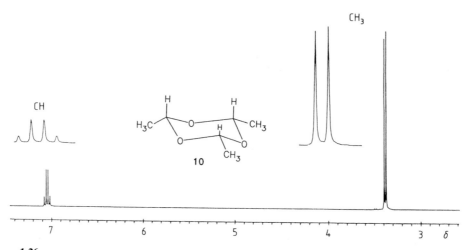

Figure 1-26.
250 MHz ^{1}H NMR spectrum of paraldehyde (**10**). The quartet and doublet are shown expanded, both by the same factor.

1.6.2.5 Multiplicity Rules

The number of lines in a multiplet, which is called the multiplicity M, can be calculated from Equation (1-24):

$$M = 2nI + 1 \qquad (1\text{-}24)$$

Here n is the number of equivalent neighbor nuclei. For nuclei with $I = 1/2$, which includes those that most concern us, Equation (1-24) simplifies to:

$$M = n + 1 \qquad (1\text{-}25)$$

For couplings to nuclei with $I = 1/2$ the signal intensities within each multiplet correspond to the coefficients of the binomial series, which can be obtained from Pascal's Triangle:

$n = 0$ 1

$n = 1$ 1   1

$n = 2$ 1   2   1

$n = 3$ 1   3   3   1

$n = 4$ 1   4   6   4   1

$\vdots$ $\vdots$

We can now understand the splitting patterns for ethyl acetate (**7**) in Figure 1-20. The two methylene protons are coupled to the three equivalent CH_3 protons of the ethyl group, and in accordance with Equation (1-25) they give a quartet ($\delta \approx 4.1$). The three methyl protons are coupled to the two equivalent protons of the methylene group, which gives a triplet ($\delta = 1.4$). Such a combination of a quartet and a triplet in the intensity ratio $2:3$ always enables one to recognize immediately that the compound being examined contains an ethyl group. Finally, the singlet at $\delta = 2$ in the spectrum of **7** originates from the three methyl protons of the acetyl group.

1.6.2.6 Couplings between Three Non-equivalent Nuclei (AMX Spin System)

Figure 1-27 shows the spectrum of styrene (**11**). Here we are only interested in the signals of the three mutually coupled nonequivalent vinyl protons H^A, H^M and H^X (three-spin AMX system).

In the spectrum we find for each proton four signals of nearly equal intensities. The splitting schemes shown above the expanded regions of the spectrum in Figure 1-27 show how the multiplets for each of the protons can be reconstructed and analyzed. One begins by considering the spectrum without coupling, which would consist of three resonance lines at ν_A, ν_M and ν_X. Then one restores one of the couplings for each nucleus – preferably the largest in each case – and splits each line into a doublet corresponding to the coupling constant. This is repeated for the second, smaller, coupling constant, so that each line of the first doublet is further split into a doublet. This leads to a

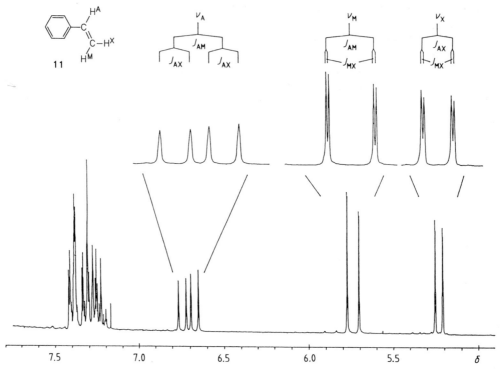

Figure 1-27.

250 MHz ^{1}H NMR spectrum of styrene (**11**) in $CDCl_3$. The protons H^A, H^M and H^X each give a doublet of doublets, shown expanded by the same factor in each case.
$J_{AM} = 17.6$ Hz, $J_{AX} = 10.9$ Hz, $J_{MX} = 1.0$ Hz.

doublet of doublets for each of the protons **H^A**, **H^M** and **H^X**, whose centers (v_A, v_M and v_X) correspond to the δ-values.

1.6.2.7 Couplings between Equivalent Nuclei (A_n Spin Systems)

Why is it that in the ^{1}H NMR spectrum of an isolated methyl group we find only one signal, even though each proton has as its neighbors two other protons of the methyl group, thereby providing the conditions for coupling to occur? Why does one only find one signal for the six protons of benzene, despite the fact that in benzene derivatives the protons are seen to be coupled to each other?

Exact answers to these questions are provided by quantum-mechanical calculations. Without going into the details of the theory, we shall here merely express this result in the form of a general rule:

Couplings between equivalent nuclei cannot be observed in the spectrum!

In the next section it will become necessary to qualify this rule to some extent, as it applies only to first-order spectra.

In the methyl and methylene groups considered up to now, and in benzene, the protons are equivalent. Consequently the couplings within them cannot be observed, and the spectra are simple and easy to interpret.

1.6.2.8 The Order of a Spectrum

A spectrum which contains only singlets is said to be a *zero-order* spectrum. Most ^{13}C NMR spectra belong to this class as a result of the method of recording which is used (^{1}H broad-band decoupling; see Section 5.3).

If the multiplets can be analyzed according to the rules already given, one is dealing with a *first order* spectrum. Such spectra are to be expected in all cases where the frequency interval Δv between the coupled nuclei is large compared with the coupling constants, i. e. $\Delta v \gg J$. If this condition is not satisfied the intensity ratios within the multiplets alter, and additional lines may appear. Such a spectrum is described as being of *higher order*. These effects will be considered in more detail in Chapter 4. In higher-order spectra the couplings between equivalent nuclei also become evident. The analysis of such spectra is more complicated, and often it is only possible by using a computer. In the course of such an analysis it also emerges that the coupling constants can have either positive or negative sign. However, we shall see (in Section 4.3.1) that the appearance of spectra of the first-order type is not affected by the signs.

1.6.2.9 Couplings between Protons and other Nuclei; ^{13}C Satellite Spectra

In the ^{1}H NMR spectra of organic molecules one normally only sees H,H couplings. However, for molecules containing fluorine, phosphorus or other nuclei which have a magnetic moment, the couplings to these nuclei are also seen. The same rules apply as for H,H couplings. For these heteronuclear couplings there is the additional simplification that $\Delta v \gg |J|$, and thus the condition for first-order spectra is nearly always fulfilled.

At this point special mention must be made of the couplings between protons and ^{13}C nuclei. These C,H couplings make themselves apparent in the ^1H NMR spectrum by the ^{13}C satellite signals. What are ^{13}C satellites? Figure 1-28 shows the ^1H NMR spectrum of chloroform. It consists of a main peak at $\delta = 7.24$, with two small peaks to the right and left of it (in the figure these are shown greatly amplified). The smaller peaks are caused by the 1.1 % of chloroform molecules that contain a ^{13}C nucleus, i. e. by $^{13}CHCl_3$. The proton is coupled to the ^{13}C nucleus, and this gives a doublet in the ^1H NMR spectrum, the ^{13}C *satellites*. The interval between the two satellite peaks is equal to $J(C,H) = 209$ Hz. The intensity of each satellite is 0.55 % of that of the main peak. (A corresponding doublet with the separation $J(C,H)$ is also observed in the ^{13}C NMR spectrum of chloroform.) The ^{13}C satellite signals are not always so easy to interpret as in the chloroform spectrum (see Section 4.7).

More information on the couplings of protons to "other" nuclei, on heteronuclear couplings that do not involve protons, and on couplings involving nuclei with a spin I greater than 1/2, can be found in Section 3.7.

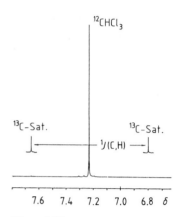

Figure 1-28.
250 MHz ^1H NMR spectrum of chloroform; the ^{13}C satellites are shown with a 15-times amplification; $^1J(C,H) = 209$ Hz.

1.6.3 The Intensities of the Resonance Signals

1.6.3.1 ^1H Signal Intensities

The area under the signal curve is referred to as the *intensity* or the *integral* of the signal. Comparing these intensities in a spectrum directly gives the ratios of the protons in the molecule. In the case of multiplets one must, of course, integrate over the whole group of peaks. An example of an integration is shown in Figure 1-29.

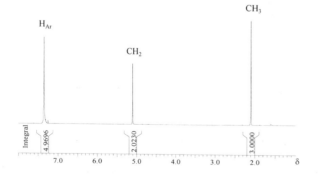

Figure 1-29.
300 MHz ^1H NMR spectrum of benzyl acetate (**12**) with integrated intensities shown below.

34

In the ^1H NMR spectrum of benzyl acetate **(12)** we find three singlets corresponding to C_6H_5, CH_2 and CH_3. Integration gives the result $5:2:3$ for the ratio of their areas, and from this all the signals can be assigned.

Signal intensities are next in importance to chemical shifts and indirect spin-spin coupling constants as aids to structure determination; they also make possible the *quantitative analysis* of mixtures.

1.6.3.2 ^{13}C Signal Intensities

In principle it should also be possible from the signal intensities in ^{13}C NMR spectra to reach conclusions about the numbers of carbon atoms present in the molecule. However, because of the low natural abundance and sensitivity compared with protons, detection methods are used which have the undesirable side-effect of distorting the integrals. For this reason it is not usual to give integral curves in ^{13}C NMR spectroscopy. The detailed causes of this are as follows:

- The amplitudes of the different frequency components of the pulse become smaller with increasing distance from the transmitter frequency v_1. Consequently nuclei with different resonance frequencies are excited to varying extents (Section 1.5.1 and Fig. 1-8).
- A resonance peak is not stored in the computer as a complete continuous curve but as a relatively small number of points (Fig. 1-30). In the integration one determines the area bounded by the straight lines joining these points. The closer the spacing of the points the more exact is the integral. The number of data points used in recording the spectrum is usually determined by the time available for the recording. Of the two curves shown in Figure 1-30 one – the broken curve – was recorded with 32 K data points, and the other – the solid curve – was recorded with 2 K data points. Whereas in the broken curve the points are at intervals of about 0.01 Hz, the intervals in the solid curve are more than 0.2 Hz. One can clearly see that the latter curve does not correctly reproduce the line shape. The amplitude is too small, the half-height width is too great, and consequently the integral must also be incorrect. Furthermore, the positions of the maxima in the two curves, from which the δ-values are calculated, differ by about 0.1 Hz. In practice, however, this error is negligible.
- The time interval between two successive pulses during the accumulation is usually so short that the spin system cannot return to equilibrium by relaxation. This leads to incorrect

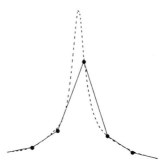

Figure 1-30.
Comparison of the same ^1H NMR signal recorded with 32 *K* data points (dashed curve) and with 2 *K* data points (solid line). The dashed curve gives the true line shape. The solid line gives only a distorted peak, with errors in height, width, area and position of maximum.

integrals, the effects being much greater for nuclei with long relaxation times T_1 than for those with a short T_1.

- ^{13}C NMR spectra are normally recorded with ^{1}H broad-band decoupling (Section 5.3). Under these conditions the signals are amplified by the nuclear Overhauser effect (NOE, Chapter 10). The increase in intensity depends on the number of directly bonded hydrogen atoms, and on other factors influencing the relaxation times (Chapter 7).

Every ^{13}C NMR spectrum is affected by all these four sources of error. It is not possible to generalize about the magnitudes of the different effects and of the total integration error – every measurement therefore involves an individual compromise between accuracy and time spent in recording. The causes of incorrect intensity measurement can be completely, or at least partially, avoided by taking precautions with the spectrometer and the recording technique – although at the cost of increased recording times. For *accurate determination of ^{13}C intensities* the following measures are essential:

- The pulse must be of sufficiently high power to ensure that the fall-off of intensity in the frequency components over the full width of the spectrum is negligible. This condition is usually satisfied for ^{1}H spectra, but not for ^{13}C spectra. It is even more critical for other nuclides – ^{31}P for example – for which the spectral width is greater than for ^{13}C.
- Where the spectral width is large and the lines are narrow a computer with a large memory capacity is needed. If a spectrum with a width of 5000 Hz is recorded using 4 K ($= 4096$) data points the resulting digital resolution is only 1.25 Hz per data point. However, the line widths are usually less than this, so one needs to use 8 K, 16 K or 32 K data points, or alternatively to record spectra of a smaller width, i.e. selected regions of the spectrum.

Errors caused by different relaxation times T_1 and different NOES are more difficult to eliminate; furthermore, these are the largest sources of error. The following methods are available:

- Errors caused by too high a pulse repetition rate can be avoided by inserting a delay equal to 5 T_1 between successive pulses. A spin system needs this amount of time to undergo nearly complete relaxation after a 90° pulse. For relaxation times as long as 100 s, such as are found for quaternary carbon nuclei, this would require a waiting time of 8 to 10 minutes between pulses! In practice such an experiment would be unrealistic, and in these circumstances it is usual to do without intensity measurements.
- If all the sources of error discussed up to now have been eliminated by a suitable choice of experimental conditions,

it becomes worthwhile to suppress the NOE for each quantitative measurement. This can be achieved in two ways:

– Adding paramagnetic ions to the sample solution shortens the relaxation times T_1 (and T_2). Chelate complexes of chromium, such as $Cr(acac)_3$, are generally used for this purpose, but too high a concentration must be avoided as it would broaden the lines (Section 7.3.3). This method is not usually employed, and is especially unsuitable if the sample is needed for further experiments. Another method has therefore been developed, viz.

– A pulse experiment which is the inverse of the gated decoupling described in Section 5.3.2. In this the broadband (BB) decoupler is switched on only during the observing pulse and the subsequent data accumulation. One obtains a decoupled ^{13}C NMR spectrum with the correct intensities, since the NOE is unable to build up during this short time. When the FID has been stored and the BB decoupler switched off, the system must relax again before the next pulse. However, the time needed for this is shorter than 5 T_1.

To summarize, quantitative ^{13}C NMR measurements are possible if the following conditions are ensured:

• high pulse power and small spectral width
• high digital resolution
• a pulse repetition rate that is not too high
• suppression of the NOE.

All the methods described can be carried out without difficulty, but they are very costly in measurement time, and are therefore only used in special circumstances.

1.6.4 Summary

Three types of spectral parameters are obtainable from NMR spectra: chemical shifts, indirect spin-spin couplings, and intensities.

• In this chapter we learned that the chemical shift is caused by the magnetic shielding of the nuclei by their surroundings, mainly by the electrons. The resonance frequencies depend on the magnetic flux density, and for this reason absolute line positions are never specified. Instead we define a dimensionless quantity, the δ-value, which gives the position of the signal relative to that of a reference compound, and also relates it to the measurement frequency. Consequently δ-values are always given in ppm independent of the spectrometer used and can be directly compared. The reference compound generally used in ^{1}H and ^{13}C NMR spectroscopy is tetramethylsilane (TMS).

- The interaction between neighboring nuclear dipoles leads to a fine structure. The strength of this interaction is given by the spin-spin coupling constant J. Since the coupling occurs through chemical bonds, it is called the indirect spin-spin coupling. The splitting patterns and intensity distribution of the multiplets can be predicted using simple rules. The indirect spin-spin coupling is independent of the external field, and the coupling constants J are therefore given in Hz. Couplings are observed not only between nuclei of the same species but also between different nuclei (heteronuclear couplings). For our purposes the most important are the H,H and C,H couplings.

- In ^{1}H NMR spectroscopy the signal intensities are also determined for each spectrum, but in routine ^{13}C NMR spectra the intensities cannot be measured.

The relationships between chemical shifts and molecular structure will be examined in detail in Chapter 2, and those between coupling constants and molecular structure in Chapter 3.

1.7 "Other" Nuclides [5, 6]

Up to now we have been concerned almost exclusively with the properties of ^{1}H and ^{13}C nuclei and their NMR spectra. However, it is possible to obtain NMR spectra of nearly all elements, although not always from observations on the isotope with the highest natural abundance, as can be seen from the examples of carbon and oxygen. For reasons connected with the historical development of NMR spectroscopy, nuclei of all species other than ^{1}H are referred to as *heteronuclei*.

The procedures for recording spectra of heteronuclides often differ considerably from those for ^{1}H and ^{13}C, since it is necessary, even for routine measurements, to adjust the experimental conditions to suit the special properties of the nuclei to be observed. For example, the spin–lattice relaxation times for some nuclides, such as ^{15}N and ^{57}Fe, are very long, whereas for others (especially those with an electric quadrupole moment) they are very short. Also the spectra observed for some nuclides contain interfering signals caused by other materials present, for example the glass of the sample tube (^{11}B, ^{29}Si), the spectrometer probe unit (^{27}Al), or the transmitter/receiver coil. For many nuclides the sample temperature and its constancy are important factors; for example, quadrupolar nuclides such as ^{17}O give narrower signals when the temperature is increased. For some nuclides such as ^{195}Pt and ^{59}Co, and also for ^{31}P, the chemical shifts are found to be strongly temperature-dependent, with coefficients that can exceed 1 ppm K^{-1}. Special care is also needed in the choice of a suitable chemical shift standard [1] and of an appropriate spectral range, since often the spectrum may consist of only one peak, and mistakes can easily arise in determining the resonance frequency (for example, where back-folding occurs). In cases where the resonance frequency is low (see Table 1-1) it is often found that the base-line of the spectra is not constant, which causes problems especially in detecting broad signals of insensitive nuclides. In cases such as these it is often necessary to carry out several experiments under different operating conditions to arrive at unambiguous assignments and yield correct spectral data. Nuclides with spin $I = 1/2$ and those with greater spin will be treated separately, as there are fundamental differences between these two cases.

1.7.1 Nuclides with Spin $I = 1/2$

The behavior in a magnetic field of any nucleus with spin $I = 1/2$ is similar to that of ^1H and ^{13}C. This group includes ^3H, ^{15}N ^{19}F, ^{29}Si, ^{31}P, ^{57}Fe , ^{119}Sn, ^{195}Pt, and many others. Some of these nuclides, such as ^3H, ^{19}F, and ^{31}P, are easy to observe (*sensitive* nuclides) , as they have a large magnetogyric ratio γ and a large magnetic moment μ, whereas others, such as ^{15}N and ^{57}Fe, are not so favorable. These insensitive nuclides suffer from the additional disadvantage of low natural abundances (e. g., 0.37 % for ^{15}N and 2.12 % for ^{57}Fe; see Table 1-1). For nuclides such as these the pulsed method of observation is essential, as in the case of ^{13}C. However, this often presents technical problems owing to the fact that the range of chemical shifts for nuclei that differ in their substituents or coordination is usually very large, requiring a correspondingly large spectral width. For example, the range of chemical shifts for ^{195}Pt can be as large as 8000 ppm, and with a magnetic flux density B_0 of 2.35 T (resonance frequency 21.497 MHz for ^{195}Pt or 100 MHz for ^1H) this corresponds to a spectral width of about 1.7 x 10^5 Hz!

A few spin 1/2 nuclides, such as ^{57}Fe, have extremely long spin–lattice relaxation times, which requires long delay times between successive pulses during the accumulation of the FID. In such cases one can add paramagnetic compounds such as $Cr(acac)_3$ to reduce the accumulation time to an acceptable value (see Sections 1.5.5 and 1.6.3.2).

1.7.2 Nuclides with Spin $I > 1/2$

By far the majority of heteronuclides belong to the group with $I > 1/2$. A small selection of these are listed in Table 1-1. All such nuclides have an electric quadrupole moment Q, and they usually give broad NMR signals due to the shortening of the relaxation times through the interaction of the quadrupole moment with local electric field gradients (Section 7.3.3). Often this means that one is unable to observe any multiplet splittings due to couplings with other nuclei, or even to resolve chemical shift differences. Exceptions to this are those nuclides that have a relatively small quadrupole moment, such as deuterium, ^2H. Other exceptions occur when the quadrupolar nucleus is in a symmetrical environment; a typical example is the ^{14}N resonance of the symmetrical ammonium ion $NH_4^\oplus$ (Section 3.7).

For many quadrupolar nuclides, especially the heavier ones, the chemical shifts between non-equivalent nuclei can be extremely large, making it possible to observe separate signals despite the broadness of the resonances (see Section 2.5). For example, the range of chemical shifts for ^{59}Co is about 20 000 ppm, which at $B_0 = 2.35$ T (^{59}Co resonance frequency 23.727 MHz) corresponds to a spectral width of the order of 4×10^5 Hz! However, in addition to the usual technical problems associated with such a large spectral width (Section 1.7.1), there is the difficulty that for very broad signals the pulse duration is of the same order as the relaxation times. In favorable cases, of which ^{59}Co is an example, one can resort to the older CW method of recording (Section 1.4.2), but this is not possible when the natural abundance is so low as to make the pulsed FT method mandatory.

Many nuclides with $I > 1/2$ belong to metals and transition metals that have important functions in biochemistry. In most cases this involves ions in aqueous solutions. Some of the most important ion species are those of the alkali and alkaline earth metals, such as $^{23}Na^\oplus$, $^{39}K^\oplus$, $^{25}Mg^{2\oplus}$, and $^{43}Ca^{2\oplus}$. Despite the great advances in measurement methods that have occurred, the low detection sensitivity for some of these ions means that one still has to resort to using materials enriched in the relevant isotopes, which is laborious, time-consuming, and costly. This applies particularly to the ions $^{25}Mg^{2\oplus}$ and $^{43}Ca^{2\oplus}$, and also to $^{57}Fe^{2\oplus}$. Another nuclide with a spin greater than 1/2 is ^{17}O ($I = 5/2$), with only 0.038 % natural abundance and a moderately large electric quadrupole moment (Table 1-1), which typically results in line-widths between 20 and 300 Hz.

In conclusion it should be mentioned that for many nuclides of both these classes, including ^{15}N, ^{17}O, ^{25}Mg, and ^{29}Si, the magnetogyric ratio γ is negative (Table 1-1). In terms of the classical picture this means that the magnetic moment $\boldsymbol{\mu}$ and the angular momentum vector $\boldsymbol{P}$ lie in opposite directions (see Section 1.2). This becomes important especially in experiments that involve the nuclear Overhauser effect (NOE, Section 10.2.2).

1.8 Bibliography for Chapter 1

[1]   R. K. Harris, E. D.Becker, S. M. Cabral de Menezes, R. Goodfellow and P. Granger. *Nuclear Spin Properties and Conventions for Chemical Shifts (IUPAC Recommendations 2001)*. In: Encyclopedia of Nuclear Magnetic Resonance, Vol. 9. Chichester: John Wiley & Sons, 2002, p. 5.

[2]   R. R. Ernst and W. A. Anderson, *Rev. Sci. Instrum. 37* (1966) 93.

[3]   R. R. Ernst: Nuclear Magnetic Resonance – Fourier Transform Spectroscopy in *Angew. Chem. Int. Ed. Engl. 31* (1992) 805.

[4]   P. Bigler: *NMR Spectroscopy: Processing Strategies*. Weinheim: Wiley-VCH, 2002, 2nd Edition.

[5]   S. Berger, S. Braun, H.-O. Kalinowski: *NMR Spectroscopy of the Non-Metallic Elements*. Chichester: John Wiley & Sons, 1997.

[6]   J. Mason (Ed.): *Multinuclear NMR*. New York: Plenum Press, 1987.

Additional and More Advanced Reading

M. H. Levitt: Spin Dynamics – Basics of Nuclear Magnetic Resonance. Chichester: John Wiley & Sons, 2001.

R. K. Harris: *Nuclear Magnetic Resonance Spectroscopy. A Physicochemical View*. London: Pitman, 1983.

F. J. M. van de Ven: *Multidimensional NMR in Liquids. Basic Principles and Experimental Methods*. New York: VCH Publishers, 1995.

A. E. Derome: *Modern NMR Techniques for Chemistry Research*. Oxford: Pergamon Press, 1987.

2 The Chemical Shift

2.1 Introduction

In molecules the nuclei are magnetically shielded, i. e. the effective field at the position of a nucleus is weaker than the externally applied field. This effect is measured by the shielding constant σ.

There have been many attempts at calculating the shielding constants σ from theory, but none of the theoretical approaches has yielded exact values. If such calculations were possible, the spectrum could be exactly predicted. Theory and experiment lead to the conclusion that the reduction of the field B_0 and of the associated resonance frequency is determined mainly by the distribution of electron density in the molecule. The chemical shifts are therefore considerably affected by substituents which specifically influence this electron distribution.

Whereas inductive and mesomeric substituent effects are transmitted through chemical bonds, interactions through space are also possible – for example if the observed nuclei have magnetically anisotropic neighbors such as a carbonyl group, a CC double or triple bond, or a phenyl ring. Intermolecular interactions also contribute to the shielding.

In the following we shall be mainly concerned with the shielding of ^1H and ^{13}C nuclei, which are the most important ones – at least for organic chemistry. Later (in Section 2.5) we will look at some of the characteristic features of the chemical shifts of "other" nuclides.

So as to get a general idea of the positions of the NMR signals for different classes of compounds, some data are summarized in Figures 2-1 and 2-2. It is seen that the resonances of hydrogen or carbon nuclei in similar bonding situations are grouped in characteristic regions. Because of this fact the chemist can reach conclusions from the signal positions about the structure of the molecule under investigation, or at least about certain of its structural components. Before discussing specific examples, we must first consider the causes of the difference between the shieldings of ^1H and ^{13}C nuclei, and of the differences between chemical shifts in ^1H and ^{13}C NMR spectroscopy.

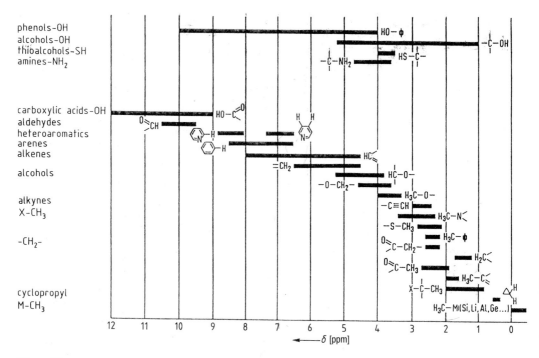

Figure 2-1.
Chemical shifts of ^{1}H nuclei in organic compounds.

2.1.1 Influence of the Charge Density on the Shielding

As has already been mentioned, the magnetic shielding of a nucleus is determined mainly by the shell (or shells) of electrons. The effect is explained by assuming that the magnetic field B_0 induces an electron current in the electron shell. This produces an opposing field at the position of the nucleus, which reduces the effective B_0.

For the hydrogen atom with one electron, and for other nuclei with a spherically symmetric charge distribution, this contribution to the shielding, called the *diamagnetic shielding term* σ_{dia}, can be calculated using the simple classical model of an electron in a circular orbit (Lamb formula). For the hydrogen atom this gives a value of 17.8×10^{-6} for σ_{dia}. This value is small, but as the number of electrons increases σ_{dia} increases rapidly. Thus, for ^{13}C σ_{dia} is already 260.7×10^{-6}, and for ^{31}P it is 961.1×10^{-6} [1]. However, in practical NMR spectroscopy these absolute shielding constants are unimportant. The simple classical model fails for molecules, as the charge distribution here is not usually spherically symmetric. Attempts were never-

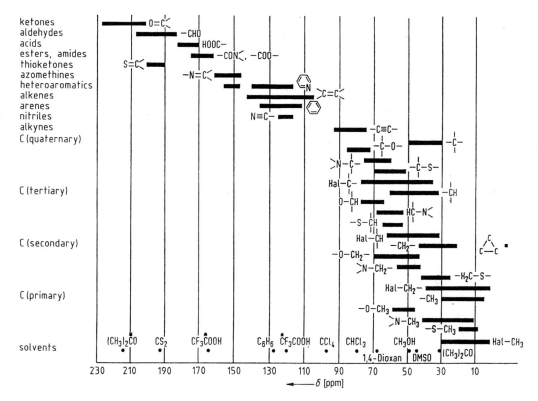

Figure 2-2.
Chemical shifts of ^{13}C nuclei in organic compounds.

theless made to apply it to small molecules, but it was found to give values that were too large. For the hydrogen molecule H_2, for example, the calculation gave a value of $\sigma_{dia} = 32.1 \times 10^{-6}$, compared with the observed value of 26.6×10^{-6}.

A second term, called the *paramagnetic shielding term* σ_{para}, aims to correct these discrepancies by taking into account the effect of the non-spherical charge distribution. (The name "paramagnetic" comes from the fact that σ_{para} has the opposite sign to σ_{dia}.) In order to be able to calculate the paramagnetic shielding term σ_{para} one needs to know, amongst other things, the wave functions of excited states, but these are only known in a few exceptional cases. Exact calculations have therefore only been possible up to now for small molecules such as H_2 and LiH. However, the theory provides an important result which enables us to understand some basic differences between ^1H and ^{13}C NMR spectroscopies, namely that σ_{para} is inversely proportional to ΔE, the average electronic excitation energy:

$$\sigma_{para} \propto \Delta E^{-1} \qquad (2-1)$$

In other words this means that the smaller ΔE is, the greater is the contribution of σ_{para} to the shielding. For the hydrogen atom – even when it is bound – ΔE is very large, and consequently σ_{para} plays only a minor part in ^1H NMR spectroscopy. For heavy atoms, and even for ^{13}C, the situation is very different. These atoms have low-lying excited states, so that ΔE is small and σ_{para} becomes an important factor. However, σ_{para} is never greater in magnitude than σ_{dia}, and the net shielding therefore remains positive.

The charge distribution, and therefore also the shielding, are greatly influenced by substituents. Electronegative substituents reduce the shielding owing to their – I effect, while electropositive substituents cause a corresponding increase in shielding. Such substituent effects, and also mesomeric effects, will be examined in detail later for individual classes of compounds.

The shielding constants σ – and therefore also the chemical shifts – are in general *anisotropic quantities,* since the distribution of electron density in a molecule is not in general spherically symmetrical, and consequently σ (and also, therefore, the resonance frequency of the nucleus being observed) depends on the orientation of the molecule relative to the external magnetic field $\boldsymbol{B}_0$. However, as the molecules in a solution are in rapid tumbling motion (Brownian motion), this *chemical shift anisotropy* is effectively averaged out, and in the spectra one always obtains narrow peaks and averaged signal positions.

The situation in solids is quite different. The NMR peaks observed for solids are typically many kilohertz in width, and a significant proportion of this broadening is caused by chemical shift anisotropy. However, it can be shown theoretically that this effect is averaged to zero if the solid sample (e. g., in powder form) is rotated at a very high frequency (up to 50 kHz!) about an axis inclined at 54.7° to the direction of the external field $\boldsymbol{B}_0$. This angle of 54.7° ($=\cos^{-1}1/\sqrt{3}$) is called the "magic angle", and the technique is known as magic angle spinning (MAS). Provided that certain additional favorable conditions are satisfied (see the cited literature for details), this method yields quite narrow NMR signals, aided by the fact that the anisotropic dipole-dipole couplings between the nuclei are also averaged to zero, or at least reduced, by MAS. Thus one can, under favorable circumstances, obtain "high-resolution" NMR spectra of solids. More about the latest developments in solid-state NMR spectroscopy can be found in the review cited in Sections 2.6 and 13.4 under "Solid-State NMR".

2.1.2 Effects of Neighboring Groups

From the foregoing discussion we have:

$$\sigma = \sigma_{dia} + \sigma_{para} \qquad (2\text{-}2)$$

However, the two shielding terms σ_{dia} and σ_{para} are not sufficient to bring the calculated and measured shielding constants into agreement for large molecules. Further terms must be included to take into account the effects of neighboring groups in the molecule and the intermolecular contributions. The most important of these are found to be:

- the contribution from the magnetic anisotropy of neighboring groups (σ_N)
- the ring current effect in arenes (σ_R)
- the electric field effect (σ_e)
- effects of intermolecular interactions (σ_i), e.g. hydrogen bonding and solvent effects.

Equation (2-2) is therefore replaced by:

$$\sigma = \sigma_{dia}^{local} + \sigma_{para}^{local} + \sigma_N + \sigma_R + \sigma_e + \sigma_i \qquad (2\text{-}3)$$

σ_{dia}^{local} and σ_{para}^{local} are essentially the same as σ_{dia} and σ_{para} in Equation (2-2), but they specifically exclude contributions other than those from electrons in the immediate vicinity of the nucleus under observation.

In ^1H NMR spectroscopy the most important additional contributions are σ_N and σ_R.

2.1.2.1 Magnetic Anisotropy of Neighboring Groups

Chemical bonds are in general magnetically anisotropic; they have different susceptibilities along the three directions (Cartesian axes) in space. Consequently the magnetic moments induced by an external magnetic field B_0 are not equal for different directions, and the shielding of a nucleus depends on its geometrical position in relation to the rest of the molecule.

For groups with an axially symmetric charge distribution the derivation of the theoretical relationships is straightforward. In this case there are two susceptibilities $\chi_\perp$ and $\chi_{||}$, perpendicular to and parallel to the bond axis respectively. The effect on the shielding for this case has been calculated by McConnell (Eq. (2-4)). He made the simplifying assumption of treating the induced magnetic moment in the field as a point dipole; he also took into account the fact that the dissolved molecules change their orientations.

$$\bar{\sigma}_N = \frac{1}{3 \, r^3 \, 4\pi} \, (\chi_\parallel - \chi_\perp) \, (1 - 3 \cos^2 \Theta) \qquad (2\text{-}4)$$

This gives $\bar{\sigma}_N$, the averaged contribution to the shielding of a nucleus, as a function of
- the distance r to the point dipole's center Z, and
- the angle Θ between the line joining this center to the observed nucleus and the axis A, which is the direction of the induced magnetic moment.

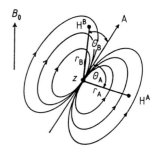

Figure 2-3.
Differential shielding of H^A and H^B in the dipolar field of a magnetically anisotropic neighboring group (McConnell model).

Figure 2-3 shows two hydrogen atoms H^A and H^B in a dipolar field. Whereas the external field at H^A is reinforced by the induced magnetic moment in the magnetically anisotropic neighboring group, the field at H^B is weakened.

From Equation (2-4) we can recognize two important facts:
- $\bar{\sigma}_N$ depends only on the geometry and the susceptibilities, and not on the nuclide being observed! Therefore the magnitude of the effect is the same for ^{1}H and ^{13}C nuclei.
- $\bar{\sigma}_N$ becomes zero when the angle $\Theta = 54.7°$! Thus we can define a double cone around the magnetically anisotropic group separating regions of positive and negative shielding contributions. At the surface of the double cone $\bar{\sigma}_N = 0$.

The classical example of a magnetically anisotropic group is acetylene. Figure 2-4a shows the double cone for this case and the signs of the shielding contributions.

Here the hydrogen atom lies on the molecular axis within the region of positive sign, and is additionally shielded. Consequently the signal of this proton in the ^{1}H NMR spectrum is found at relatively high field, at $\delta = 2.88$. On the basis of the electron density distribution alone, one would have expected a smaller shielding than for the protons in ethylene, which have a δ value of 5.28. Thus, the magnetic anisotropy of the CC triple bond explains one of the more striking observations of ^{1}H NMR spectroscopy (Figure 2-1).

Although the electron density distribution in the CC and CO double bonds is no longer axially symmetric, anisotropy cones have also been calculated for these groups (Figure 2-4 b and c). The magnetic anisotropy of the carbonyl group explains qualitatively why aldehydic protons ($\delta \approx 9$ to 10) are so weakly shielded: the aldehydic proton is in the region of the cone for which the shielding contribution is negative.

For cyclohexane derivatives with a sterically fixed chair conformation, chemical shift differences of 0.1 to 0.7 ppm between the axial and equatorial protons are found, the axial protons being more strongly shielded than the equatorial. On the basis of observations such as these a magnetic anisotropy is also ascribed to the CC single bond (Figure 2-4 d). It can be seen from Figure 2-4 that in all these types of molecules, with the exception of acetylene, the regions with a negative shielding contribution lie along the direction of the bond axis.

The anisotropy effect is independent of the nuclide being observed, as we have already seen above. The shifts for protons and ^{13}C nuclei are therefore equal in magnitude; they amount to several ppm. However, since the total range of chemical shifts in ^{13}C NMR spectroscopy is much greater than for protons, the fraction attributable to magnetic anisotropy is smaller – although not negligible.

The ^{13}C resonances of the triple-bonded carbon atoms in acetylene and its derivatives are found – as in the ^{1}H NMR spectra – between those of the alkanes and the alkenes. The magnetic anisotropy of the triple bond is, of course, only partly responsible for the large shielding in alkynes; a significant proportion of it may be attributed to the higher excitation energy ΔE in alkynes compared with alkenes (see Section 2.1.1). Carbon atoms directly bonded to a triply-bonded carbon atom show an additional shielding. This corresponds to similar effects found in the ^{1}H NMR spectra.

2.1.2.2 Ring Current Effects

The ^{1}H NMR signal of the protons in benzene is found at $\delta = 7.27$, and that of ethylene at $\delta = 5.28$. Furthermore this is not an isolated case, but applies quite generally: protons in arenes are less shielded than those in alkenes.

This effect can be explained in terms of an induced *ring current*, which is set up when the molecule with its delocalized π-electrons is placed in a magnetic field. The ring current in turn generates an additional magnetic field, whose lines of force at the center of the arene ring are in the opposite direction to the external magnetic field B_0 (Figure 2-5). This again leads to regions of increased and reduced shielding in the vicinity of the arene molecule. Hydrogen atoms directly attached to the arene ring are in a position where the lines of force increase the B_0 field, i. e. in the region of reduced shielding.

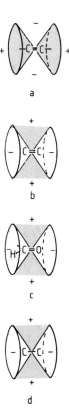

Figure 2-4.
Shielding contribution due to the magnetic anisotropy of the CC triple bond (a), the CC and CO double bonds (b and c), and the CC single bond (d).
+: increased shielding zone;
−: reduced shielding zone.

Figure 2-5.
Sketch showing the ring current effect in arenes, with zones of increased (+) and reduced (−) shielding.

The ring current effect is greatest when the plane of the benzene ring is perpendicular to the field direction, as shown in Figure 2-5. It is zero when the molecule is orientated with one of its in-plane axes parallel to the field, i. e. when the magnetic field lines do not pass through the ring. Experimentally one always measures averaged values, as the molecules in solution are in rapid motion. (The ring current model is only one of the possible models for interpreting the experimental results [2].)

A few examples from ^1H NMR spectroscopy will show how large the ring current effect on chemical shifts can be:

○ In 1,4-decamethylenebenzene (**1**) it is found that the CH_2 protons in the middle of the chain are more strongly shielded ($\delta \approx 0.8$) than those directly adjacent to the benzene ring ($\delta = 2.6$). It is true that the chemical shift of the latter is influenced by the inductive effect of the ring, but this alone cannot account for the difference.

1

In large unsaturated ring systems where the number of π-electrons satisfies the Hückel rules ($4n + 2$), effects are found which again indicate the existence of a ring current. Two examples of this are:

○ In the [18]-annulene (**2**) the six inner hydrogen atoms project into the region where the shielding contribution is positive, whereas the twelve outer ones are in the region of negative shielding contribution. In the spectrum (Figure 11-5, Chapter 11) we find one signal at $\delta = -1.8$ and a second signal of twice the intensity at $\delta = 8.9$. Comparing these values with the δ-value of 5.7 for the protons of the non-planar, non-aromatic molecule cyclooctatetraene (**3**), the effect of the ring current is clearly seen.

○ In *trans*-10b,10c-dimethyldihydropyrene (**4**) the signal of the methyl protons is found at $\delta = -4.25$! In the absence of the specific shielding effect of the ring current one would have expected a signal at about $\delta \approx 1$.

There are many examples showing how useful the ring current model is. Attempts have even been made to use it as a means of defining aromaticity.

In ^{13}C NMR spectroscopy the ring current effect is less important, as it contributes only a few percent to the total shielding in this case. For benzene this can be explained, though in a greatly over-simplified way, by the fact that the carbon atoms are on the "current loop", which is just where the induced field is zero.

2

$\delta(H) = 5.7$

3

$\delta(CH_3) = -4.25$

4

2.1.2.3 Electric Field Effects

In a molecule containing polar groups such as a carbonyl or nitro group an intramolecular electric field exists. This field influences the electron density distribution in the molecule and therefore the magnetic shielding of the nuclei. From this point of view the shifts of ^{1}H and ^{13}C resonances which occur on protonation – e.g. of amines – are also a result of electric field effects.

2.1.2.4 Intermolecular Interactions – Hydrogen Bonding and Solvent Effects

Hydrogen bonding: Protons attached to oxygen by hydrogen bonds have particularly low shielding values, the δ-values being greater than 10 in many cases. However, exact chemical shift values cannot be specified, as the signal positions depend on the temperature and the concentration. This is also the case for exchangeable protons in NH and SH groups.

The protons bonded to the oxygen atom in enols have extremely low shieldings. For acetylacetone in the enol form with chloroform as solvent the signal of the OH proton is found at $\delta = 15.5$ (!). This extreme downfield shift is probably caused by the electron withdrawing effect of the two oxygen atoms (see Section 11.3.6 and Figure 11-6).

Solvent effects: If the dissolved substance interacts with the solvent this shows up as a shift in the positions of the signals. Such effects are clearly seen when the spectrum of a substance is measured in turn in a non-polar solvent (CCl_4), in a polar solvent ([D$_6$]-DMSO), and in a magnetically anisotropic solvent ([D$_6$]-benzene). However, it is not possible to make generalizations.

On the basis of such effects, solvents can be used as an aid to assigning signals. This topic will be treated in Sections 6.2.4 and 6.3.5. The special interaction effects between chiral solutes and chiral solvents will be discussed in detail in Chapter 12.

2.1.2.5 Isotope Effects [3]

Figure 1-28 (Section 1.6.2.9) shows the ^{1}H resonances of ^{12}CHCl$_3$ and ^{13}CHCl$_3$: a singlet at $\delta = 7.24$ for ^{12}CHCl$_3$ and a doublet – the so-called ^{13}C satellites – for ^{13}CHCl$_3$. The center of the doublet is shifted by about 0.7 Hz to the right (i.e. to higher shielding) relative to the main signal. This shift is caused by the *isotope effect.*

Replacing ^{12}C by ^{13}C causes only a small shift of the ^1H resonances. Effects that are much greater and more commonly observed occur in the ^{13}C NMR spectra of deuterated compounds. Figure 2-6 shows the effect on the ^{13}C resonances of benzene when all the hydrogen atoms are replaced by deuterium. The figure shows the $^{13}C\{^1H\}$ NMR spectrum of a mixture consisting of about 10 % C_6H_6 and 90 % C_6D_6. For C_6H_6 we obtain a singlet at $\delta = 128.53$, and for C_6D_6 a triplet, as the C,D coupling is not eliminated by the ^1H broad-band decoupling. The middle line of the triplet is shifted relative to the singlet by 33.3 Hz (with an observing frequency of 62.89 MHz), or 0.53 ppm! The average shift caused by a directly bonded deuterium atom is about 0.25 ppm, and that caused by a geminal deuterium atom is about 0.1 ppm.

Here again, as in chloroform, replacement of the lighter isotope by the heavier isotope results in an increased shielding.

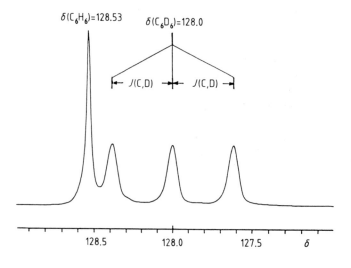

$\delta(C_6H_6)=128.53$ $\delta(C_6D_6)=128.0$

$J(C,D)$ $J(C,D)$

128.5   128.0   127.5 δ

Figure 2-6.
62.9 MHz ^{13}C {^1H} NMR spectrum of a mixture of about 10 % C_6H_6 and 90 % C_6D_6. For C_6D_6 a triplet is obtained because of the C,D coupling. Isotope shift: 33.3 Hz (0.53 ppm); $^1J(C,D) = 24.55$ Hz.

2.1.3 Summary

^1H NMR: The chemical shifts of protons are determined mainly by the diamagnetic shielding term σ_{dia}. The paramagnetic shielding term σ_{para} is only a correction term of relatively minor importance. Substituents, magnetic anisotropy of neighboring groups, ring current effects and electric field effects,

hydrogen bonding, and intermolecular interactions with solvents and complexing agents influence the shielding in specific ways. All these effects contribute to the differences that are found between the resonance frequencies of protons with different bonding.

^{13}C NMR: The positions of the resonances for ^{13}C and heavier nuclides are determined mainly by σ_{para}. Intra- and intermolecular effects, expressed in ppm, are similar in magnitude to those in ^{1}H NMR spectroscopy, and when considered in relation to the total shifts they are therefore less important.

The ^{1}H and ^{13}C chemical shifts of important classes of compounds will first be described, followed by a discussion of how spectra are affected by symmetry, equivalence and chirality. This chapter will conclude with a look at the chemical shifts of some "other" nuclides.

2.2 ^{1}H Chemical Shifts of Organic Compounds

What relationships exist between the chemical shifts and the structure of a molecule?

As we saw in Section 2.1, exact theoretical predictions are rarely possible. The spectroscopist must therefore proceed with interpretation in an empirical way, relying either on his own experience or on published results – textbooks, published tables, spectra catalogs, and more recently on computerized spectral data. As an aid to getting started, an account will be given below of the chemical shifts of some classes of compounds and – so far as possible – of their special characteristics.

Experience teaches that the signals of more than 95 % of the protons in organic molecules lie within the narrow range of $\delta = 0$ to 10. Characteristic ranges for different functional groups are shown in Figure 2-1, on a δ scale relative to the tetramethylsilane reference signal (δ (TMS) = 0). Definite limits for the individual groups cannot be given, however. Also the ranges often overlap. For example, signals of olefinic protons occasionally appear in the range for aromatic signals, or aliphatic proton signals in the range for olefinic protons. Although the overall spread of about 10 ppm for the proton resonances of nearly the entire range of different types of molecules may seem very small, one must appreciate that the chemical shifts (δ-values) can be determined with a precision of 0.01 ppm or better.

2.2.1 Alkanes and Cycloalkanes

n-Alkanes: The main influence on the chemical shifts of protons in alkanes is that due to substituents. The signals of substituted alkanes are distributed over a wide range. Table 2-1 lists a few values for methane derivatives. Of the protons in methyl halides, the most strongly shielded are those of CH_3I, while the least shielded are those of CH_3F. One may suspect that there is a relationship here between the chemical shift and the electronegativity of the substituent. This idea is supported by the values for the series $C-CH_3$, $N-CH_3$, $O-CH_3$: with increasing electronegativity of the substituents the shielding becomes smaller, i. e. the δ-value increases. The strongly electronegative nitro group shifts the methyl signal down as far as $\delta = 4.33$. In organometallic compounds such as $LiCH_3$, the protons are particularly strongly shielded as a result of the electropositive character of this element. The reference substance TMS also belongs to this class of compounds.

All these examples are generally considered in terms of *inductive substituent effects.* The effects decrease with the distance between the substituents and the observed protons, as is shown by the following sequence:

	CH_3Cl	$CH_3\text{-}CH_2\text{-}Cl$	$CH_3\text{-}CH_2\text{-}CH_2\text{-}Cl$
δ:	3.05	1.42	1.04

For multiple substitution the influence of each additional substituent is slightly less:

	CH_4	CH_3Cl	CH_2Cl_2	$CHCl_3$
δ:	0.23	3.05	5.33	7.26
$\Delta\delta$:		2.82	2.28	1.93

For assigning ^1H NMR signals, δ-values are in practice often estimated by using empirical rules. Out of the many such rules that exist, only two will be mentioned here (for some others see Section 6.2.2):

- For a given substituent X, methyl protons are usually more strongly shielded than methylene protons, and these in turn are more strongly shielded than methine protons, e. g.

	$(CH_3)_2CHCl$	CH_3CH_2Cl	CH_3Cl
δ:	4.13	3.51	3.05

- For methylene group protons $X\text{-}CH_2\text{-}Y$ with two substituents X and Y the chemical shift can be estimated to a good approximation using *Shoolery's rule:*

$$\delta = 0.23 + S_x + S_y \qquad (2\text{-}5)$$

Table 2-1.
^1H chemical shifts δ [ppm] of methyl protons for different substituents X.

X	$\delta(X-CH_3)$	$E_X^{a)}$
Li	−1	1.0
R_3Si	0	1.8 (Si)
H	0.4	2.1
CH_3	0.8	2.5 (C)
NH_2	2.36	3.0 (N)
OH	3.38	3.5 (O)
I	2.16	2.5
Br	2.70	2.8
Cl	3.05	3.0
F	4.25	4.0
COOH	2.08	
NO_2	4.33	

[a] E_X: electronegativities according to Pauling [4]

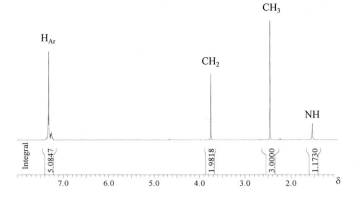

Figure 2-7.
300 MHz ^{1}H NMR spectrum of
N-benzylmethylamine (**5**) in CDCl$_3$
with integrated intensities shown
below.

Here S_x and S_y are effective shielding constants whose values for some substituents are listed in Table 6-1 (Section 6.2.2.1).

○ As an example we will assume that it is desired to calculate the chemical shift of the methylene protons in N-benzylmethylamine (**5**):

$$S_{\text{phenyl}} = 1.83$$
$$\underline{S_{\text{NHR}} = 1.57}$$
$$\Sigma S = 3.40$$

$$\delta\,(CH_2) = 0.23 + 3.40 = 3.63$$

Comparing this with the observed δ-value of about 3.75 (Figure 2-7) we find good agreement between the estimate and the experimental result. The remaining three signals in Figure 2-7 can be easily assigned on the basis of the signal intensity ratio (5:2:3:1).

Shoolery's rule can also be applied to trisubstituted methanes provided that appropriate care is taken.

Cycloalkanes: In cycloalkanes the chemical shifts of the protons depend on the size of the ring, on the conformational mobility, and on steric effects. In alkyl substituted cycloalkanes the steric effect predominates over all others. The δ-values for a few small-ring cycloalkanes are given in Table 2-2.

The most striking feature is the high shielding of the protons in cyclopropane, with the value $\delta = 0.22$! This is accounted for by the diamagnetic anisotropy of the cyclopropane ring and is therefore also the case for substituted cyclopropanes.

When the spectrum of an unknown sample is found to contain signals in the neighborhood of the TMS signal, a three-membered ring is immediately suspected. Other compounds with signals in this region, such as metal alkyl compounds, can in most cases be easily ruled out on the basis of the sample's chemical history.

From cycloheptane up to the larger rings the δ-values are nearly constant, varying only by tenths of a ppm.

Table 2-2.
^{1}H chemical shifts δ [ppm] of cycloalkanes.

Compound	δ
Cyclopropane	0.22
Cyclobutane	1.94
Cyclopentane	1.51
Cyclohexane	1.44
Cycloheptane	1.54
Cyclooctane	1.54

2.2.2 Alkenes

Table 2-3 lists δ-values for the olefinic protons in ethylene derivatives, which range from $\delta = 4$ to 7.5. The effects of substituents can be inductive, mesomeric or steric. A method for estimating the chemical shifts using empirically determined substituent increments will be described later, in Section 6.2.2.2, by giving examples of its application.

2.2.3 Arenes

In aromatic compounds the shielding is determined mainly by the mesomeric effects of the substituents. Thus in aniline (**6**) the protons in the *ortho* and *para* positions are more strongly shielded than the *meta* (Figure 2-8). Scheme I illustrates this by showing the mesomeric limiting structures.

Furthermore, all the protons here are more strongly shielded than those in benzene ($\delta = 7.27$). Evidently the amino group increases the electron density in the ring through the $+ M$ effect – especially at the *ortho* and *para* positions in this case.

For nitrobenzene (**7**) on the other hand, all the signals appear at greater δ-values; the protons are less shielded than the

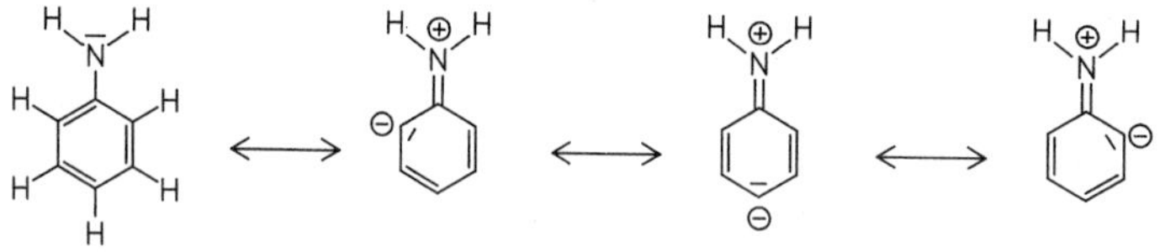

Table 2-3.
^{1}H chemical shifts δ [ppm] of monosubstituted ethylenes; δ (ethylene) = 5.28 ppm.

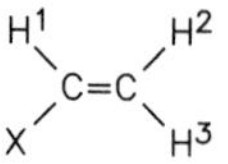

$$\begin{array}{c} H^1 \qquad H^2 \\ \diagdown C=C \diagup \\ X \diagup \qquad \diagdown H^3 \end{array}$$

X	δ (H^1) (gem)	δ (H^2) (trans)	δ (H^3) (cis)
CH$_3$	5.73	4.88	4.97
C$_6$H$_5$	6.72	5.20	5.72
F	6.17	4.03	4.37
Cl	6.26	5.39	5.48
Br	6.44	5.97	5.84
I	6.53	6.23	6.57
OCH$_3$	6.44	3.88	4.03
OCOCH$_3$	7.28	4.56	4.88
NO$_2$	7.12	5.87	6.55

6

Scheme I

H–2,6

H–3,5

H–4

7.2   7.0   6.8   6.6   6.4 δ

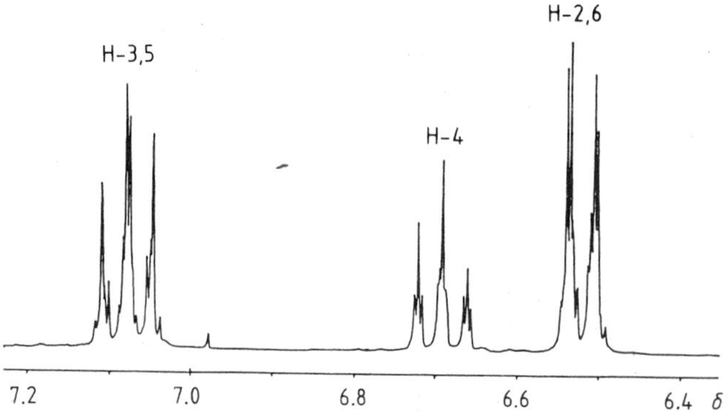

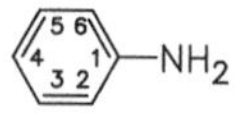

6

Figure 2-8.
Portion of the 250 MHz ^{1}H NMR spectrum of aniline (**6**) in CDCl$_3$ (δ (NH$_2$) = 3.45 ppm).

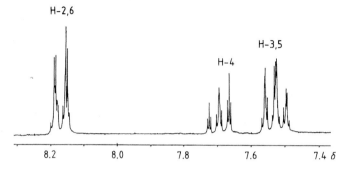

7

Scheme II

H-2,6

H-3,5

H-4

8.2   8.0   7.8   7.6   7.4 δ

Figure 2-9.
250 MHz ^{1}H NMR spectrum of nitrobenzene (**7**) in CDCl$_3$.

protons in benzene, as the nitro group withdraws electrons from the ring (− M effect). Those most strongly affected are again the *ortho* protons, with a δ-value of 8.17. These are followed by the *para* proton at δ = 7.69 and the *meta* protons at δ = 7.53 (Figure 2-9). The mesomeric limiting structures (Scheme II) again give a qualitatively correct prediction of this result.

It has been found experimentally that for multiple substitution the substituents give nearly constant additive contributions to the chemical shifts of the remaining ring protons. For many substituents these contributions have been determined from the experimental data in the form of incremental constants. These can be used, together with the δ-value for benzene (7.27), to estimate the chemical shifts of the aromatic protons for different benzene derivatives. The necessary formula and incremental constants are given in Section 6.2.2.3 together with an example of their application.

2.2.4 Alkynes

The special position of acetylene has already been discussed in detail in Section 2.1.2.1, where the cause of the unexpectedly high shielding was explained. Unfortunately the region where the chemical shifts of acetylenic protons are found (δ ≈ 2 to 3)

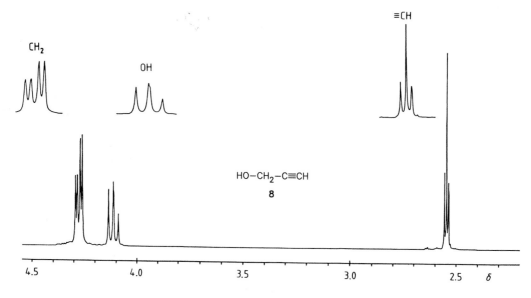

Figure 2-10.
250 MHz ^{1}H NMR spectrum of propynol (**8**). The multiplets are shown expanded by the same factor in each case. ^4J(H,H) = 2.4 Hz; ^3J(H,OH) = 5.8 Hz.

overlaps with those of many other types of protons, especially that of substituted alkanes. Consequently a signal in this region does not provide unambiguous evidence for protons next to a triple bond. The coupling multiplet pattern and the magnitudes of the coupling constants may serve as further aids to assignment, since acetylenic proton signals can only be split by long-range couplings. An example is provided by the spectrum of propynol (**8**) (Figure 2-10).

The chemical shifts of acetylenic protons in individual cases depend on the electronegativities of the substituents, on the conjugation, and – to an especially large extent in alkynes – on the solvent.

For example, an alkyl group increases the shielding, while an aryl group reduces it:

$$HC \equiv CH \qquad HC \equiv C - CH_3 \qquad HC \equiv C - C_6H_5$$
$$\delta: \quad 2.36 \qquad\qquad 1.8 \qquad\qquad\qquad 3.0$$

2.2.5 Aldehydes

The signals of aldehydic protons R**CHO** can be recognized immediately from their characteristic position in the region $\delta = 9$ to 11. Whenever signals consisting of simple multiplets

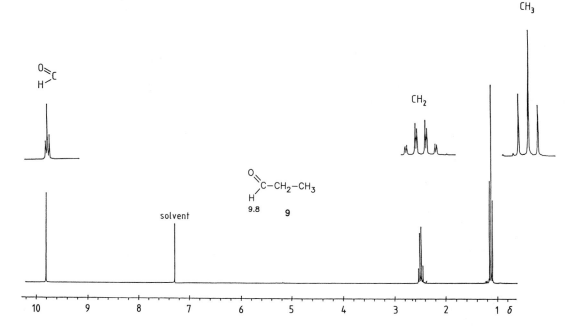

Figure 2-11.
250 MHz ^{1}H NMR spectrum of propionaldehyde (**9**) in CDCl$_3$. The multiplets are shown expanded by the same factor in each case. $J(H^1,H^2) = 1.4$ Hz.

appear in this part of the spectrum of an unknown substance, an aldehyde is at once suspected.

For propionaldehyde (**9**), for example, we find a triplet at $\delta \approx 9.8$ (Figure 2-11).

The substituent effects in this case are not large. Thus, the signal of the aldehydic proton of acetaldehyde (**10**) is at $\delta = 9.8$, while that of crotonaldehyde (**11**) is at $\delta = 9.48$, and that of benzaldehyde (**12**) at $\delta = 10.0$. Even conjugation with a CC double bond or with a phenyl ring does not cause any great shifts, as these examples show.

2.2.6 OH, SH, NH

The positions of the resonances of protons in OH, SH and NH groups are subject to considerable variation. These hydrogen atoms can form hydrogen bonds, they can undergo exchange, and they have varying degrees of acidic character. Their chemical shifts are further influenced by concentration, temperature, solvent, and by impurities such as water. The

59

measured δ-values are only reproducible under well-defined experimental conditions. Consequently the chemical shift ranges listed in Table 2-4 for a few particular classes of compounds can be used for signal assignment only with great care. Some generalizations are possible from these values, however: OH signals may be found anywhere in the spectrum, whereas SH and NH resonances usually fall within narrower regions.

The spectra of two alcohols have already been encountered: that of benzyl alcohol in Figure 1-24, and that of propynol in Figure 2-10. For benzyl alcohol under the measurement conditions used the OH signal is at $\delta = 1.8$; it is split into a triplet by its coupling with the methylene protons. The OH signal of propynol is found at $\delta = 4.11$. This signal too is split into a triplet by coupling with the methylene protons. OH signals are more usually singlets, however, and these are often broadened. This is caused by the exchange of the hydrogen atoms under observation (Section 11.3.7). Where several exchangeable hydrogen atoms are present in the molecule, intra- and intermolecular exchange processes lead to averaged signals.

In amides the NH protons do not usually exchange so readily, with the result that even couplings to vicinal protons can be observed.

The exchange of hydrogen atoms with deuterium is of great practical importance, as the corresponding signals in the ^{1}H NMR spectrum disappear following the H-D exchange. This effect is used as an aid to assignment (Section 6.2.5).

Table 2-4.
Chemical shift ranges for OH, NH and SH protons.

Compound types		δ (H) [ppm]
– OH:	Alcohols	1 – 5
	Phenols	4 – 10
	Acids	9 – 13
	Enols	10 – 17
– NH:	Amines	1 – 5
	Amides	5 – 6.5
	Amido groups in peptides	7 – 10
– SH:	Thiols	
	aliph.	1 – 2.5
	arom.	3 – 4

2.3 ^{13}C Chemical Shifts of Organic Compounds

Figure 2-2 gives an impression of the regions in the spectrum where one can expect to find the ^{13}C resonance signals of the carbon atoms in organic molecules. As in ^{1}H NMR spectroscopy, the reference substance is TMS. It should be noticed that ^{13}C resonances extend over a range of 200 ppm, which is about twenty times greater than that for ^{1}H resonances. Consequently, for the same line width one can expect to obtain better separation of the signals in ^{13}C spectroscopy than in ^{1}H spectroscopy. Owing to the method of observation used, the chemical shifts are usually the only parameters that can be obtained from the ^{13}C NMR spectrum. A knowledge of the relationships between chemical shifts and molecular structure is therefore even more important in ^{13}C spectroscopy than in ^{1}H spectroscopy.

Often the spectrum contains simply *one* signal for each carbon atom in the molecule, and it is thus necessary to assign each signal to the correct carbon atom. In the following sections the chemical shifts of some selected classes of compounds will be discussed by considering a few examples. More detailed information on aids to assignment and special techniques will be given in Section 6.3. There we shall also discuss how chemical shifts can be predicted using empirical correlations.

The δ-values listed in Tables 2-5 to 2-14 are taken from the extensive collection of data published by Kalinowski, Berger and Braun [5].

2.3.1 Alkanes and Cycloalkanes

Whereas the ^{1}H NMR spectra of hydrocarbons consist of broad bands or multiplets in the narrow range $\delta = 0.8$ to 2, which are often difficult or impossible to assign, their ^{13}C resonances extend over 50–60 ppm, with the result that usually a separate signal appears for each carbon atom, unless the molecule has a symmetrical structure. These characteristics can be seen on comparing the 300 MHz ^{1}H NMR spectrum and the

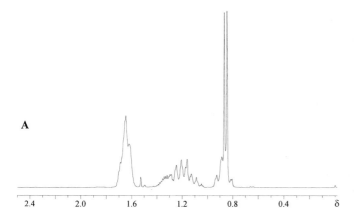

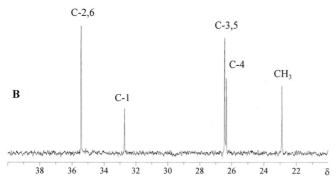

Figure 2-12.
A: 300 MHz ^{1}H NMR spectrum of methylcyclohexane (**13**) in CDCl$_3$.
B: 75 MHz ^{13}C NMR spectrum of **13**.

61

75 MHz ^{13}C NMR spectrum of methylcyclohexane (**13**) in Figure 2-12. In Chapters 8 and 9 we will learn about methods that can be used to arrive at unambiguous assignments of the signals in both the ^{1}H and the ^{13}C spectra.

Alkanes: In alkanes the chemical shift of a particular ^{13}C nucleus depends on the number of neighboring carbon atoms at the α and β positions and on the degree of branching. Some representative δ-values are given in Table 2-5.

The chemical shifts can be predicted very satisfactorily using the relationships derived by Grant and Paul and by Lindeman and Adams (Section 6.3.2).

Substituents have a considerable influence on the chemical shifts. This is shown in Table 2-6 where a number of propane derivatives are taken as examples. It is seen that merely replacing a hydrogen atom by a methyl group reduces the shieldings of the α- and β-^{13}C nuclei by 8.8 and 8.6 ppm respectively, while in contrast the shielding of the γ-^{13}C nucleus is increased by 3 ppm (referred to as the α-, β- and γ-effects respectively). The α-effect increases with increasing electronegativity of the substituent (Table 2-6) – fluorine induces a shift of nearly 70 ppm! For chloro- and bromopropane too there is a clear relationship to the electronegativity. The effect of iodine is an exception to the pattern. Here the large number of electrons in the iodine atom and the spin-orbit coupling [6] apparently influence the diamagnetic shielding term of the directly bonded ^{13}C nucleus; this "heavy atom effect" increases where there is multiple substitution. Thus for CI$_4$ we find $\delta = -292.2$, which is up to now the largest upfield shift found for a neutral molecule.

The β-effect is much smaller, and is always a deshielding effect. However, there appears to be no direct relationship to the electronegativity. The γ-effect causes the ^{13}C nuclei concerned – those of the CH$_3$ group in our example – to be more strongly shielded. Such an increase in shielding is found for almost all substituents in acyclic alkanes. This γ-effect is attributed to steric interactions.

In Section 6.3.2.1 we shall see how substituent increments can be used to predict the chemical shifts of the ^{13}C nuclei at the α-, β- and γ-positions:

Cycloalkanes: Table 2-7 gives the δ-values for cyclopropane up to cycloheptane. The ^{13}C nuclei in cyclopropane have an exceptionally large shielding, which is an effect similar to that found in ^{1}H NMR spectroscopy. For all other rings from the five-membered up to the largest the ^{13}C resonances are within a very narrow range between about $\delta = 24$ and $\delta = 29$. Only for the strained ring cyclobutane does the shielding remain somewhat greater than that for the larger rings.

Substituent effects have been studied in particular detail for the cyclohexane system, especially for alkyl substitution [7].

Table 2-5.
^{13}C chemical shifts δ [ppm] of alkanes.

Compound	$\delta\,(C^1)$	$\delta\,(C^2)$
CH$_4$	-2.3	
H$_3$C$-$CH$_3$	6.5	
CH$_2$(CH$_3$)$_2$	16.1	16.3
(H$_3$C$-$CH$_2$)$_2$	13.1	24.9
CH(CH$_3$)$_3$	24.6	23.3
C(CH$_3$)$_4$	27.4	31.4

Table 2-6.
^{13}C chemical shifts δ [ppm] of propane derivatives
$XC^\alpha H_2-C^\beta H_2-C^\gamma H_3$

X	$\delta\,(C^\alpha)$	$\delta\,(C^\beta)$	$\delta\,(C^\gamma)$
H	16.1	16.3	16.1
CH$_3$	24.9	24.9	13.1
NH$_2$	44.6	27.4	11.5
OH	64.9	26.9	11.8
NO$_2$	77.4	21.2	10.8
F	85.2	23.6	9.2
Cl	46.7	26.0	11.5
Br	35.4	26.1	12.7
I	9.0	26.8	15.2

Table 2-7.
^{13}C chemical shifts δ [ppm] of cycloalkanes.

Compound	δ
Cyclopropane	-2.8
Cyclobutane	22.4
Cyclopentane	25.8
Cyclohexane	27.0
Cycloheptane	28.7

2.3.2 Alkenes

The ^{13}C resonances of double-bonded carbon nuclei are found in a broad range from $\delta \approx 100$ to 150. The chemical shifts given in Tables 2-8 and 2-9 for some ethylene derivatives show how these are affected by alkyl substituents and by substituents with widely differing inductive and mesomeric properties.

From the few values given in Table 2-8 we can already see that for alkyl-substituted ethylenes the olefinic ^{13}C nuclei carrying the substituents are less shielded ($\delta \approx 120$ to 140) than those at the terminal position ($\delta \approx 105$ to 120). The presence of a second alkyl group in geminal dialkyl derivatives further increases this effect ($\delta > 140$ in most cases). Scarcely any difference is found between E and Z isomers.

Table 2-8.

^{13}C chemical shifts δ [ppm] of alkenes.

Compound	δ (C^1)	δ (C^2)	δ (C^3)
$H_2C^1 = C^2H_2$	123.5		
$H_3C^3C^1H = C^2H_2$	133.4	115.9	19.9
$H_3CCH = CHCH_3$ *(cis)*	124.2		11.4
$H_3CCH = CHCH_3$ *(trans)*	125.4		16.8
$(H_3C)_2C = CH_2$	141.8	111.3	24.2
Cyclohex-1-ene	127.4		25.4
			(C^4: 23.0)

The large effects that substituents can cause are seen in Table 2-9. With the exceptions of Br, CN and I, the substituent effects reduce the C^1 shielding and increase the C^2 shielding compared with ethylene. Thus, in contrast to the alkanes the substituent effect alternates in this case. The large influence on the C^1 shielding can be qualitatively accounted for by the inductive substituent effect. In addition one expects a contribution from the mesomeric effect.

The very large β-effects of CH$_3$O and CH$_3$COO as substituents can best be understood by considering the mesomeric limiting structures (**14**). These show that the mesomeric effect increases the electron density at the C^2 atom and therefore also the nuclear shielding.

$$H_3C - \overset{\oplus}{\underline{O}} - C^1H = C^2H_2 \longleftrightarrow H_3C - \underline{O} = C^1H - \overset{\ominus}{\underline{C}^2H_2}$$

δ: 52.5   153.2   84.1

14

Table 2-9.

^{13}C chemical shifts δ [ppm] of monosubstituted ethylenes.

$$\begin{array}{c} H \\ \diagdown \\ X \end{array} \overset{1}{C} = \overset{2}{C} \begin{array}{c} H \\ \diagup \\ \diagdown H \end{array}$$

X	δ (C^1)	δ (C^2)
H	123.5	123.5
CH$_3$	133.4	115.9
CH = CH$_2$	137.2	116.6
C$_6$H$_5$	137.0	113.2
F	148.2	89.0
Cl	125.9	117.2
Br	115.6	122.1
I	85.2	130.3
OCH$_3$	153.2	84.1
OCOCH$_3$	141.7	96.4
NO$_2$	145.6	122.4
CN	108.2	137.5

Iodine behaves anomalously as a substituent owing to the heavy atom effect.

In Section 6.3.2.2 we shall see how empirically determined substituent increments can be used to predict the chemical shifts of substituted olefins.

2.3.3 Arenes

The ^{13}C signals of benzene, alkyl-substituted benzenes, fused ring arenes and annulenes are found within the comparatively narrow range of $\delta \approx 120$ to 140. This is illustrated by the examples shown in Scheme III.

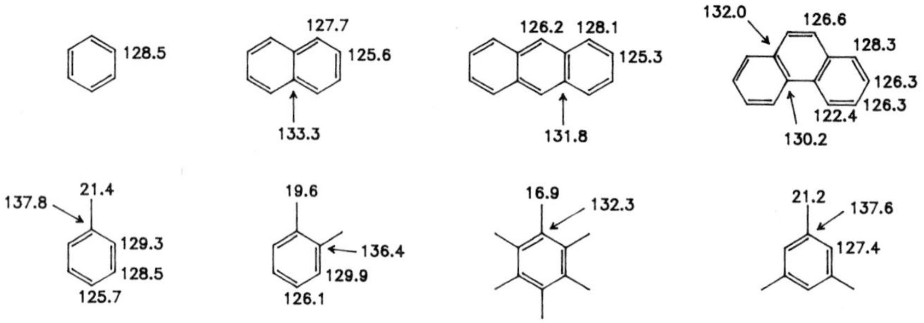

Scheme III

Introducing substituents extends the range of chemical shifts, and signals of substituted arenes can be expected anywhere within the range $\delta = 100$ to 150. As we have already seen, the ^{13}C resonances of alkenes also lie within this region, and this can sometimes cause problems in assignments. Table 2-10 lists the ^{13}C resonances of the ring carbons for some monosubstituted benzenes.

Table 2-10.
^{13}C chemical shifts δ [ppm] of monosubstituted benzenes.

X	$\delta\,(C^1)$	$\delta\,(C^2)$	$\delta\,(C^3)$	$\delta\,(C^4)$
H	128.5			
Li	186.6	143.7	124.7	133.9
CH$_3$	137.7	129.2	128.4	125.4
COOH	130.6	130.1	128.4	133.7
F	163.3	115.5	131.1	124.1
OH	155.4	115.7	129.9	121.1
NH$_2$	146.7	115.1	129.3	118.5
NO$_2$	148.4	123.6	129.4	134.6
I	94.4	137.4	131.1	127.4

Comparing these with the δ-value for benzene, the extreme cases in the table are phenyllithium ($\delta = 186.6$) with a very large downfield shift of the C^1 resonance relative to benzene, and iodobenzene ($\delta = 94.4$) with a large upfield shift of C^1. In all such compounds the ^{13}C resonance most strongly affected is that of the carbon atom bearing the substituent, followed by the $C^{2,6}$ (*ortho*) and C^4 (*para*) resonances; the $C^{3,5}$ (*meta*) resonances show the smallest effects. The shifts depend on the inductive and mesomeric properties of the substituents. (For substituents with appreciable $+ M$ or $- M$ effects, such as OH, NH_2 or NO_2, it is helpful to draw the mesomeric limiting structures.) In addition steric effects or anisotropy effects may play a part, and for iodine the heavy atom effect must again be taken into account.

It should be mentioned here that there have been many attempts, with varying degrees of success, to correlate such chemical shifts with theoretically calculated charge densities, or with substituent constants such as the Hammett and Taft constants [8, 9].

The large amount of experimental data that are available on benzene derivatives with two or more substituents have led to the conclusion that the substituent effects are, to a good approximation, additive. Based on this a convenient incremental system has been developed for predicting chemical shifts, and will be described in Section 6.3.2.3 together with a table of increments and examples of its use. Experience shows that good agreement is obtained between calculated and measured values, provided that there are not too many substituents and that they are not in positions *ortho* to each other.

Heteroaromatic compounds: In aromatic heterocycles the shieldings of the ring carbon nuclei are essentially determined by the heteroatom. Pyridine (**15**) will be considered as a typical representative of this class of compounds. Here the ^{13}C nuclei at the α and γ positions are less shielded than the carbons of benzene, whereas those at the β position are slightly more shielded. These observations are qualitatively accounted for by the electron density distribution and the mesomeric limiting structures shown in Scheme IV.

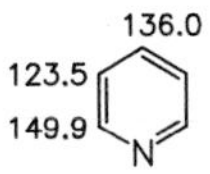

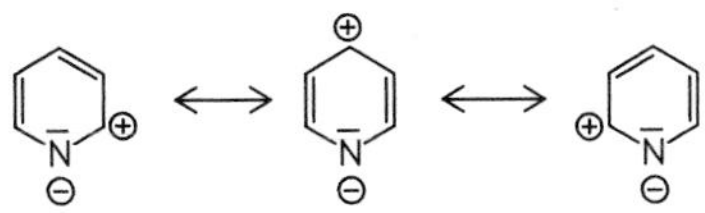

Scheme IV

For pyridine derivatives the chemical shifts can again be predicted using substituent increments, taking as the starting δ-values for the individual ^{13}C nuclei those measured for pyridine itself.

2.3.4 Alkynes

Comparing the measured δ-values for acetylene (71.9) and ethylene (123.5) we see that the ^{13}C nuclei in acetylene are remarkably highly shielded. The magnetic anisotropy of the CC triple bond, which accounted for the large shielding of the protons in alkynes (Section 2.2.4), is insufficient to explain the ^{13}C shielding. To find the main cause we must look at the paramagnetic shielding term σ_{para}, as it is chiefly this term which determines the shieldings of ^{13}C nuclei (Section 2.1.1). Since ΔE is greater for acetylene than for ethylene, this means that σ_{para} is smaller in magnitude for acetylene than for ethylene. However, σ_{para} is a negative correction term in Equation (2-2), and consequently the ^{13}C nuclei in acetylene are more highly shielded than those in ethylene. The same argument also holds for substituted alkynes (Table 2-11). The substituent effects are surprisingly large in some cases, e. g. for the ethoxy group.

In halogen-substituted acetylenes the heavy atom effect is particularly noticeable when the series is compared. In $C_4H_9-C^2 \equiv C^1-I$ the C^1 signal is found at $\delta = -3.3$, while that of C^2 is at 96.8.

Table 2-11.
^{13}C chemical shifts δ [ppm] of monosubstituted acetylenes.

$$H - \overset{1}{C} \equiv \overset{2}{C} - X$$

X	$\delta\,(C^1)$	$\delta\,(C^2)$
H-	71.9	
Alkyl-	68.6	84
H-C ≡ C-	64.7	68.8
Phenyl-	77.2	83.6
CH_3CH_2O-	23.4	89.6

2.3.5 Allenes

In allene (**16**) the most striking feature is the low shielding of the central carbon nucleus (C^2) with $\delta = 212.6$. The same effect is also observed in allene derivatives, whose central carbon nuclei have δ-values between 195 and 215. Conversely the two outer ^{13}C nuclei ($C^{1,3}$) have shieldings greater than those of the olefinic carbons in alkenes. The shielding values are influenced by substituents; replacing all four hydrogen atoms by methoxy groups (**17**) even reverses the relationship between the chemical shifts.

16

17

2.3.6 Carbonyl and Carboxy Compounds

The ^{13}C resonances of carbonyl groups (in ketones and aldehydes) and of carboxyl groups and derivatives thereof (in anhydrides, esters, acyl halides and amides) are found in the range from $\delta = 160$ to 220 (Fig. 2-2). In most cases these are quaternary carbon atoms, which in general give very low signal intensities owing to their long spin-lattice relaxation times T_1

(see Chapter 7). The low shielding values of the ^{13}C nuclei in these functional groups are mainly attributable to the paramagnetic shielding term.

2.3.6.1 Aldehydes and Ketones

The ^{13}C chemical shifts for a number of aldehydes and ketones are listed in Table 2-12. These examples show that the shielding values of the carbonyl ^{13}C nuclei in aldehydes and ketones are among the smallest found, being in the range from $\delta = 190$ to 220.

With increasing alkyl substitution the carbonyl shieldings are further reduced (i.e. δ increases). Di-t-butyl ketone with $\delta = 218.0$ and hexachloroacetone with $\delta = 175.5$ are two opposite extreme cases. Where there is conjugation with an unsaturated moiety such as a vinyl or phenyl group the shielding is increased. However, the substituent effects are much smaller than for alkanes.

Table 2-12.
^{13}C chemical shifts δ [ppm] of aldehydes and ketones.

Compound	δ (C^1)	δ (C^2)	δ (C^3)	δ (C^4)
H$_3$C^2-C^{1}HO	200.5	31.2		
H$_3$C-CH$_2$-CHO	202.7	36.7	5.2	
(CH$_3$)$_2$CH-CHO	204.6	41.1	15.5	
(CH$_3$)$_3$C-CHO	205.6	42.4	23.4	
H$_2$C = CH-CHO	193.3	136.0	136.4	
C$_6$H$_5$-CHO	191.0			
CH$_3$C^{2}OC^{1}H$_3$	30.7	206.7		
C^{4}H$_3$C^{3}H$_2$C^{2}OC^{1}H$_3$	27.5	206.3	35.2	7.0
(CH$_3$)$_2$CHCOCH$_3$	27.5	212.5	41.6	18.2
(CH$_3$)$_3$CCOCH$_3$	24.5	212.8	44.3	26.5
(CH$_3$)$_3$CC^{3}OC^2(CH$_3$)$_3$	28.6	45.6	218.0	
Ph-CO-Ph	195.2			
Cl$_3$C^{1}C^{2}OCCl$_3$	90.2	175.5		
H$_2$C^1 = C^{2}H-C^{3}OCH$_3$	128.0	137.1	197.5	25.7

The chemical shift ranges for aldehydes and ketones overlap, and in many cases a clear decision between the two possibilities cannot be made on this basis alone. Such a problem can easily be resolved, however, by recording the ^{13}C spectrum with off-resonance decoupling or by a DEPT experiment (Sections 5.3.3 and 8.6). The signal of the carbonyl ^{13}C nucleus in a ketone remains a singlet in the off-resonance spectrum, whereas that in an aldehyde becomes a doublet; the DEPT spectrum too allows one to distinguish easily between these two carbonyl ^{13}C situations (quaternary and CH).

In 1,3-diketones the δ-values are about the same as in monoketones. For example, the carbonyl signal of acetylacetone in the keto form **18a** is at $\delta = 201.1$, differing considerably from the signal of the corresponding carbon in the enol form **18b** (Scheme V).

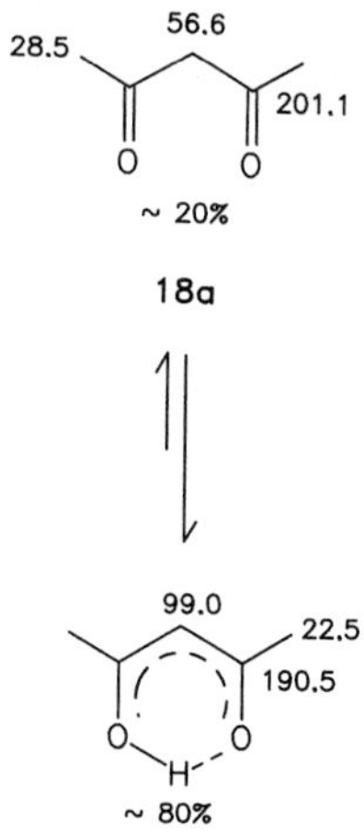

28.5   56.6   201.1   O   O
~ 20%

18a

99.0   22.5   190.5   O   H   O
~ 80%

18b

Scheme V

2.3.6.2 Carboxylic Acids and Derivatives

The shieldings of the carboxyl group ^{13}C nuclei in monocarboxylic acids are greater than those for the carbonyl group in ketones and aldehydes. These resonances are found in the region from $\delta \approx 160$ to 180. Table 2-13 lists chemical shifts for several compounds that are derived, either directly or formally, from acetic acid; the corresponding substituted derivatives of other carboxylic acids show similar trends in their δ-values.

Table 2-13.
^{13}C chemical shifts δ [ppm] of acetic acid derivatives.

Compound	δ (C^1)	δ (C^2)	
C^{2}H$_3$C^{1}OOH	176.9	20.8	(pD 1.5)[a]
CH$_3$COO$^\ominus$	182.6	24.5	(pD 8)[a]
CH$_3$CON(CH$_3$)$_2$	170.4	21.5	CH$_3$: 35 and 38.0
CH$_3$COCl	170.4	33.6	
CH$_3$COOCH$_3$	171.3	20.6	OCH$_3$: 51.5
CH$_3$COOCH=CH$_2$	167.9	20.5	=CH: 141.5
			=CH$_2$: 97.5
(CH$_3$CO)$_2$O	167.4	21.8	
CH$_3$COSH	194.5	32.6	

[a] Solvent: D$_2$O.

The transition from the acid to the carboxylate ion which occurs in alkaline solutions causes a reduction in the shielding of the carboxyl ^{13}C nucleus, and also in those of the carbon nuclei at the α, β and γ positions. On the other hand the shielding in amides, acyl halides, esters and anhydrides is in all cases greater than in the parent acids; in general the δ-values decrease in the sequence stated.

Table 2-14.
^{13}C chemical shifts δ [ppm] of α-substituted acetic acids.

Compound	$\delta\,(C^1)$	$\delta\,(C^2)$	$\delta\,(C^3)$
H−C^{2}H$_2$C^{1}OOH	175.7	20.3	
C^{3}H$_3$-C^{2}H$_2$-C^{1}OOH	179.8	27.6	9.0
(CH$_3$)$_2$CH−COOH	184.1	34.1	18.1
(CH$_3$)$_3$C−COOH	185.9	38.7	27.1
H$_2$N−CH$_2$−COOH (D$_2$O)			
pD 0.45	171.2	41.5	
pD 12.05	182.7	46.0	
HO−CH$_2$−COOH (D$_2$O)	177.2	60.4	
ClCH$_2$−COOH	173.7	40.7	
Cl$_3$C−COOH	167.0	88.9	
H$_2$C^3 = C^{2}H−C^{1}OOH	168.9	129.2	130.8
C$_6$H$_5$−COOH	168.0		

If the hydrogen atoms of the CH$_3$ group in acetic acid are formally replaced by methyl groups, the carboxyl ^{13}C signal appears at a higher δ-value (Table 2-14).

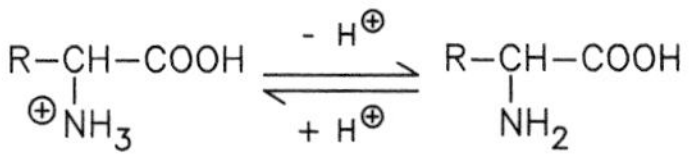

The chemical shifts of amino acids are strongly pH-dependent, as expected from their amphoteric character (Scheme VI).

Carboxylic acids form dimers by hydrogen bonding. Dimers and monomers coexist in solution, the equilibrium being dependent on concentration, temperature and, above all, on the choice of solvent. Chemical shift changes of several ppm due to effects of this kind are generally found.

Where the carboxyl group undergoes conjugation with an unsaturated group, as in acrylic acid (δ = 168.9) or benzoic acid (δ = 168.0), the shielding is increased, i.e. the δ-value is smaller than that for acetic acid.

In the unsaturated dicarboxylic acids maleic acid (**19**) and fumaric acid (**20**), the effect of the two different carboxyl group configurations on the carboxyl ^{13}C resonances is less noticeable than that on the olefinic resonances.

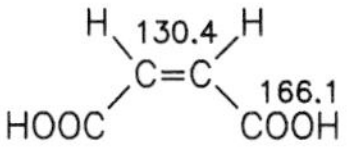

When the oxygen atom in carbonyl compounds is replaced by sulfur, the signals for all these classes of compounds are shifted 20–40 ppm towards higher δ-values, i.e. the ^{13}C nuclei are less shielded. Finally, it should be mentioned that for carboxylic acids and amino acids too there exist empirical correlations for predicting the chemical shifts as functions of the types and positions of substituents [10].

$$R-CH-COOH \atop {}^{\oplus}NH_3 \quad \underset{+ H^\oplus}{\overset{- H^\oplus}{\rightleftharpoons}} \quad R-CH-COOH \atop NH_2$$

Scheme VI

H, 130.4, H
C=C 166.1
HOOC, COOH
19

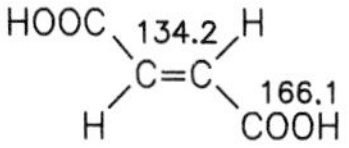

HOOC, 134.2, H
C=C 166.1
H, COOH
20

69

2.4 Relationships between the Spectrum and the Molecular Structure

2.4.1 Equivalence, Symmetry and Chirality

In Chapter 1 we learned the following two rules:
- equivalent nuclei have the same resonance frequency, and
- couplings between equivalent nuclei cannot be observed in first-order spectra.

From these rules we can, by a purely qualitative argument, conclude further that the greater the extent of equivalence between the nuclei in a molecule, the simpler will be its spectrum and the fewer the lines in it.

Equivalence can result not only from molecular symmetry, but also from conformational mobilities such as rotations or inversions (Chapter 11). The effects of *equivalence* and *symmetry* are seen especially clearly in the ^{13}C NMR spectrum, where, due to the large chemical shift differences, a singlet is usually found for each chemically different type of carbon atom in the molecule, provided that C,H couplings are eliminated by ^1H broad-band decoupling. The number of signals is reduced according to the extent of equivalence between the ^{13}C nuclei.

Examples to illustrate effects of equivalence:
- *Methyl groups* play an important role in ^1H NMR spectroscopy, as the three hydrogen nuclei are always equivalent. In methylene groups also the two hydrogen nuclei are often equivalent; thus it is usually easy to recognize the signals of an *ethyl group* – a triplet and a quartet.
- In *benzene*, as a result of the high degree of symmetry in the molecule, all the ^1H nuclei and all the ^{13}C nuclei are equivalent, giving only *one* line in each spectrum.
- In a *monosubstituted benzene* the symmetry gives three sorts of hydrogen nuclei, the *ortho* pair, the *meta* pair, and the single *para* proton. The spectrum can be very complex, as in the case of nitrobenzene (**7**)' (Figure 2-9), but despite this one can recognize the three groups of multiplets. The ^{13}C spectrum of **7** shows the molecular symmetry very clearly, with only four signals for the six carbon nuclei (Table 2-10).
- Figure 2-13 shows the ^1H NMR spectra of three isomeric *disubstituted benzenes*. The three spectra can be correctly assigned at once from symmetry considerations: *p*-dichlorobenzene (**21**) gives only one signal; *o*-dichlorobenzene (**22**) gives a symmetrical spectrum of the AA'BB' or [AB]$_2$ type (Sect. 4.5.2); *m*-dichlorobenzene (**23**) has three chemically different types of hydrogen nuclei, and accordingly gives a more complex spectrum (AB$_2$C type). The ^{13}C

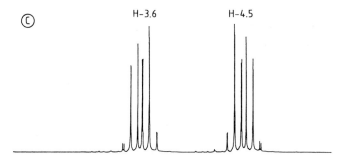

H–3.6   H–4.5

© C

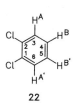

22

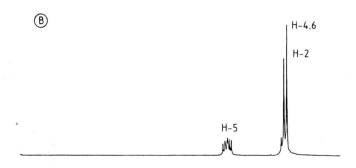

® B

H–4.6

H–2

H–5

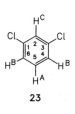

23

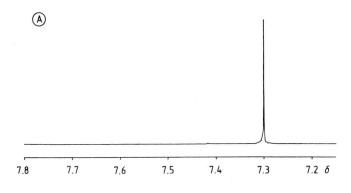

Ⓐ A

21

Figure 2-13.
250 MHz ^{1}H NMR spectra of
A: *p*-dichlorobenzene (**21**) in
CDCl₃
B: *m*-dichlorobenzene (**23**) in
CDCl₃
C: *o*-dichlorobenzene (**22**) in
acetone/CDCl₃.

7.8   7.7   7.6   7.5   7.4   7.3   7.2 δ

spectra can be assigned even more easily: for *para* substitution only two signals are obtained, for *meta* substitution four, and for *ortho* substitution three!

○ There are three isomeric dichlorocyclopropanes, namely the 1,1 (**24**), *cis*-1,2 (**25**) and *trans*-1,2 (**26**) isomers.
Here the method of assigning the spectra by simply counting the ^{13}C NMR signals fails, as all three compounds give two singlets.

71

However, in this case the ^{1}H NMR spectra clearly show the symmetry. In 1,1-dichlorocyclopropane (**24**) the four protons are equivalent, giving only one signal. The *trans* isomer (**26**) has a C_2 axis of symmetry, and the molecule therefore has two sorts of protons. The *cis* isomer (**25**) has a plane of symmetry, giving three sorts of protons, and therefore the most complex of the three spectra.

○ Next we consider the ^{1}H NMR spectra of allyl alcohol (**27**), propylene oxide (methyloxirane, **28**), and trimethylene oxide (oxetane, **29**), which are shown in Figure 2-14. These three compounds – together with propionaldehyde (**9**), whose spectrum is shown in Figure 2-11 – are constitutional isomers with the common molecular formula C_3H_6O. Although the spectra are very complex in parts, they can be assigned by analyzing the structural formulas in terms of symmetry elements and equivalent protons.

The reader should note that many of the finer details of the spectra shown in this section will be better understood after studying the next two chapters.

To end this section we consider the spectra of *chiral* molecules, i. e. those that can occur in two enantiomeric forms. The NMR spectra of the two forms are exactly the same. Nevertheless, it is possible to distinguish between them by preparing diastereomers, which have different chemical and physical properties, and also different NMR spectra (see Chapter 12).

If a molecule contains a diastereotopic group, evidence of chirality can be obtained even without preparing diastereomers. In practice such studies are confined to ^{1}H NMR spectroscopy, the commonest diastereotopic groups being CH_2 and $C(CH_3)_2$ groups. Since experience has shown that difficulties are often encountered with the analysis and interpretation of such data, we will have a closer look in the next section at the problems posed by homotopic, enantiotopic and diastereotopic groups.

24

25

26

Scheme VII

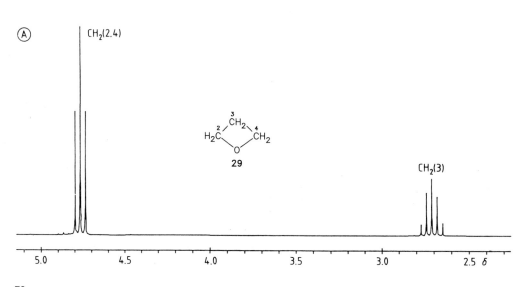

29

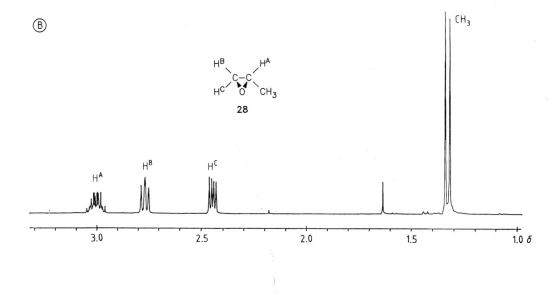

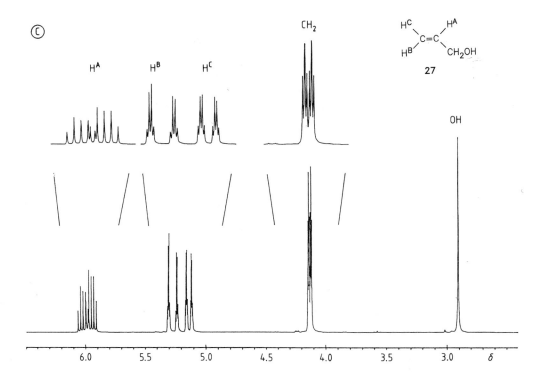

Figure 2-14.
250 MHz ^{1}H NMR spectra of A: trimethylene oxide (**29**), B: propylene oxide (**28**), C: allyl alcohol (**27**) in CDCl$_3$.

2.4.2 Homotopic, Enantiotopic and Diastereotopic Groups

If we consider a methylene group CH_2 or an isopropyl group $C(CH_3)_2$, are the two constitutionally identical hydrogen atoms or methyl groups equivalent or not? The answer to this question is of great importance for solving stereochemical problems.

If we first confine our attention to the methylene group, we need to answer the following questions: firstly, do the two methylene protons give a singlet or a four-line AB type spectrum (Section 4.3.2), and secondly, what does this tell us?

In stereochemistry it is usual to distinguish three possible relationships between the two hydrogen atoms of a CH_2 group; according to Mislow the pair can be *homotopic, enantiotopic* or *diastereotopic*.

Homotopic protons are equivalent, and therefore give *one* signal in the spectrum, if we are considering an isolated CH_2 group. An example is methylene chloride, CH_2Cl_2 (**30**), in which the two hydrogen atoms are homotopic. The molecule has a *two-fold (C_2) axis of symmetry;* this is both a necessary and a sufficient condition for the equivalence. The methylene protons in most ethyl groups are also equivalent like those in methylene chloride, as a result of rapid rotation.

However, there are some methylene groups in which the protons only appear to be equivalent, even though they would exchange positions if reflected in a plane (*mirror symmetry plane*). An example is bromochloromethane (**31**). If we imagine looking from H^1 towards the carbon atom we see that the other substituents Br, Cl and H are arranged in a clockwise (pro-R) configuration, whereas viewed from H^2 the order of the substituents is reversed, being in the anticlockwise (pro-S) configuration. Or, putting this in another way, if we were to replace one of the hydrogen atoms by deuterium, two different enantiomers would be formed. Hydrogen atoms with this property are called *enantiotopic*. Compound **31** is said to be *prochiral*, not chiral. In the allene derivative **32** the two methylene protons are again enantiotopic. In the ^1H NMR spectrum we find that *enantiotopic protons are indistinguishable*. If we assume for the moment that there are no spin-spin couplings to other neighboring nuclei, such a pair gives a singlet.

If the two hydrogen atoms of a methylene group cannot be imagined to exchange positions either by rotation about an axis of symmetry or by reflection in a plane of symmetry, they are said to be *diastereotopic*. As an example we will consider the two methylene protons H^A and H^B in 1,2-propanediol (**33**). If one of these hydrogen atoms were replaced by a substituent R, two different diastereomers would be obtained – hence the

30

31

32

term diastereotopic. The two atoms H^A and H^B are attached to a prochiral center, and are therefore not equivalent; furthermore, even rapid rotation around the C^1-C^2 bond does not make them equivalent. Consequently H^A and H^B always give separate signals, except in cases where the resonance frequencies are accidentally equal (*isochronous*). The simplest way of explaining this is by considering the Newman projection formulas for the three rotational isomers as shown in Scheme VIII.

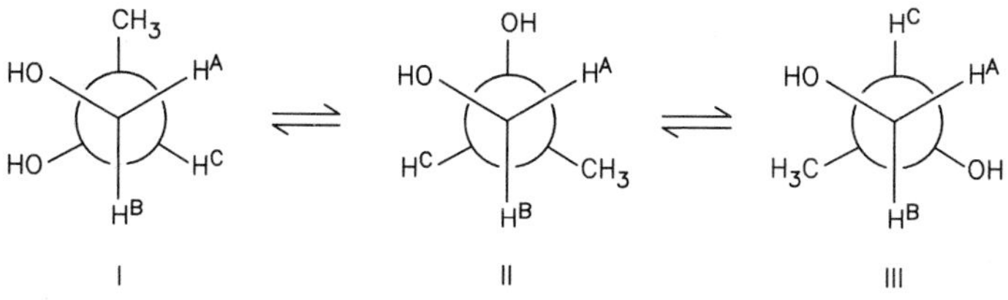

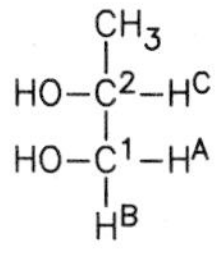

Scheme VIII

How do the respective environments, and thus the magnetic shieldings, appear to the protons H^A and H^B in the three conformers I, II and III? If we list only the nearest and second nearest neighbors, we obtain the following expressions:

for H^A:

OH,CH$_3$/H,H(δ_1)   OH,OH/H,CH$_3$(δ_2)   OH,H/H,OH(δ_3)

for H^B:

H,H/OH,OH(δ_4)   H,CH$_3$/OH,H(δ_5)   H,OH/OH,CH$_3$(δ_6)

All six environments are different. If the rotation around the central C-C bond were frozen out the chemical shifts observed for H^A would be δ_1 in rotamer I, δ_2 in rotamer II, and δ_3 in rotamer III, while for H^B the corresponding δ-values would be δ_4, δ_5 and δ_6. However, as very fast rotation occurs at room temperature, these δ-values become averaged:

$$\bar\delta_A = x_I\,\delta_1 + x_{II}\,\delta_2 + x_{III}\,\delta_3$$
$$\bar\delta_B = x_I\,\delta_4 + x_{II}\,\delta_5 + x_{III}\,\delta_6 \tag{2-6}$$

x_I, x_{II} and x_{III} are the weightings (i. e. mole fractions) which must be applied to the δ-values to take into account the relative probabilities of the three rotamers I to III (see Chapter 11). It follows directly from (2-6) that, even in the case when all three rotamers are present in equal amounts ($x_I = x_{II} = x_{III} = 1/3$), $\bar\delta_A$ and $\bar\delta_B$ are not equal, unless they coincide accidentally! The same argument applies whether one considers the enantiomer or the racemate.

CH$_2$ groups or C(CH$_3$)$_2$ groups are always non-equivalent when the molecule contains an asymmetrical atom (which may be a carbon, trivalent phosphorus or sterically fixed trivalent nitrogen atom), or, expressing this more generally, when the molecule is *chiral*.

Examples:

○ In the spectrum of valine (**34**, Figure 2-15) the methyl protons give four signals near $\delta \approx 1$. This is explained by the fact that C^2 is an asymmetrical carbon atom, and the two methyl groups are therefore diastereotopic; consequently they give separate resonances. Due to the coupling with H^3 we find two doublets.

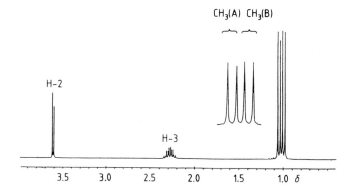

Figure 2-15.
250 MHz ^{1}H NMR spectrum of valine (**34**) in D_2O. The methyl region is shown expanded.

○ For the biphenyl derivate **35** (Figure 2-16) we find two signals at $\delta = 1.42$ and 1.65 for the two methyl groups of the isopropyl moiety. In this case the molecule is chiral owing to the hindered rotation around the central CC bond (i. e. it exhibits atropisomerism, or has conformational enantiomers).

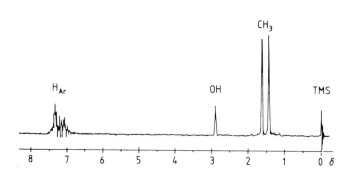

Figure 2-16.
60 MHz ^{1}H NMR spectrum of the biphenyl derivative **35** in $CDCl_3$. Adding a trace of CF_3COOH causes the signal at $\delta \approx 3$ to disappear (exchange of the OH proton, see Section 11.3.7).

There are some CH_2 groups in which the two hydrogen atoms are diastereotopic even though the molecule as a whole is achiral. The classical example is acetaldehyde diethyl acetal (**36**), in which the two methylene protons of each ethyl group are diastereotopic. The ethyl group component of the spectrum is much more complex than usual (Figure 2-17), and is of the ABX_3 type (see Chapter 4). In this case C^1 is a prochiral center.

76

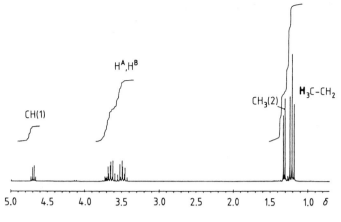

HᴬHᴮ

CH(1)

CH₃(2)

H₃C-CH₂

5.0 4.5 4.0 3.5 3.0 2.5 2.0 1.5 1.0 δ

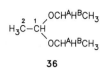

OCHᴬHᴮCH₃

H₃C–CH

OCHᴬHᴮCH₃

36

Figure 2-17.
250 MHz ¹H NMR spectrum of
acetaldehyde diethyl acetal (**36**) in
CDCl₃, with integrated curve.

An analogous situation is found in the partly deuterated citric acid (**37**, Figure 2-18). Here too the two methylene protons are diastereotopic, and they give an AB type spectrum (see Section 4.3).

COOD
|
Hᴬ–C–Hᴮ
|
DOOC–C–OD **37**
|
Hᴬ–C–Hᴮ
|
COOD

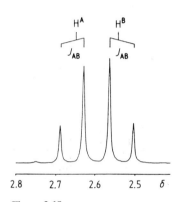

Hᴬ   Hᴮ

Jᴀʙ   Jᴀʙ

2.8 2.7 2.6 2.5 δ

Figure 2-18.
250 MHz ¹H NMR spectrum of
citric acid (**37**) in D₂O.

Disregarding for the moment the last two cases of molecules with a prochiral center, one can formulate the following general rule: if the methylene protons or the two methyl groups of an isopropyl moiety are non-equivalent, the molecule is chiral. On the other hand, if the spectrum indicates that they are equivalent the converse is not necessarily true, since it is possible for the resonance frequencies to coincide accidentally.

2.4.3 Summary

Molecular symmetry has the effect of simplifying the spectrum, since *equivalent* nuclei have the same resonance frequency, i. e. they are *isochronous*. Enantiomers are indistinguishable in the NMR spectrum. If a methylene or isopropyl group is attached to a chiral group, the geminal hydrogen atoms or methyl groups are diastereotopic; they have different magnetic shieldings and therefore give separate signals.

2.5 Chemical Shifts of "Other" Nuclides [11]

For heteronuclides the chemical shift is usually the only spectral parameter used in the analysis (Section 1.7). The range of chemical shifts is generally much greater than for ^1H and ^{13}C; spectral widths of many thousands of ppm or several hundred kHz are not uncommon.

Many elements allow the possibility of observing two or more different isotopes by NMR spectroscopy; thus, all three hydrogen isotopes ^1H, ^2H, and ^3H are observable, while for boron both ^{10}B and ^{11}B are available, for nitrogen ^{14}N and ^{15}N, and so on. In principle it does not matter which isotope one chooses to measure the chemical shifts, since the values of δ (as defined in Section 1.6.1.2) are the same except for some very small isotope effects. Therefore one always chooses to make measurements on the isotope that is most favorable from the standpoints of technical simplicity and useful nuclear properties. If the element has an isotope with a nuclear spin of 1/2 this is usually preferred, even if it has a low natural abundance, as in the case of ^{15}N.

The chemical shifts δ for heteronuclides are measured relative to an appropriately defined standard [12], in exactly the same way as described in Section 1.6.1.2 for ^1H and ^{13}C. The sign convention is the same as for ^1H and ^{13}C, i.e. with positive values of δ to the left of the standard and negative values to the right, in accordance with Equation (1-22). However, for practical purposes it is important to note that this convention is not always observed, especially in the older literature, and in such cases one must reverse the signs. With regard to the choice of reference substances, it turns out that many compounds that appear suitable on chemical shift grounds cannot be simply used as *internal standards* in the same way as TMS because of their chemical reactivity. In such cases one can use an *external standard*, by inserting into the sample tube a capillary containing the reference compound. Also sometimes the chemical shifts are related to a standard frequency generated within the spectrometer [13, 14].

Deuterium (^2H) and tritium (^3H)

The hydrogen isotope deuterium, ^2H, plays an important indirect role in ^1H NMR spectroscopy. Firstly, deuteration or H–D exchange can cause signals to disappear from the ^1H NMR spectrum, thereby helping with assignments. Secondly, this replacement of some of the protons in the molecule by ^{2}H simplifies the spectrum of the remaining protons (Section 5.2.1), since the H,D coupling constants are only about one-sixth of the corresponding H,H coupling constants, and are usually only detectable as a slight broadening of the resonance lines. Thus we will be encountering deuterium in many different sections of this book. It is unnecessary to discuss ^{2}H chemical shifts at this point, as the δ-values are identical to those for ^{1}H except for the small isotope effects discussed earlier (Section 2.1.2.5).

The nuclear properties of the heaviest hydrogen isotope, tritium, make it the most ideally suited of all known nuclides from the NMR standpoint; however, it has the great disadvantage that it is unstable, being a weak β-emitter. Consequently one can only measure tritium resonances in laboratories with special safety provisions.

Boron (^{10}B and ^{11}B)

Of the two boron isotopes ^{10}B ($I = 3$) and ^{11}B ($I = 3/2$) the ^{11}B isotope is usually chosen for NMR measurements, as it has a larger magnetic moment and also a higher natural abundance (80.1 % compared with 19.9 %). The total range of chemical shifts is about 200 ppm. The reference compound usually chosen is $BF_3(OEt_2)$, which is used as an external standard and defines the zero of the scale ($\delta[BF_3(OEt_2)] = 0$).

The chemical shift is greatly influenced by the substituents attached directly to the boron atom. For example, in boron-oxygen compounds the number of oxygen moieties OR attached to the boron atom can be easily determined from the chemical shift, as can be seen from the following examples:

$$\delta[B(CH_3)_3] = +86.3 \qquad \delta[B(CH_3)_2(OCH_3)] = +53.8$$
$$\delta[B(CH_3)(OCH_3)_2] = +32.1 \qquad \delta[B(OCH_3)_3] = +18.3$$

In these four examples the shielding increases with each additional OR group. On the other hand, the nature of the group R at the β-position has scarcely any effect.

From the position of the signal one can determine the coordination number of the boron atom, since the resonances of fourfold coordinated boron are found to be in the approximate range $\delta = +20$ to -128, whereas those of threefold coordinated boron are typically in the range $\delta = +92$ to -8.

Nitrogen (^{14}N and ^{15}N)

The two isotopes ^{14}N ($I = 1$) and ^{15}N ($I = 1/2$) both have only small values of γ, and thus belong to the class of insensitive nuclides. Although the electric quadrupole moment of ^{14}N is relatively small and the signals are therefore not very greatly broadened, the overwhelming majority of nitrogen NMR studies are now performed on ^{15}N using unenriched samples, despite the low natural abundance. The spread of chemical shifts for the widest possible variety of compounds is about 900–1000 ppm. Nitrogen nuclei in amines are the most strongly shielded, while those in nitroso compounds are the least shielded. Liquid nitromethane is used as standard, the ^{15}N chemical shifts then being determined relative to the δ-value for this compound. Another frequently used standard is an aqueous solution of NH_4NO_3, assigning a δ-value of zero for the $^{15}NH_4^{\oplus}$ ion in this compound.

Oxygen (^{17}O)

^{17}O is the only oxygen isotope that gives NMR signals. With a nuclear spin of 5/2, a natural abundance of only 0.038%, and an electric quadrupole moment (even though this is not very large), ^{17}O is not a very favorable nuclide for NMR measurements. Nevertheless, owing to the great importance of oxygen numerous studies have been carried out, mostly confined to the measurement of chemical shifts. The most strongly shielded ^{17}O nuclei are those in compounds with singly-bonded oxygen, such as alcohols and ethers ($\delta = -50$ to $+100$, referred to $\delta[H_2O] = 0$). The least shielded are those in nitrites ($\delta \approx 800$) and nitro compounds ($\delta \approx 600$), where the oxygen atom is doubly bonded. A carboxy group $OC=O$ gives only one signal, showing that the two oxygen nuclei are isochronous (see Section 4.2.2).

Applications of ^{17}O NMR spectroscopy can be found in the publications cited in Section 2.6 (see under "Other" Nuclides).

Fluorine (^{19}F)

^{19}F, with a spin of 1/2 and a natural abundance of 100%, ranks alongside ^{1}H as one of the easiest nuclides to observe. However, compared with ^{1}H it has the additional advantage of a considerably wider range of chemical shifts, which means that the spectra are in general simpler than those of the analogous hydrogen compounds, as there is less overlapping between the groups of peaks. As fluorine forms compounds with nearly all other elements, there exists a wealth of highly informative experimental data. Between the fluorine resonance of the compound ClF, with the highest known shielding, and that of FOOF with the smallest shielding, there is an interval of 1313 ppm; at $B_0 = 2.35$ T (^{19}F resonance frequency 94.094 MHz) this corresponds to more than 120 kHz!

With regard to ^{19}F chemical shifts in organic molecules, it is found that nuclei in saturated compounds are more strongly shielded than those in unsaturated compounds; however, unlike the situation for ^{1}H resonances, there is severe overlapping between the ^{19}F resonances in olefines and in arenes. The effects of neighboring groups that are significant in ^{1}H NMR spectroscopy (magnetic anisotropy and ring current effects, etc.) are of little importance for ^{19}F. On the other hand, large substituent effects are found, but these are complex and difficult to interpret.

The usual reference for ^{19}F chemical shifts is liquid CCl_3F.

Phosphorus (^{31}P)

Phosphorus resonances, like those of ^{1}H and ^{19}F, can be observed without difficulty, and consequently a large amount of experimental data exists. The region containing most of the chemical shifts extends over more than 300 ppm, and even small structural differences affect the spectra markedly. For example, the resonances for $P(CH_3)_3$ and $P(C_2H_5)_3$ are separated by 40 ppm. The least shielded phosphorus nuclei are those of the trihalides PBr_3 ($\delta = 227$; external standard: 85% H_3PO_4), PCl_3 ($\delta = 219$), and PI_3 ($\delta = 178$). In PF_3 the deshielding effect on the phosphorus nucleus is considerably smaller ($\delta = 97$). These observations have been interpreted as arising from the interplay of two opposing effects, namely the ionic character on the one hand, and the double bond character of the P-X bond on the other. For Br, Cl, and I the dominant effect is the ionicity, whereas for F it is the double bond character of the P–F bond. Pentavalent phosphorus generally shows higher shielding values than trivalent phosphorus; for example, $\delta(PBr_5) = -101$, $\delta(PCl_5) = -80$, $\delta(PF_5) = -80$. In Chapter 14 we will meet some examples of *in vivo* studies using ^{31}P resonances.

Alkali and alkaline earth metals

All alkali and alkaline earth metals have nuclei with a quadrupole moment and a small magnetogyric ratio γ, and they are therefore not very suitable for NMR studies. Most measurements have been carried out on the ions of these elements in aqueous solution. The results show that the chemical shifts are greatly affected by ion–ion interactions, and they therefore depend on the nature of the counterions (e. g., $Cl^{\ominus}$, $Br^{\ominus}$, $I^{\ominus}$) and on concentration and temperature. It is therefore necessary to use as the reference an external standard consisting of a salt solution of defined concentration [12].

Studies on $^{23}Na^{\oplus}$, $^{39}K^{\oplus}$, $^{25}Mg^{2\oplus}$, and $^{43}Ca^{2\oplus}$ are of interest in relation to biochemical problems, but of these only $^{23}Na^{\oplus}$ is easy to observe. For $^{25}Mg^{2\oplus}$ and $^{43}Ca^{2\oplus}$ it is usually necessary to use enriched samples.

Silicon (^{29}Si) and tin (^{119}Sn)

In addition to carbon the two other most important Group IV elements that have been studied by NMR spectroscopy are silicon and tin. The relevant isotopes are ^{29}Si and ^{119}Sn; both have the favorable nuclear spin value of 1/2, and their natural abundances (4.68 % and 8.59 %, respectively) are sufficient to allow measurements on unenriched samples. The range of chemical shifts is about 600 ppm for silicon and about 3000 ppm for tin, and accordingly the effects of such variables as substituents, types of bonding, bond angles, and number of coordinated ligands are quite large. These two elements form a wide variety of compounds of importance in organometallic and inorganic chemistry, and consequently there is a wealth of experimental NMR data. The reference compounds mainly used, usually in the form of external standards, are the tetramethyl compounds $Si(CH_3)_4$ and $Sn(CH_3)_4$.

The transition metals: iron (^{57}Fe) and platinum (^{195}Pt)

Iron is an important element in catalysts, in the form of its carbonyl and other complexes. In this context the ^{57}Fe resonances of a large number of complexes with 1,3-dienes or cyclopentadiene as ligands have been intensively studied. It is found that the chemical shifts of ^{57}Fe are strongly influenced by stereoelectronic effects, by substituents, and by the formal charges of the metal complexes, resulting in an overall spread of about 2000 ppm. For example, it is found that introducing a methyl group at one of the *cis* positions in the [$Fe(CO)_3$(1,3-butadiene)] complex reduces the shielding of the iron nucleus by about 105 ppm compared with the unsubstituted complex, whereas the effect of a *trans*-methyl group is only about 30 ppm [15].

The most thoroughly studied of the transition metal nuclides is ^{195}Pt, with spin $I = 1/2$ and 33.83 % natural abundance. The range of chemical shifts covers many thousands of ppm. Two extreme values found are those for analogous platinum(IV) complex ions: $\delta = 11847$ for $[PtF_6]^{2\ominus}$ (in CH_2Cl_2) and $\delta = -1545$ for $[PtI_6]^{2\ominus}$ (in H_2O). In the broad range between these extremes are found the values for complexes with mixed, including neutral, ligands (H_2O, NH_3, CO, etc.), and platinum(II) complexes.

Chemical shifts for metal nuclides are nowadays often determined not in relation to a reference substance but relative to a reference frequency generated by the computer from the ^{2}H resonance of the solvent or the ^{1}H resonance of TMS [12].

2.6 Bibliography for Chapter 2

[1]   R. K. Harris: *Nuclear Magnetic Resonance Spectroscopy. A Physico-chemical View.* London: Pitman, 1983.

[2]   C. W. Haigh and R. B. Mallion: Ring Current Theories in Nuclear Magnetic Resonance. In: *Prog. Nucl. Magn. Reson. Spectrosc. 13* (1980) 303.

[3]   P. E. Hansen: Isotope Effects on Nuclear Shielding. In: *Annual Reports on NMR Spectroscopy*, G. A. Webb (Ed.): 15 (1983) 105.

[4]   L. Pauling: *The Nature of the Chemical Bond and the Structure of Molecules and Crystals*, 3rd ed., Cornell University Press 1960.

[5]   H.-O. Kalinowski, S. Berger, S. Braun: *Carbon-13 NMR Spectroscopy.* Chichester: John Wiley & Sons, 1988.

[6]   A. A. Cheremisin and P. V Schastner, *J. Magn. Reson. 40* (1980) 459.

[7]   Ref. [5], p. 118.

[8]   Ref. [5], p. 104 ff.

[9]   D. J. Craik: Substituent Effects on Nuclear Shielding. In: *Annual Reports on NMR Spectroscopy*, G. A. Webb (Ed.), 15 (1983) 2.

[10] Ref. [5], p. 231ff.

[11] J. Mason (Ed.): *Multinuclear NMR.* New York: Plenum Press, 1987.

[12] R. K. Harris, E. D. Becker, S. M. Cabral de Menezes, R. Goodfellow and P. Granger. *Nuclear Spin Properties and Conventions for Chemical Shifts (IUPAC Recommendations 2001).* In: Encyclopedia of Nuclear Magnetic Resonance, Vol. 9. Chichester: John Wiley & Sons, 2002, p. 5.

[13] Ref. [11], p. 533.

[14] M. L. Martin, J.-J. Delpuech and G. J. Martin: *Practical NMR Spectroscopy.* London: Heyden, 1980, p. 177.

[15] C. M. Adams, G. Cerioni, A. Hafner, H. Kalchhauser, W. v. Philipsborn, P. Prewo and A. Schwenk, *Helv. Chim. Acta 71* (1988). 1116.

Additional and More Advanced Reading

E. Breitmaier and W. Voelter: *Carbon-13 NMR spectroscopy. High Resolution Methods and Applications in Organic Chemistry and Biochemistry*, 3rd Edition Weinheim: VCH Verlagsgesellschaft, 1989.

"Other" Nuclides

S. Berger, S. Braun, H.-O. Kalinowski: *NMR Spectroscopy of the Non-Metallic Elements.* Chichester: John Wiley & Sons, 1997.

C. Brevard and P. Granger: *Handbook of High Resolution Multinuclear NMR.* New York: John Wiley & Sons, 1981.

R. K. Harris and B. E. Mann (Eds.): *NMR and the Periodical Table.* London: Academic Press, 1978.

J. Mason (Ed.): *Multinuclear NMR.* New York: Plenum Press, 1987.

Solid-State NMR

N. J. Clayden: Developments in Solid State NMR. In: *Annual Reports on NMR Spectroscopy*, G. A. Webb (Ed.), 24 (1992) 1.

E. O. Stejskal and J. D. Memory. *High Resolution NMR in the Solid State. Fundamentals of CP/MAS.* New York, Oxford: Oxford University Press, 1994.

3 Indirect Spin-Spin Coupling

3.1 Introduction

How are the magnitudes of the coupling related to chemical structure? Are H,H and C,H coupling constants influenced in the same way by substituents? What is the significance of the signs of coupling constants? Can coupling constants be calculated from theory? What is the underlying mechanism of the coupling?

In this chapter we will attempt to answer these questions. The main emphasis will be on H,H and C,H couplings, but C,C and H,D couplings and some of those involving "other" nuclides will also be discussed.

Coupling constants between protons will be denoted by $J(H,H)$, those between ^{13}C nuclei by $J(C,C)$, between protons and deuterons by $J(H,D)$ and between protons and ^{13}C nuclei by $J(C,H)$. The number of bonds between the coupled nuclei will be indicated by a superscript preceding the J; thus ^1J denotes a coupling between the nuclei of atoms directly bonded to each other, ^2J a geminal coupling, ^3J a vicinal coupling, and ^{3+n}J a long-range coupling.

In practice the most important types of couplings are those between protons, since J-values occur as spectral parameters in every ^1H NMR spectrum. In contrast C,H coupling constants are not usually evident from the ^{13}C NMR spectrum, as their effects are eliminated by ^1H broad-band decoupling (Section 5.3). Whether or not it is worthwhile to undertake a separate experiment to determine these couplings depends on the nature of the problem. With regard to determining C,C coupling constants, the question of time and effort in relation to the value of the information arises even more forcefully.

Table 3-1 gives an approximate indication of the ranges of values that can be expected for the different types of coupling constants, extreme cases being here ignored.

The values of ^1J, ^2J and ^3J vary over quite wide ranges. Evidently, therefore, the internuclear distances alone cannot account for the values. Analysis of the known data has shown that the molecular structure is of crucial importance. For this reason coupling constants are very interesting spectral parameters from the chemist's point of view.

Table 3-1.

General summary of the orders of magnitude and signs of H,H,
C,H and C,C coupling constants.

	J(H,H) [Hz]	Sign	J(C,H) [Hz]	Sign	J(C,C) [Hz]	Sign[b]
^1J	276[a]	positive	125–250	positive	30–80	positive
^2J	0–30	usually neg.	−10 to +20	pos./neg.	< 20	pos./neg.
^3J	0–18	positive	1–10	positive	0–5	positive
^{3+n}J	0– 7	pos./neg.	< 1	pos./neg.	< 1	pos-neg.

[a] For H_2. [b] Determined in only a few cases.

The factors influencing constants include:
- the hybridization of the atoms involved in the coupling
- bond angles and torsional angles
- bond lengths
- the presence of neighboring π-bonds
- efects of neighboring electron lone-pairs
- substituent effects.

The coupling constant is therefore not an easy quantity to interpret; furthermore, theoretical predictions of values have only been successful in a few special cases. It is nearly always a question of proceeding the other way round, i. e. trying to account theoretically for the experimentally determined coupling constants.

The majority of spectra are analyzed by first order methods. This gives the absolute magnitudes of the coupling constants, which is quite sufficient for most problems. However, certain effects can only be understood by introducing the sign into the discussion.

The following sections will show some of the ways in which H,H, C,H and C,C coupling constants are influenced by the chemical structures of the molecules under investigation. Couplings between the ^1H or ^{13}C nuclei and other species such as ^{14}N, ^{15}N, ^{19}F, ^{29}Si, ^{31}P or ^{195}Pt will only be mentioned occasionally; however, couplings to deuterium will be discussed in some detail, as they are of considerable practical and theoretical importance. The final part of the chapter will give a simplified discussion of the theory of the coupling mechanism, and will consider some other related questions such as how one can measure coupling constants when the coupled nuclei are equivalent (Section 3.6.2).

3.2 H,H Coupling Constants and Chemical Structure

3.2.1 Geminal Couplings $^2J(H,H)$

A geminal coupling can be observed between the protons of a CH_2 group, provided that they are not chemically equivalent (Section 1.6.2). They are non-equivalent if, for example, the CH_2 group forms part of a rigid molecular structure, or, more generally, if the two protons are diastereotopic (Section 2.4.2).

The results from analyzing a large amount of experimental data have shown that the magnitude of the geminal coupling depends on:
- the $H-C-H$ bond angle
- the hybridization of the carbon atom, and especially
- the substituents.

Geminal coupling constants are usually negative, i.e. $^2J(H,H) < 0$.

3.2.1.1 Dependence on Bond Angle

The dependence of geminal coupling constants $^2J(H,H)$ on the bond angle is illustrated by the values for methane (**1**), cyclopropane (**2**) and ethylene (**3**) (Figure 3-1).

^2J [Hz]	−11 to −14	−2 to −5	+3 to −3
Φ	109°	120°	120°
Examples:	methane:	cyclopropane:	ethylene:
	−12.4 Hz	−4.5 Hz	+2.5 Hz
	1	2	3

Figure 3-1.
Dependence of geminal coupling constants on the bond angle Φ.

From these and many other examples the following correlation is observed: the greater the bond angle Φ between the coupled nuclei, the more positive is $^2J(H,H)$.

The coupling constant of −4.5 Hz for cyclopropane lies between the values for methane and ethylene. Thus we again find that cyclopropane and its derivatives occupy a special position, as was the case for the ^1H and ^{13}C chemical shifts (Sections 2.2.1 and 2.3.1).

3.2.1.2 Substituent Effects

Substituted alkanes and cycloalkanes: In saturated compounds an electronegative substituent at the α position introduces a positive contribution to the geminal coupling. The coupling constants listed in Table 3-2 show this effect taking a number of methane derivatives as examples. For multiple substitution the substituent effects are additive.

In three-membered rings (Table 3-3) $^2J(H,H)$ again becomes more positive with increasing electronegativity of the substituent.

It is seen from these examples that the signs of the coupling constants must be taken into account. If the signs were ignored the values for the three-membered rings would not show a consistent trend.

In cycloalkanes the geminal coupling constants generally have values of -10 to -15 Hz. The only exception is cyclopropane with a value of -4.5 Hz.

Formaldehyde (**4**) has a coupling constant of $+41$ Hz, which is the largest so far known. This is probably due to a combination of several different factors, namely the hybridization, the electronegativity of the substituent, the proximity to a π-bond, and the electron lone pair on the oxygen atom.

Substituted ethylenes: In the case of ethylene derivatives an electronegative substituent (e.g. fluorine) at the β position causes a negative contribution to the geminal coupling, whereas an electropositive substituent such as lithium causes a positive contribution. Values for a few compounds are listed in Table 3-4.

3.2.1.3 Effects of Neighboring π-Electrons

Neighboring π-electrons generally cause a negative contribution to the geminal coupling. Since $^2J(H,H)$ usually has negative sign, the absolute magnitude increases. This effect is especially large when the line joining the two coupled protons is parallel to the neighboring p- or π-orbital (see sketch).

Such a stereochemical configuration is present in cyclopentene-1,4-dione (**5**), for which $^2J(H,H) = -22$ Hz! The effect is also apparent in toluene (**6**), in which the methyl protons have a geminal coupling constant of -14.4 Hz. It is true that this value is only 2 Hz more negative than that found for methane, but here one must take into account the fact that the measured value is an average; the rapid rotation of the methyl group does not result in the optimal relative orientation.

Table 3-2.
Geminal H,H coupling constants in substituted methanes.

Conpound	^2J [Hz]
CH$_4$	-12.4
CH$_3$OH	-10.8
CH$_3$Cl	-10.8
CH$_3$F	-9.6
CH$_2$Cl$_2$	-7.5
CH$_2$O	$+41.0$

4

Table 3-3.
Geminal H,H coupling constants in three-membered rings.

X	^2J [Hz]	E_X[a]
CH$_2$	-4.5	2.5
S	$(\pm)0.4$	2.5
NR	$+2.0$	3.0
O	$+5.5$	3.5

[a] Electronegativities according to Pauling.

5

$^2J = -22$ Hz

6

$^2J = -14.4$ Hz

3.2.2 Vicinal Couplings $^3J(H,H)$

Vicinal proton-proton couplings in saturated compounds have been very thoroughly studied, both experimentally and theoretically, and the factors influencing $^3J(H,H)$ are therefore very well understood. They are:

- the torsional or dihedral angle
- the substituents
- the distance between the two carbon atoms concerned
- the H−C−C bond angle.

An initial impression of the values and trends is obtained from Table 3-5, which summarizes the expected ranges and typical values for different classes of compounds, the sign being positive throughout.

In the following discussion the main emphasis will be concerned with the dependence of vicinal coupling constants on the dihedral angle and on substituents.

Table 3-4.
Geminal H,H coupling constants in monosubstituted ethylenes.

X	$^2J(H,H)$[a] [Hz]	E_X[b]
Li	+7.1	1.0
H	+2.5	2.2
Cl	−1.4	3.0
OCH$_3$	−2.0	3.5
F	−3.2	4.0

[a] Values from [1] p. 384.
[b] Electronegativities according to Pauling.

Table 3-5.
Ranges and typical values of vicinal H,H coupling constants.

Compound		$^3J(H,H)$ [Hz]	
		Range[b]	Typical value[b]
Cyclopropane	cis	6–10	8
	trans	3– 6	5
Cyclobutane	cis	6–10	–
	trans	5– 9	–
Cyclohexane	a, a	6–14	9
	a, e	3– 5	3
	e, e	0– 5	3
Benzene	ortho	6–10	9
Pyridine	2,3	5– 6	5
	3,4	7– 9	8
H−C−C−H		0–12	7
=CH−CH=		9–13	10
−CH=CH$_2$	cis	5–14	10
	trans	11–19	16
>CH−CHO		1– 3	3
=CH−CHO		5– 8	6
CH−NH[a]		4– 8	5
CH−OH[a]		4–10	5
CH–SH[a]		6– 8	7

[a] Not exchanging. [b] All values are positive.

3.2.2.1 Dependence on the Dihedral Angle

Karplus curves: An important contribution to the understanding or vicinal couplings in saturated systems was made by
M. Karplus [2]. On the basis of calculations he stated the form
of the dependence of the vicinal coupling constant on the
dihedral angle Φ. The lower curve of Figure 3-2 corresponds
approximately to the theoretical Karplus curve. The hatched
area indicates the range within which $^3J(H,H)$ is found to vary
in practice.

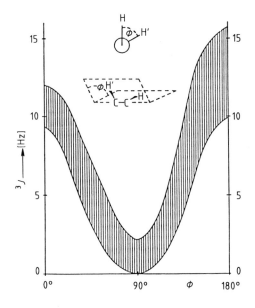

Figure 3-2.
Range of observed vicinal coupling
constants for different values of the
dihedral angle Φ (Karplus curve).

Evidently the coupling constants are largest for $\Phi = 0°$ or
$180°$, and smallest for $\Phi = 90°$.

The most important uses of the Karplus relationship are in
determining the conformations and configurations of ethane
derivatives and saturated six-membered rings. It is also applied
when the coupling is transmitted through nitrogen, oxygen or
sulfur atoms, provided that no exchange of protons attached to
the heteroatom occurs. The dependence of the vicinal coupling
constant on the dihedral angle, as formulated by Karplus, is
without doubt one of the most important relationships in conformational analysis, possibly more so than any other.

Ethane derivatives: The vicinal coupling constants $^3J(H,H)$
for ethane derivatives, e.g. in ethyl groups, are usually about
7 Hz. This typical value corresponds to an averaged coupling,
since at room temperature there is rapid exchange between the
different rotamers (Scheme I).

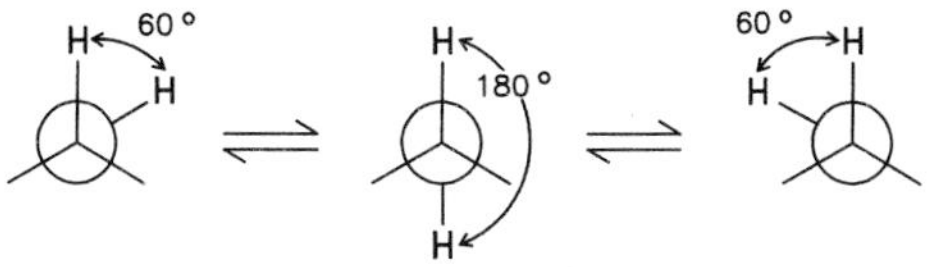

Scheme I

In rotamers I and III there is a *gauche* coupling (^3J_g) with $\Phi = 60°$, while in rotamer II there is a *trans* coupling (^3J_t) with $\Phi = 180°$. From Figure 3-2 we read off the following values for the coupling constants at these angles: $^3J_g \approx 3$–5 Hz, $^3J_t \approx 10$–16 Hz. If all three rotamers are involved in the equilibrium in equal amounts, fast rotation results in a vicinal coupling constant which is the arithmetic average as given by Equation (3-1):

$$^3J = 1/3\,(2\,^3J_g + \,^3J_t) \approx 7 \text{ Hz} \qquad (3\text{-}1)$$

However, if the three rotamers have different energies the position of the equilibrium is determined by the energy differences between them. The observed coupling constant is then given by:

$$^3J = x_\mathrm{I}\,J_g + x_\mathrm{II}\,J_t + x_\mathrm{III}\,J_g \qquad (3\text{-}2)$$

where x represents the mole fraction of each rotamer. If the spectrum is recorded at different temperatures the equilibrium ratios are shifted, and in such cases the observed coupling constant ^3J is temperature-dependent.

Example:

○ For 2-methyl-3-dimethylamino-3-phenylpropionic acid ethyl ester (**7**) the vicinal coupling constant is found to be 11 Hz, and is unaffected by whether the molecules are in the *threo* or *erythro* form. This value is significantly larger than the 7 Hz average that one would expect from Equation (3-1) if we had $x_\mathrm{I} = x_\mathrm{II} = x_\mathrm{III} = 1/3$. From this we can conclude that in both forms the two coupled protons occur preferentially in the *trans* conformation, i. e. the equilibrium in each case is shifted in favor of conformer II (see Scheme II).

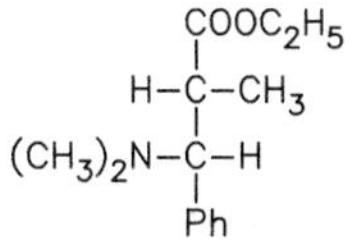

7

Six-membered rings, cycloalkanes: The conformational analysis of ring compounds, especially cyclohexane derivatives and the stereochemically similar carbohydrates of the pyranose type, presents a rich field for the application of the Karplus curve. The preferred conformation of all these six-membered rings – e.g. cyclohexane (**8**) – is the chair form. In this conformation a distinction is made between the axial (a) and equatorial (e) positions of the hydrogen atoms. Depending on the relative positions of the coupled hydrogen nuclei three different vicinal couplings are thus possible: $^3J_{aa}$, $^3J_{ae}$ and $^3J_{ee}$. For the coupling

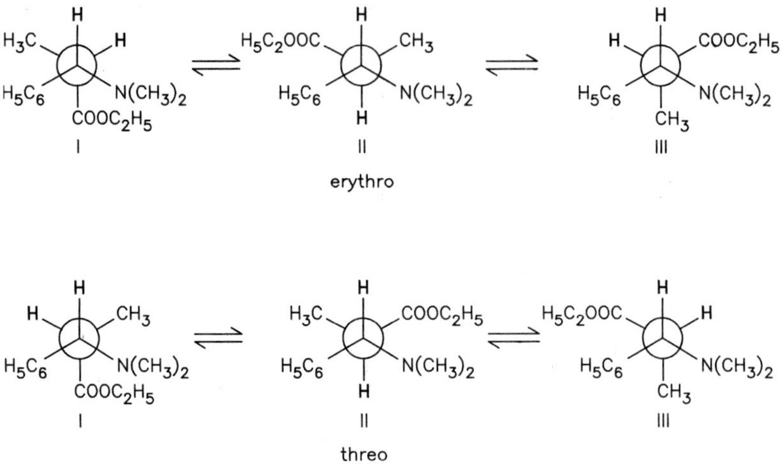

H₃C, H
H₃C—[ ]—H
H₅C₆ N(CH₃)₂
COOC₂H₅

I

H₅C₂OOC, CH₃
H₅C₂OOC—[ ]—CH₃
H₅C₆ N(CH₃)₂
H

II

H, COOC₂H₅
H—[ ]—COOC₂H₅
H₅C₆ N(CH₃)₂
CH₃

III

erythro

H, CH₃
H—[ ]—CH₃
H₅C₆ N(CH₃)₂
COOC₂H₅

I

H₃C, COOC₂H₅
H₃C—[ ]—COOC₂H₅
H₅C₆ N(CH₃)₂
H

II

H₅C₂OOC, H
H₅C₂OOC—[ ]—H
H₅C₆ N(CH₃)₂
CH₃

III

threo

Scheme II

$^3J_{aa}$ the dihedral angle is approximately 180°, and a coupling constant of 10–16 Hz is therefore expected. The values found experimentally are about 7–9 Hz. For protons in the ee or ae conformations the angle is only about 60°, which should give coupling constants of about 3–5 Hz. In practice values of about 2–5 Hz are found.

As the two ranges are quite well separated it is nearly always possible to distinguish between aa couplings on the one hand and ae or ee couplings on the other hand. Having determined the relative positions of the hydrogen atoms from the vicinal H,H coupling, those of the substituents follow automatically. By this means it has been possible to elucidate the structures of many sugars.

$$^3J_{aa} \approx 7\text{–}9 \text{ Hz}$$
$$^3J_{ae} \approx {}^3J_{ee} \approx 2\text{–}5 \text{ Hz}$$

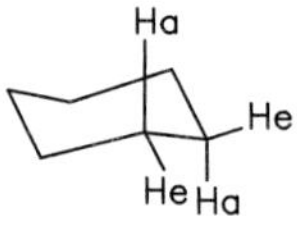

Ha

He

He Ha

8

Example:

○ Figure 3-3 shows the 250 MHz ¹H NMR spectrum of glucose in D₂O. The solution contains glucose in the α and β forms (**9 α** and **9 β**).

We wish to find the relative proportions of these two anomers. For this we first need to know which signals come from α-glucose and which from β-glucose. The simplest method is to look for the resonance of the proton on C¹. This anomeric proton, H-1, ist the least shielded of the ring protons because of the two α oxygen atoms. Furthermore, it is coupled to only one other proton, H-2, and is therefore the only proton in the molecule which gives a doublet. We find one such doublet at δ = 4.65 and the other at δ = 5.24. These have vicinal coupling constants ³J(H-1,H-2) of 7.9 Hz and 3.7 Hz respectively. The larger value is characteristic of an axial-axial coupling, which one expects for β-glucose where the OH group is in the equatorial configuration. The smaller coupling is therefore identified as the axial-equatorial coupling in α-glucose. The integrated areas of the two doublets give proportions of about 40 % α-glucose and 60 % β-glucose.

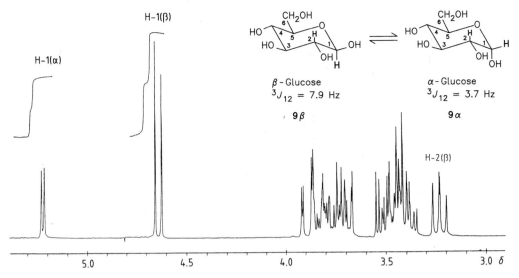

Figure 3-3.
250 MHz ^{1}H NMR spectrum of glucose (**9**) in D$_2$O. The residual HDO signal of the solvent was suppressed (see Section 7.2.4 and Ref. [32] of Chap. 11).

Cyclopropane (**2**) and its derivatives are the only cyclo-alkanes – apart from six-membered ring compounds with a rigid chair conformation – for which stereochemically distinct vicinal coupling constants are found. The ^3J values for *cis* proton pairs are usually between 6 and 10 Hz, while those for *trans* proton pairs are between 3 and 6 Hz (see sketch).

Other examples of the application of the Karplus relationship are described in Section 13.3.1.2.

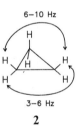

3.2.2.2 Substituent Effects

Saturated compounds, alkanes: From the coupling constants for a few ethane derivatives as listed in Table 3-6 a general tendency can be seen: electronegative substituents reduce the coupling, but the effect is not very great.

The relationship between coupling constants and the differences between the electronegativities E_X of the substituents and the value E_H for hydrogen can be described by a simple empirical equation:

$$^3J(H, H) = 8.0 - 0.8 (E_X - E_H) \qquad (3\text{-}3)$$

However, we must bear in mind that usually only an averaged vicinal coupling is observed owing to rotation about the CC bond.

Table 3-6.
Vicinal H,H coupling constants in monosubstituted ethanes.

$$X-CH_2-CH_3$$

X	$^3J(H,H)$ [Hz]	E_X[a]
Li	8.4	1.0
H	8.0	2.2
CH$_3$	7.3	2.5
Cl	7.2	3.0
OR	7.0	3.5

[a] Electronegativities according to Pauling.

93

Ethylene derivatives: In ethylene and ethylene derivatives the coupling between *cis* proton pairs is smaller than that between *trans* proton pairs.

$^3J_{cis}$ = 6–14 Hz (usually 10)　　　$^3J_{trans}$ = 14–20 Hz (usually 16)

Scheme III

As is evident from the data in Table 3-7, both these types of couplings are greatly affected by substituents, and become smaller as the electronegativities of the substituents increase.

Table 3-7.
Vicinal H,H coupling constants in monosubstituted ethylenes.

X	$^3J_{cis}$ [Hz][a)	$^3J_{trans}$ [Hz][a)	E_X[b)
Li	19.3	23.9	1.0
H	11.6	19.1	2.2
Cl	7.3	14.6	3.0
OCH$_3$	7.1	15.2	3.5
F	4.7	12.8	4.0

[a) Values from [1] p. 384.
[b) Electronegativities according to Pauling.

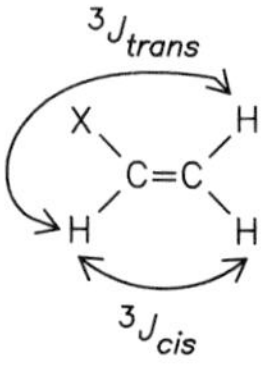

The dependence of the *cis* and *trans* coupling constants on substituent electronegativity is empirically described by Equations (3-4):

$$^3J_{cis} = 11.7 - 4.7\,(E_X - E_H)$$
$$^3J_{trans} = 19.0 - 3.3\,(E_X - E_H)$$

(3-4)

The resulting ranges of values for $^3J_{cis}$ and $^3J_{trans}$ do not overlap except in a few cases (Li and F), and the measured coupling constants can therefore be used to determine the configuration at the double bond.

Cycloalkenes: In small strained rings the *cis* vicinal couplings remain significantly smaller than the normally expected value of about 10 Hz (Table 3-8), and only when we reach the seven-membered ring does the coupling increase to 10 Hz. Evidently the H−C−C bond angle has an important effect here.

Table 3-8.
Vicinal H,H *cis* coupling constants in cycloalkenes.

Compound	$^3J_{cis}$ [Hz]
Cyclopropene	1.3
Cyclobutene	3.0
Cyclopentene	5.0
Cyclohexene	9.0
Cycloheptene	10.0

94

Aldehydes: The vicinal coupling to an aldehydic proton is relatively small. For acetaldehyde (**10**) ^3J is 2.9 Hz, and for propionaldehyde (**11**, see Figure 2-11) it is 1.4 Hz.

10

3.2.3 H,H Couplings in Aromatic Compounds

11

Benzene derivatives: In benzene and its derivatives the *ortho*, *meta* and *para* couplings are different (Table 3-9), and by analyzing the aromatic region of the proton spectrum (δ = 6 to 9) one can therefore determine the arrangement of the substituents. It is often possible to analyze the spectrum by first-order methods, particularly if the spectra have been recorded at a high resonance frequency. A good review of this topic is given in the article by M. Zanger [3].

For naphthalene (**12**) two different *ortho* coupling constants are found: $^3J_{12}$ = 8.3 Hz and $^3J_{23}$ = 6.9 Hz (J_{13} = 1.2 Hz and J_{14} = 0.7 Hz). The difference between the *ortho* couplings is probably due mainly to the different bond lengths.

Heteroaromatics: In heteroaromatics the coupling constants depend on the electronegativity of the heteroatom, the bond lengths, and the charge distribution in the molecule.

Table 3-9.
H,H coupling constants in benzene and benzene derivatives.

J [Hz]	Benzene	Derivatives
J_o	7.5	7–9
J_m	1.4	1–3
J_p	0.7	< 1

12

Table 3-10.
H,H coupling constants in pyridine and pyridine derivatives.

13

| | | J(H,H) [Hz] | | |
		Pyridine	Derivatives
ortho		2.3	4.9	5–6
		3.4	7.7	7–9
meta		2.4	1.2	1–2
		3.5	1.4	1–2
		2.6	–0.1	0–1
para		2.5	1.0	0–1

In pyridine (**13**) and its derivatives the *ortho* coupling is, as in benzene, considerably larger than the *meta* and *para* couplings, and thus it is usually possible from these and the chemical shifts to completely determine the pattern of substitution.

In five-membered aromatic heterocycles such as furan, thiophene and pyrrole, on the other hand, the differences between the *ortho*, *meta* and *para* couplings are much smaller, and consequently they do not always lead to structural assignments.

3.2.4 Long-range Couplings

In saturated systems couplings through more than three bonds are often less than 1 Hz. However, they can be particularly large for molecules in which the bonds linking the coupled nuclei are in a sterically fixed "W" configuration, as shown by the examples **14** to **18** in Scheme IV.

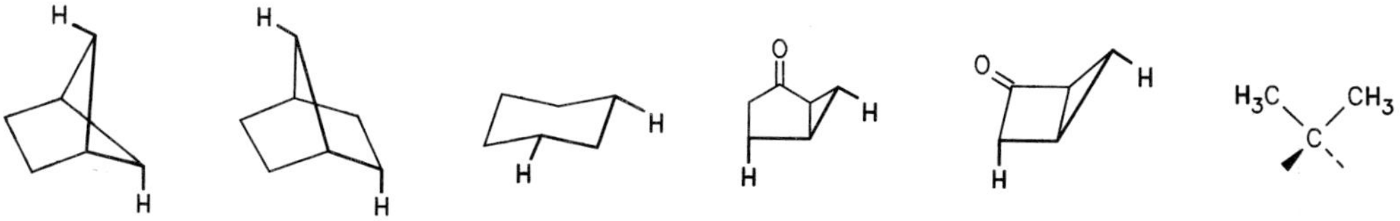

^4J(Hz) +7	+3 to +4	+1.1	+0.9	+1.2	<0.5
14	**15**	**16**	**17**	**18**	**19**

Scheme IV

Coupling constants ^4J(H,H) in W configurations always have positive sign.

In non-fixed systems such as propane (**19**) and its derivatives, ^4J(H,H) values are less than 0.5 Hz, and are therefore only observable under conditions of good resolution.

Couplings through four bonds between protons in allylic compounds can be quite large. These couplings are strongly dependent on the angle Φ between the C-H bond and the axis of the π-orbital in the double bond (see sketch). The sign of the allylic coupling is always negative.

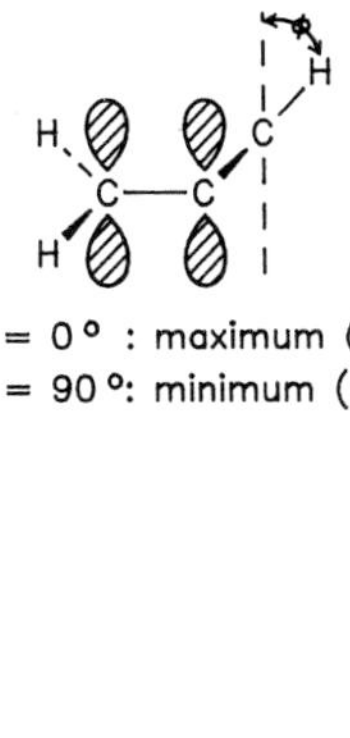

Couplings through five or more bonds can only rarely be observed. Exceptions to this are found for couplings through zig-zag bond systems in unsaturated compounds, e.g. in naphthalene (**20**), benzahldehyde (**21**), and in allenes (**22**) and alkynes (**23**) (see Scheme V).

Very large values are often found for homo-allylic proton pairs. In unsaturated five-membered heterocycles (**24**) a ^5J coupling through four single bonds and one double bond is observed.

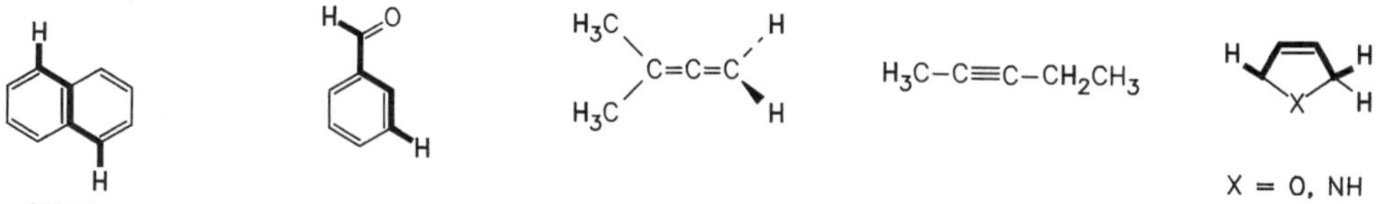

$\Phi = 0°$: maximum (−3 Hz)
$\Phi = 90°$: minimum (−0.5 Hz)

0.8 Hz	0.4 Hz	3.0 Hz	2.6 Hz	X = O, NH up to 7.0 Hz
20	**21**	**22**	**23**	**24**

Scheme V

3.3 C,H Coupling Constants and Chemical Structure

3.3.1 C,H Couplings through One Bond ^1J(C,H)

3.3.1.1 Dependence on the s-Fraction

The C,H coupling constants measured for ethane, ethylene and acetylene (Table 3-11) suggest that there may be a relationship between ^1J(C,H) and the hybridization of the carbon atom involved. These observations are in fact described by an empirical correlation between the s-fraction (denoted by s) and the coupling constant:

$$^1J(C,H) = 500\ s \qquad (3\text{-}5)$$

The quantity s can assume values which range from 0.25 to 0.5 as the hybridization changes from sp^3 to sp.

Table 3-11.
Coupling constants ^1J(C,H) in ethane, ethylene, benzene and acetylene with hybridization and fraction of s-character in the hybrid orbitals.

	H$_3$C–CH$_3$	H$_2$C=CH$_2$	C$_6$H$_6$	HC≡CH
^1J(C,H) [Hz]	124.9	156.4	158.4	249.0
Hybridization	sp^3	sp^2	sp^2	sp
s-fraction	0.25	0.33	0.33	0.5

Equation (3-5) is valid to a good approximation for acyclic hydrocarbons and for both saturated and unsaturated cyclic hydrocarbons with six or more carbon atoms in the ring (Tables 3-12 and 3-13). For the strained three- and four-membered rings, however, larger deviations from the equation are found.

Table 3-12.
Coupling constants ^1J(C,H) in cycloalkanes [4].

Compound	^1J(C,H) [Hz]
Cyclopropane	160.3
Cyclobutane	133.6
Cyclopentane	128.5
Cyclohexane	125.1
Cyclodecane	124.3

Table 3-13.
Coupling constants ^1J(C,H) in cycloalkenes.

Compound	^1J(=C,H) [Hz]
Cyclopropene	228.2
Cyclobutene	168.6
Cyclopentene	161.6
Cyclohexene	158.4
C$_n$H$_{2n-2}$ ($n>6$)	≈ 156

3.3.1.2 Substituent Effects

Substituents have a considerable influence on C,H coupling constants. If, for example, one of the hydrogen atoms of methane is replaced by the strongly electronegative substituent fluorine, the coupling constant increases from 125 Hz to 149.1 Hz, whereas lithium, being an electropositive substituent, reduces it to 98 Hz. It seems likely that the changes in $^1J(C,H)$ with these highly polar substituents are due to their inductive effects rather than to differences in hybridization. Some further data are listed in Table 3-14.

The effects found for methane derivatives can also be observed for other saturated compounds, and this leads to the following general rule: *electronegative substituents increase $^1J(C,H)$ while electropositive substituents reduce it.*

In ethylene derivatives too, $^1J(C,H)$ is strongly affected by a geminal substituent; for example, $^1J(C,H)$ for the a-carbon in fluoroethylene (**25**) is 200.2 Hz compared with 156.4 Hz in ethylene (Table 3-11)!

The value of $^1J(C,H)$ for benzene is 158.4 Hz (Table 3-11). In a monosubstituted benzene there are three different $^1J(C,H)$ values, namely ^1J(C-2, H-2), ^1J(C-3, H-3) and ^1J(C-4, H-4), and the effect of the substituent diminishes with increasing distance; the largest deviations from the benzene coupling constant are found to be about 10 Hz.

$^1J(C,H)$ coupling constants are *always positive!*

Table 3-14.
Coupling constants $^1J(C,H)$ in monosubstituted methanes.

$^{13}CH_3-X$	$^1J(C,H)$ [Hz]
F	149.1
Cl	150.0
OH	141.0
H	125.0
CH$_3$	124.9
Li	98.0

25

3.3.2 C,H Couplings through Two or More Bonds

3.3.2.1 Geminal Couplings (i. e. $^2J(C,H)$ in $H-C-^{13}C$)

The magnitudes of geminal coupling constants depend on the molecular system in question, as is evident from the values listed in Table 3-15. If we exclude the extreme values found for acetylene and its derivatives, the rest of the geminal coupling constants lie within the range -10 to $+20$ Hz.

From a chemical standpoint the structural element $H-C-^{13}C$ can undergo many different possible variations in making up a molecule: substitution at one or both carbon atoms, incorporation into chains or rings, single or double bonding between the carbon atoms, etc. All these variations affect the coupling constant in different ways [6]. The couplings in arenes will be discussed in Section 3.3.3.

Table 3-15.
Geminal C,H coupling constants ($^2J(C,H)$) in ethane, ethylene, benzene and acetylene.

Compound	$^2J(C,H)$ [Hz]
Ethane	$-$ 4.5
Ethylene	$-$ 2.4
Benzene	$+$ 1.1
Acetylene	$+49.6$

3.3.2.2 Vicinal Couplings
(i. e. ^3J(C,H) in H−C−C−^{13}C)

Since the coupling between vicinal protons is of such great importance for elucidating the stereochemistry of organic molecules (Section 3.2.2), it is appropriate to ask whether this also applies to ^3J(C,H) couplings. Theoretical studies on propane [7, 8] do in fact lead to a relationship between these vicinal coupling constants and the dihedral angle, which is identical to that shown for H,H couplings in Figure 3-2. The shape of the curve has been confirmed by measurements on nucleosides and carbohydrates. The largest couplings of 7–9 Hz were found for $\Phi = 0°$ and 180°, while for $\Phi = 90°$ the coupling was near zero. The relationship ^3J (180°) > ^3J (0°) was again found to hold. The vicinal couplings also depend on the CC bond length, the bond angle and the electronegativities of substituents.

In ethylene derivatives such as propene (**26**) the *trans* coupling is found to be greater than the *cis* coupling, as in case of the corresponding H,H couplings. Also in toluene (**27**), with the coupled nuclei in a *cis* configuration, ^3J(C,H) is smaller than in benzene (**28**) with a *trans* configuration. However, both these couplings are smaller than the corresponding couplings in propene (**26**). (This comparison does not take into account the different bond lengths.)

26

27

28

3.3.2.3 Long-range Couplings ^{3+n}J(C,H)

Long-range C,H couplings are not usually detectable, but a few exceptions to this rule have been found for conjugated π-bonded systems.

3.3.3 C,H Couplings in Benzene Derivatives

In benzene there are four different C,H couplings: ^1J(C,H), ^2J(C,H), ^3J(C,H) and ^4J(C,H). In a monosubstituted benzene there are 16 coupling constants, excluding any couplings to the substituent nuclei, namely three ^1J(C,H) couplings, five each of ^2J(C,H) and ^3J(C,H) couplings, and three ^4J(C,H) couplings. Table 3-16 lists a selection of these coupling constants for benzene, toluene, chlorobenzene and fluorobenzene; in each case the couplings given are those between C^1 and the ring protons. These data serve merely to indicate the approximate magnitudes of the different types of J(C,H) couplings in these compounds. (The remaining 13 coupling constants for these and many other monosubstituted benzenes can be found in [6]).

The most important result is: *couplings through three bonds are greater than those through two bonds*, i.e. ^3J(C,H) $> ^2J$(C,H).

Table 3-16.
Some C,H coupling constants for benzene, toluene, chlorobenzene and fluorobenzene.

	X	^1J(C^1,H^1)	^2J(C^1,H^2)	^3J(C^1,H^3)	^4J(C^1,H^4)
				[Hz]	
	H	158.4	+1.1	+ 7.6	−1.3
	CH$_3$		+0.5	+ 7.6	−1.4
	Cl		−3.4	+10.9	−1.8
	F		−4.9	+11.0	−1.7

This statement still holds when the coupled ^{13}C nucleus is at the 2, 3 or 4 position in the ring.

The effects of substituents on the coupling constants differ greatly, and no generalizations are possible. To give an indication of the range of variations that can be found for a single substituent, the ranges of the couplings for chlorobenzene (**29**) are listed here.

For heteroaromatic compounds such as pyridine there is not such a clear division between the ranges of values found for ^2J(C,H) and ^3J(C,H).

^1J(C,H): 161.4 to 164.9 Hz
^2J(C,H): −3.4 to +1.6 Hz
^3J(C,H):   5.0 to 11.1 Hz
^4J(C,H): −0.9 to −2.0 Hz

29

3.4 C,C Coupling Constants and Chemical Structure

Despite new measurement techniques and spectrometers with high sensitivity, measuring C,C coupling constants is still a time-consuming task; this is because the probability of two ^{13}C atoms being present together in the same molecule is low (about 10^{-4}). One must either have a very large sample (>100 mg) and a very long data accumulation time, or resort to using ^{13}C-labeled compounds with an enrichment of 80 to 90 %.

The $^1J(C,C)$ values given in Table 3-17 for ethane, ethylene and acetylene show how widely the coupling constants vary, suggesting that they have the potential to yield much useful information.

For the biochemist especially the qualitative evidence of a $^1J(C,C)$ coupling is interesting, as it can contribute to the elucidation of biochemical reaction pathways (Chapter 14).

For couplings through more than one bond see Ref. [9].

Table 3-17.
C,C coupling constants in ethane, ethylene and acetylene.

Compound	$^1J(C,C)$ [Hz]
Ethane	34.6
Ethylene	67.6
Acetylene	171.5

3.5 Correlations between C,H and H,H Coupling Constants

Having now encountered in this discussion a wide variety of C,H and H,H couplings, we need to consider whether there exist patterns which are common to both. Are C,H and H,H couplings affected in similar ways by stereochemical factors, changes in hybridization, substituents and other variables? – in other words, is there a correlation between C,H and H,H couplings?

Let us therefore compare the C,H and H,H coupling constants for molecules with the same or similar structures, first for geminal then for vicinal couplings. If such a correlation exists we should expect to find a constant value for the ratio $J(C,H):J(H,H)$.

With this aim in mind Marshall and Seiwell [10] synthesized a series of carboxylic acids with ^{13}C-labeled carboxyl groups, e.g. crotonic, benzoic and cyclohexylcarboxylic acids. As comparison compounds for measuring the H,H coupling constants they took propene, benzene and cyclohexane respectively. To explain which pairs of C,H and H,H coupling constants are to be compared, just one example is illustrated here, that of labeled crotonic acid (**30**) and propene (**31**) (Scheme VI).

The experiments gave widely differing results. The ratio $^3J(C,H):^3J(H,H)$ for the couplings through *three* bonds for alkanes, alkenes and arenes is always within the range 0.5 to 0.85, with an average value of 0.61. The variations are so small (standard deviation 0.061) that we can state the following correlation:

$$^3J(C,H) \approx 0.6 \ ^3J(H,H) \qquad (3\text{-}6)$$

In cases where H,H coupling constants can be measured Equation (3-6) can be used to estimate the corresponding C,H couplings. However, the following points must always be kept in mind:

- the comparisons are only valid for molecules with similar structures;

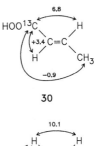

30

31

Scheme VI

- in arriving at the correlation (3-6), the compounds studied were those in which a hydrogen atom in the comparison compound was replaced by a carboxyl or methyl group;
- the calculation gives only an approximate estimate of C,H coupling constants.

A comparison of C,H and H,H couplings through *two* bonds unfortunately does not lead to any simple general relationship. For alkanes it is found that the ratio $^2J(C,H) : {}^2J(H,H)$ is less than 0.5, whereas for alkenes and arenes it varies between 1.5 and 4, and for carbonyl compounds between 0.5 and 1.0. Although these ratios can be used to give rough estimates of $^2J(C,H)$ values in each of the different types of compounds, the large differences found between the different compound types indicate that substituents and changes in hybridization in the $\mathbf{H-C-{}^{13}C}$ nad $\mathbf{H-C-H}$ structural units do not affect the two coupling constants in the same way in these cases.

3.6 Coupling Mechanisms

3.6.1 The Electron-Nuclear Interaction

In Section 1.6.2.1 we excluded the direct interaction through space from our discussion, since in solution it is averaged to zero. We assumed instead that the interaction between neighboring nuclei with magnetic moments, the *indirect spin-spin coupling*, takes place through the bonds. It depends on the magnetic moments of the coupled nuclei, and is independent of the strength of the applied magnetic field (Chapter 1). Is there a simple physical picture to describe this interaction and give us a clear understanding of its mechanism?

The only way in which information can be carried between the coupled nuclei is through the bonding electrons. However, in order for the electrons to transmit information about the orientations of neighboring nuclei in the magnetic field, there must be an interaction between the nuclei and the electrons. In the case of H,H and C,H couplings this interaction arises from the so-called *Fermi contact term*. The Fermi contact is the name used to describe the direct interaction between the magnetic moments of the nuclei and those of the bonding electrons in s-states, since these are the only ones which have a finite probability density function at the position of the nucleus. In hydrogen these are the 1s electrons, and in carbon the 2s electrons.

We can use a greatly simplified vector model (due to Dirac) to understand the coupling mechanism, taking as an example the HD molecule. The magnetogyric ratio γ is positive for both the H and the D nuclei.

According to this model the energetically preferred state is that in which the nuclear magnetic moment (e.g. that of hydrogen) and the magnetic moment of the nearest bonding electron are in an antiparallel configuration. According to the *Pauli exclusion prinicple* the second electron of the bonding pair must then orientate itself with its spin in the opposite direction. Since this bonding electron is on average closer to the second nucleus (the deuteron in our example), this nucleus in turn has a preferred orientation in which its magnetic moment is antiparallel to that of the electron. In this way each nucleus responds to the orientation of the other. Figure 3-4 shows the energetically preferred configuration of the HD molecule.

We can now understand why coupling constants often show a strong dependence on the s-fraction of the bonding orbitals as determined by the hybridization. The most striking example of this was in Equation (3-5) (Section 3.3.1), which gives the relationship between $^1J(\mathrm{C,H})$ and s-fraction.

The coupling constant is defined as having positive sign if its effect is to stabilize the state in which the nuclear spins are antiparallel (Section 4.3.1), as in our example of the HD molecule. In this case we therefore expect that, for nuclei with $\gamma > 0$, ^1J should be positive:

$$^1J > 0$$

This has been confirmed experimentally in most cases.

In coupling through two bonds there is an additional atom between the coupled nuclei. This may be a carbon atom or some other atom, e. g. oxygen as in the $\mathrm{H-O-H}$ molecule.

In the $\mathrm{CH_2}$ group the bonding electrons responsible for the coupling belong to different orbitals of the carbon atom. According to *Hund's rule* the energetically preferred state is that in which the electron spins of the two bonding electrons of the carbon atom are parallel. The preferred configuration of the remaining electron and nuclear spins follows directly from this. The resulting situation is shown in Figure 3-5.

In a $\mathrm{CH_2}$ group where this applies the favored state is that in which the spins of the protons are parallel, and according to the definition ^2J should be negative. Experiments confirm this prediction too.

These considerations lead one to expect that the sign of the coupling should change with each additional intervening bond. This is mainly true for H,H couplings, although there are many exceptions.

The coupling diminishes rapidly as the number of bonds between the coupled nuclei increases. Consequently H,H cou-

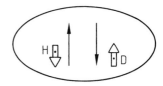

Figure 3-4.
Indirect spin-spin coupling in the HD molecule, transmitted through the bonding electrons. The sketch shows the energetically preferred configuration of the nuclear and electron spins.

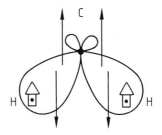

Figure 3-5.
Indirect spin-spin coupling through two bonds in a $\mathrm{CH_2}$ group. The sketch shows the energetically preferred configuration of the nuclear and electron spins.

plings through more than three bonds are only detectable in exceptional cases.

In unsaturated compounds it appears that the π-electrons are involved in the coupling, despite the fact that π-orbitals have a node at the position of the nucleus and the probability density function at this point, and therefore the contribution from the Fermi contact term, is zero. The existence of an interaction is explained by the σ-π spin polarization mechanism, which is also invoked to account for the hyperfine coupling in electron spin resonance (ESR) spectra.

To arrive at a quantitative description of C,C couplings, the Fermi contact term alone is not sufficient, even though it is the dominant factor. For F,F couplings, and to an even greater extent for heavier nuclei, the Fermi contact term plays only a minor role [11].

3.6.2 H,D Couplings

To conclude we will consider the heteronuclear coupling between hydrogen and deuterium nuclei. So far as the mechanism is concerned, we start from the fact that H,D couplings, like H,H couplings, are determined entirely by the Fermi contact term. According to theory the coupling constant is then proportional to the product of the magnetogyric ratios:

$$J(H, H) \propto \gamma_H \, \gamma_H \quad \text{and} \quad J(H, D) \propto \gamma_H \, \gamma_D \qquad (3\text{-}7)$$

Isotopic substitution does not alter the electron wave functions appearing in the calculation of the coupling, and consequently the H,H and H,D coupling constants are proportional to the magnetogyric ratios:

$$J(H, H) : J(H, D) = \gamma_H : \gamma_D \qquad (3\text{-}8)$$

Using the known values of γ_H and γ_D (Table 1-1 in Chapter 1), we have:

$$J(H, H) = 6.514 \, J(H, D) \qquad (3\text{-}9)$$

Equation (3-9) can be used to calculate either of the coupling constants from the other. This is of great practical importance since, as explained in Section 1.6.2.7, couplings between equivalent protons cannot be observed in first order spectra. The couplings between the protons of a methyl group or those in benzene were mentioned as examples. If one determines the H,D coupling constants for the compounds, the H,H couplings between the equivalent protons can be calculated from Equation (3-9).

The situation described above for H,D couplings cannot be extrapolated to C,H couplings. If only the γ-values were involved we would expect to find the ratio $J(C,H):J(H,H) = 0.25$, since γ_H is about four times γ_{13_C}. In practice, however, larger ratios are found; for example, the correlations discussed in the previous section showed that $^3J(C,H):{}^3J(H,H)$ is about 0.6. Nevertheless, the existence of correlations of this kind does at least indicate that the coupling mechanism is the same.

3.6.3 Relationship between the Coupling and the Lifetime of a Spin State

An indirect spin-spin coupling is only observed in the spectrum if the spin orientations of the coupled nuclei in the field B_0 are maintained for a certain minimum time, i.e. if the lifetime τ_1 of the spin state satisfies the condition:

$$\tau_1 > J^{-1} \ [s] \tag{3-10}$$

For a typical proton coupling constant of 10 Hz this means that the lifetime τ_1 must be greater than 0.1 s.

In ^1H and ^{13}C NMR spectroscopy condition (3-10) is usually satisfied, and the couplings can therefore be observed. The situation is different for couplings to nuclei with spins of 1 or greater, i.e. for $I \geqslant 1$. Such nuclei have, in addition to a magnetic moment μ an electric quadrupole moment Q, which results in rapid relaxation and thus in a shortening of τ_1. This causes the signal splitting associated with the coupling to disappear from the spectrum.

Thus, for example, splittings due to couplings between ^1H and ^{14}N are only rarely observed in ^1H NMR spectra. Usually only a line broadening is seen, indicating that the signal splittings have not been completely eliminated.

Occasionally it is possible to observe a N,H coupling by raising the sample temperature [12]. In compounds with a spherically symmetrical charge distribution around the nitrogen atom, such as the ammonium ion $NH_4^{\oplus}$, a splitting of the ^1H NMR signal into the characteristic $1:1:1$ triplet is found. Splittings due to N,H coupling are also often found in the spectra of isonitriles. In all these cases it appears that the relaxation process is suffciently slow or is so ineffective that the lifetimes of the nuclei in the different spin states are long enough to make the couplings visible in the spectrum.

Couplings to deuterium are nearly always observable. This is due to the fact that its electric quadrupole moment is an order of magnitude smaller than that of ^{14}N, with a correspondingly smaller contribution to the relaxation rate $1/T_1$.

3.6.4 Couplings through Space

Sometimes, in cases where an unusually large coupling is found between nuclei that are in close spatial proximity but are separated by several bonds, a through-space coupling has been invoked as an explanation. This type of coupling differs from the direct spin-spin coupling in not being averaged to zero in solution. As no definite proof exists for either the existence or non-existence of such a coupling mechanism, it will not be discussed here.

3.7 Couplings of "Other" Nuclides (Heteronuclear Couplings)

As already mentioned in Section 1.6.2.9, heteronuclear couplings can cause additional splittings in ^{1}H and ^{13}C spectra. These can be seen especially clearly in ^{13}C spectra if one eliminates the C,H couplings by broad-band decoupling.

Two examples:
○ Figure 2-6 shows the ^{13}C{^{1}H} NMR spectrum of a mixture of C_6H_6 and C_6D_6. C_6H_6 gives only a singlet, whereas C_6D_6 gives a triplet due to the C,D coupling.
○ When chloroform is used as a solvent (see Fig. 5-3 for example) the ^{13}C NMR spectrum always contains a triplet at $\delta \approx 77$. The splitting is again due to the C,D coupling.

In ^{1}H NMR spectra couplings to heteronuclei can usually be recognized by the appearance of *satellites* arranged symmetrically about the main peak. The most familiar of these are the ^{13}C satellites, which have already been discussed in Section 1.6.2.9. ^{1}H NMR spectra of nitrogen compounds also contain ^{15}N satellites, but these are more difficult to detect than the ^{13}C satellites since only 0.37 % of the molecules contain the ^{15}N isotope. The other 99.63 % of the molecules containing ^{14}N do not show a proton–nitrogen coupling, because the ^{14}N nuclei undergo rapid relaxation owing to their quadrupole moment, causing the coupling to disappear.

Couplings to other nuclides can also give satellites in routine ^{1}H or ^{13}C spectra. A typical example is the ^{1}H NMR spectrum of tetramethylsilane, TMS (SiMe$_4$), which is made up of 95.3 % ^{28}SiMe$_4$ and 4.7 % ^{29}SiMe$_4$. Whereas ^{28}Si has zero nuclear spin, that of ^{29}Si is 1/2, and consequently the signal from the molecules containing ^{29}Si is split into a doublet by the

^{29}Si, ^{1}H coupling. Close examination of the TMS signal reveals the ^{29}Si satellites at intervals of about 3 Hz to the right and left of the main peak, each with about 2.5% of the total intensity. The separation between the two satellites yields the coupling constant $^2J(^{29}$Si,^{1}H) = 6.6 Hz.

The ^{1}H NMR spectra of platinum compounds show very strong satellites, as the natural abundance of ^{195}Pt is 33.8%. Their separation corresponds to the ^{195}Pt,^{1}H coupling. In contrast, satellite spectra are difficult to detect and measure in cases where the natural abundance of the coupling isotope is less than 1%. It is then better to determine the coupling constants from the NMR spectrum of the heteronuclide itself, especially if this has a spin of 1/2, as in the cases of ^{15}N, ^{19}F, ^{29}Si, ^{31}P, ^{119}Sn, and ^{195}Pt, etc. As examples of results obtained in this way one may cite some ^{119}Sn,^{1}H and ^{195}Pt,^{1}H coupling constants. In tin compounds with methyl ligands it is found that couplings through two bonds with a carbon atom in between, denoted by $^2J(^{119}$Sn-C-^{1}H), have values in the range 40–150 Hz. This contrasts with only 6.6 Hz for the ^{29}Si,^{1}H coupling in the analogous compound TMS! In platinum complexes with methyl ligands the two-bond coupling constants $^2J(^{195}$Pt,^{1}H) are in the approximate range 40–90 Hz. For platinum(IV) hydrides values from 700 to 1370 Hz have been measured for the ^{195}Pt,^{1}H coupling constants (through one bond).

For nuclides that have a spin greater than 1/2, and therefore also an electric quadrupole moment Q, values of coupling constants are known only in exceptional cases. Usually the resonances are so broad that no splittings attributable to spin–spin couplings can be resolved, or alternatively the couplings are eliminated by fast relaxation. As already mentioned in Sections 1.7.2 and 3.6.3, exceptions to this are molecules such as the ammonium ion in which the quadrupolar nucleus is in a symmetrical environment. In the ^{14}N NMR spectrum of NH$_4^\oplus$ the coupling to the four protons results in a quintet consisting of lines that are only a few Hz in width. The value found for the N,H coupling constant $^1J(^{14}$N,^{1}H) is 52.5 Hz. In contrast to this the width of the ^{14}N signal in pyridine is about 200 Hz.

In ^{17}O NMR spectra the line widths are in general between 20 and 300 Hz, so that here too it has only been possible to determine ^{17}O,^{1}H coupling constants in a few special cases. One of these is water, for which the coupling constant $^1J(^{17}$O,^{1}H) has been measured as 83 Hz.

Relatively narrow signals are also obtained in cases where the electric quadrupole moment is very small, as in deuterium (^{2}H) and one of the two lithium isotopes (^{6}Li). The resonances of nuclei that are coupled to ^{2}H of ^{6}Li show characteristic multiplets from which one can determine the coupling constants. For these nuclides the number of lines in the multiplets is, of course, given by Equation (1-24) (Section 1.6.2.5).

A detailed discussion of the heteronuclear coupling constants that are of crucial importance in applications to molecular structure elucidation would be outside the scope of this book, and the reader is instead referred to the works cited in Section 3.8 under "Additional and More Advanced Reading".

In conclusion it should be noted that many of the one- and two-dimensional experimental procedures that have been developed in recent times are based on the existence of scalar heteronuclear couplings. Examples are the INEPT and DEPT methods, and the procedures for recording heteronuclear two- (and multi-)dimensional J-resolved and correlated (COSY) spectra. These techniques will be discussed in Chapters 8 and 9.

3.8 Bibliography for Chapter 3

[1] H. Günther: *NMR Spectroscopy*. New York: John Wiley & Sons, 1987.

[2] M. Karplus: *J. Chem. Phys. 30* (1959) 11. M. Karplus, *J. Amer. Chem. Soc. 85* (1963) 2870.

[3] M. Zanger: *Org. Magn. Reson. 4* (1972) 1.

[4] The coupling constants listed in Tables 3-12 to 3-17 are mostly taken from the extensive collection of data given by Kalinowski, Berger and Braun, Ref. [5]; a smaller proportion are from Refs. [6] and [7].

[5] H.-O. Kalinowski, S. Berger, S. Braun: *Carbon-13 NMR Spectroscopy*. Chichester: John Wiley & Sons, 1988.

[6] E. Breitmaier and W. Voelter: *^{13}C-NMR Spectroscopy*. 3rd Edition, Weinheim: VCH Verlagsgesellschaft, 1989.

[7] J. L. Marshall: *Carbon-Carbon and Carbon-Proton NMR Couplings: Applications to Organic Stereochemistry and Conformational Analysis*. Deerfield Beach: Verlag Chemie International, 1983.

[8] R. Wasylishen and T. Schäfer: *Can. J. Chem. 50* (1972) 2710.

[9] L. B. Krivdin and E. W. Della: Spin-Spin Coupling Constants between Carbons separated by more than one Bond. In: *Progress in NMR Spectroscopy*. J. W. Emsley, J. Feeney and L. H. Sutcliffe (Eds.). Oxford: Pergamon Press, 23 (1992) 304.

[10] J. L. Marshall and R. Seiwell: *Org. Magn. Res. 8* (1976) 419.

[11] R. K. Harris: *Nuclear Magnetic Resonance Spectroscopy. A Physicochemical View*. London: Pitman, 1983, p. 215.

[12] Ref. [1], p. 278 ff.

Additional and More Advanced Reading

V. M. S. Gil and W. v. Philipsborn: Effect of Electron Lone-Pairs on Nuclear Spin-Spin Coupling Constants. In: *Magn. Reson. Chem. 27* (1989) 409.

"Other" Nuclides

S. Berger, S. Braun, H.-O. Kalinowski: *NMR Spectroscopy of the Non-Metallic Elements*. Chichester: John Wiley & Sons, 1997.

R. K. Harris and B. E. Mann (Eds.): *NMR and the Periodic Table*. London: Academic Press, 1978.

J. Mason (Ed.): *Multinuclear NMR*. New York: Plenum Press, 1987.

4 Spectrum Analysis and Calculations

4.1 Introduction [1, 2]

The information obtainable from spectra consists of:
- chemical shifts δ
- intensities I
- indirect spin-spin coupling constants J
- the spectral type (symmetry)
- relaxation times T_1 and T_2
- scalar and dipolar couplings between neighboring nuclei
- line shape.

No single spectrum is capable of giving all these types of information simultaneously; special measurement techniques need to be used according to the nature of the problem. Normally the first three of these types of parameters, which can yield information as described in Chapters 2 and 3, are sufficient for an analysis. Often we determine the spectral type without consciously thinking about it, when we use the information to tell us something about the symmetry of the molecule – an important aid to solving structural problems.

In Chapter 1 we derived rules for analyzing simple first-order spectra; these are effective for a large proportion of all NMR spectrum analysis problems, attention often being limited to certain parts of the spectrum. However, there are many spectra which do not yield to such simple methods of analysis. We first came across an indication of this in the spectrum of cinnamic acid (Fig. 1-22), where the intensities within the doublets of the two-spin system assigned to the olefinic protons were not as expected. The difficulties become even more strikingly apparent when we compare the portions of the 90 MHz and 300 MHz ^{1}H NMR spectra of 2,6-dimethylaniline (**1**) which are shown in Figure 4.1.

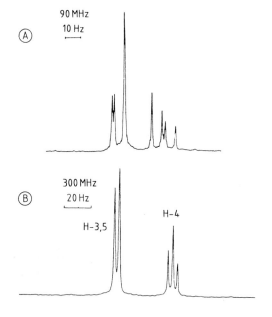

Figure 4-1.
Part of the ¹H NMR spectrum of
2,6-dimethylaniline (**1**) in CCl₄
recorded at different frequencies;
A: 90 MHz; B: 300 MHz. Only the
signals of the aromatic ring protons
are shown. In the 90 MHz spectrum
the CH₃ protons were decoupled.

In the 300 MHz spectrum (Fig. 4-1 B) there is no difficulty in applying a first-order analysis to the aromatic proton signals: the 3,5 protons give a doublet due to their coupling with H-4, and H-4 in turn gives a triplet due to its coupling with H-3 and H-5. However, quite a different situation is found in the 90 MHz spectrum (Fig. 4-1 A), where neither the number of lines nor their intensities are in accordance with the rules given in Section 1.6.2.5.

These two partial spectral serve to illustrate the transition from a first-order to a higher-order spectrum. The question of how the chemical shifts and coupling constants can be obtained from the 90 MHz spectrum will be dealt with in Section 4.4.

Spectra recorded at a higher measurement frequency are not always easier to analyze, as is shown by the example in Figure 4-2 of the 60 MHz and 500 MHz ¹H NMR spectra of ethylbenzene (**2**). The differences between the chemical shifts of the aromatic ring protons are so small that only a slightly broadened single peak is found in the 60 MHz spectrum. In the 500 MHz spectrum, however, the multiplets arising from the *ortho*, *meta* and *para* protons are clearly separated.

Such *higher-order spectra* can only be analyzed exactly by quantum mechanical methods; nevertheless, every spectrum is in principle amenable to calculation. In this chapter we shall not derive the methods for such quantum mechanical calculations, but will merely use the results from them. First it will be shown how systems of two, three and four spins can be analyzed, then the computer-aided methods of spectrum simulation and iteration used for determining the spectral parameters of large complex spin systems will be described.

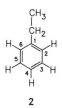

2

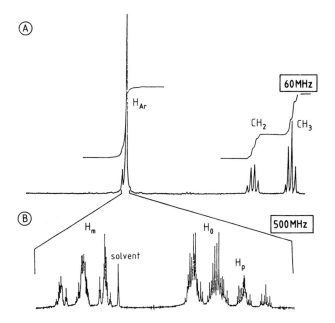

Figure 4-2.
A: 60 MHz ^{1}H NMR spectrum of ethylbenzene (**2**) in CDCl$_3$.
B: Portion of the 500 MHz ^{1}H NMR spectrum of **2**. Only the signals of the aromatic ring protons are shown.

The analysis of the spectrum yields exact values for the chemical shifts and coupling constants. However, this does not automatically solve the problem of assigning signals or groups of signals to particular nuclei in the molecule. We shall return to these problems of assignment in Chapter 6.

The methods used for dertermining relaxation times, connectivities and line shapes will be treated in detail in Chapters 7–9 and 11. Here, before we can go on to discuss the spectrum analysis itself, we need to explain some further points of nomenclature and to introduce several definitions.

The mathematical analysis of higher-order spectra is mainly of importance in ^{1}H NMR spectroscopy, and therefore the examples given in the following discussion will relate almost entirely to proton spectra (the few exceptions are in Section 4.7). Nevertheless, the theory presented here is also perfectly valid for any other spin-1/2 nuclide, although the appearance of the spectra will depend on the nuclear properties.

4.2 Nomenclature

4.2.1 Systematic Notation for Spin Systems

- Several nuclei coupled together constitutes a *spin system*, for example, the protons of an ethyl group form a five-spin system. In a case where the nuclei are not coupled, the spectrum contains only singlets; an example of such a spectrum is shown in Figure 1-18. If instead the molecule contains several spin systems which are not coupled to each other, the spectrum can be divided into partial spectra which are mutually independent. The simplest example of this kind that we have met is the spectrum of ethyl acetate (Fig. 1-20); a somewhat more complicated example is the spectrum of cinnamic acid (Fig. 1-22).

- Chemically non-equivalent nuclei are identified by different letters of the alphabet, beginning with A; where possible, depending on the type of spectrum, the labeling follows the order of their signals from left to right in the spectrum.
- Where several chemically equivalent nuclei are present they all have the same letter, and the number of equivalent nuclei is added as a subscript.
- In cases where the chemical difference Δv between the coupled nuclei is much greater than the coupling constant J (i.e. $\Delta v \gg J$), they are represented by letters that are well separated in the alphabet.

Examples:
○ The two olefinic protons of cinnamic acid (Fig. 1-22) form an AX system.
○ The partial spectrum of the three aromatic protons in 2,6-dimethyl-aniline (**1**) ist of the AX_2 type at 300 MHz, whereas at 90 MHz it is of the AB_2 type (Fig. 4-1).
○ The five spins of the ethyl group in ethyl acetate form an A_2X_3 system (Fig. 1-20).

Unfortunately, however, this ordering of the letters is not strictly observed, especially in cases where the spin system contains two or more different nuclear species. For example, where protons are coupled to a heteronucleus such as ^{13}C, the latter is always denoted by X (see also Section 4.7).

4.2.2 Chemical and Magnetic Equivalence

When we try to group the protons and classify the type of spectrum for a *para*-disubstituted benzene derivative using the rules given in Section 4.2.1, we encounter problems: the nomenclature is no longer unambiguous. This fact becomes immediately apparent if we consider not just the chemical shifts, but also the coupling constants. Let us take as an example *p*-nitrophenol (**3**). There are two pairs of chemically equivalent protons, H-2/6 and H-3/5, and each proton is therefore represented by a letter common to the pair, namely A and X respectively. However, to classify the spectrum as A_2X_2 would be incorrect.

3

Why? Let us consider the couplings of H-2 to the two X nuclei H-3 and H-5. In one case we have an *ortho* coupling, and in the other a *para* coupling! As we saw in Section 3.2.3, the *para* coupling is smaller than the *ortho* coupling. This leads to a further classification of chemically equivalent nuclei according to whether they are *magnetically equivalent* or *magnetically non-equivalent*.

The definitions of chemical and magnetic equivalence are as follows:

Chemical equivalence:
Two nuclei *i* and *k* are chemically equivalent if they have the same resonance frequency, i.e. $v_i = v_k$.

Example:
○ The pairs H-2 and H-6, or H-3 and H-5, in *p*-nitrophenol (**3**).

It is also possible for the resonance frequencies of two nuclei to coincide accidentally (isochronism). They then formally satisfy the condition for chemical equivalence, but are not equivalent in a chemical sense. However, this distinction is irrelevant for the purpose of analyzing an NMR spectrum.

Magnetic equivalence:
Two nuclei *i* and *k* are magnetically equivalent if:
• they are chemically equivalent ($v_i = v_k$), and
• for all couplings of the nuclei *i* and *k* to other nuclei such as *l* in the molecule, the relationship $J_{il} = J_{kl}$ is satisfied.

Examples:
○ In 1,2,3-trichlorobenzene (**4**) the protons at the 4 and 6 positions are chemically and magnetically equivalent, as the coupling constants ^3J(H-4, H-5) and ^3J(H-6, H-5) are equal.
○ In the case of *p*-nitrophenol (**3**) the protons H-2 and H-6 are chemically equivalent but not magnetically equivalent, since the coupling constants ^3J (H-2, H-3) and ^5J(H-6, H-3) are unequal, as also are ^3J(H-6, H-5) and ^5J(H-2, H-5).

This distinction between chemical and magnetic equivalence makes it necessary to introduce a further labeling of the nuclei. When two or more nuclei are chemically equivalent but not magnetically equivalent, the same letter is repeated with primes to show the distinction. The correct description of the spectral type for *p*-nitrophenol (**3**) is thus AA'XX'.

According to a different notation introduced by Haigh [3] this type of spectrum is written as [AX]₂. The spin systems in which the couplings are equivalent are enclosed within square brackets. In our example there are two AX systems. This method of classification has particular advantages for large spin systems in symmetrical molecules. As the following discussion deals only with relatively small spin systems, we shall adopt the notation using primes, which is somewhat simpler for these case.

4

If two or more nuclei are chemically and magnetically equivalent, they are denoted by the same letter, with the number of equivalent nuclei given by the subscript.

Examples:
○ For 1,2,3-trichlorobenzene (**4**) the spectral type is A_2B. The descriptions AX_2 or AB_2 for 2,6-dimethylaniline (**1**) were also correct (Section 4.2.1).

4.3 Two-Spin Systems

4.3.1 The AX Spin System

To derive the energy level scheme for a homonuclear two-spin system AX, we will first neglect the spin-spin coupling by simply assuming $J_{AX} = 0$! In this simple special case the total energy of the two-spin system is equal to the sum of the individual energies, which according to Equations (1-8) and (1-19) of Chapter 1 are given by:

$$E_{A\alpha,\beta} = -m_A \, \gamma \, \hbar \, (1 - \sigma_A) \, B_0$$
$$E_{X\alpha,\beta} = -m_X \, \gamma \, \hbar \, (1 - \sigma_X) \, B_0 \qquad (4\text{-}1)$$

Corresponding to the *m*-values of $+1/2$ and $-1/2$ for the A and X nuclei in the α and β states respectively, we obtain the four energy values:

$$\alpha\alpha: E_1 = E_{A\alpha} + E_{X\alpha} = -\frac{1}{2} \, \gamma \, \hbar \, (2 - \sigma_A - \sigma_X) \, B_0$$

$$\alpha\beta: E_2 = E_{A\alpha} + E_{X\beta} = -\frac{1}{2} \, \gamma \, \hbar \, (\sigma_X - \sigma_A) \, B_0$$

$$\beta\alpha: E_3 = E_{A\beta} + E_{X\alpha} = +\frac{1}{2} \, \gamma \, \hbar \, (\sigma_X - \sigma_A) \, B_0 \qquad (4\text{-}2)$$

$$\beta\beta: E_4 = E_{A\beta} + E_{X\beta} = +\frac{1}{2} \, \gamma \, \hbar \, (2 - \sigma_A - \sigma_X) \, B_0$$

These energy values (expressed in the form E/h) are shown on the left-hand side of Figure 4-3. If now the A and X nuclei are assumed to be coupled, we must also take into account the energy E_{SS} of the spin-spin coupling, which is given by:

$$E_{SS} = J_{AX} \, m_A \, m_X \, h \qquad (4\text{-}3)$$

Thus for nuclei with $I = 1/2$:

$$E_{SS} = \pm \frac{1}{4} J_{AX} \, h$$

From this we obtain four new energy values:

$$E_1 + \frac{1}{4} J_{AX} h$$

$$E_2 - \frac{1}{4} J_{AX} h$$

$$E_3 - \frac{1}{4} J_{AX} h \qquad (4\text{-}4)$$

$$E_4 + \frac{1}{4} J_{AX} h$$

The energy values for the states with parallel nuclear spins ($\alpha\alpha$ and $\beta\beta$) are both raised by an amount $J/4$ (or lowered if J is negative), while those with antiparallel spins ($\alpha\beta$ and $\beta\alpha$) are correspondingly lowered (or raised). Whether the interaction energy E_{SS} is positive or negative is determined by the sign of the coupling constant J_{AX}.

Definition: The coupling constant J_{AX} is positive if the energy levels 1 and 4 of the states with parallel nuclear spins are raised and those of 2 and 3 are lowered (Fig. 4-3, center). For a negative J-value the situation is completely reversed (Fig. 4-3, right).

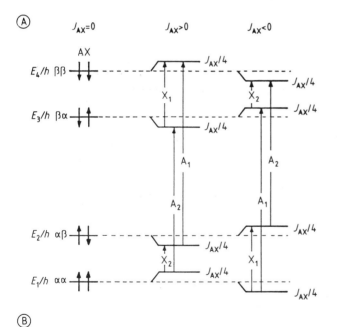

Figure 4-3.
A: Energy level scheme for a two-spin AX system for the cases $J_{AX} = 0$, $J_{AX} > 0$ and $J_{AX} < 0$. The arrows indicate the spin orientations (z-components) A_1, A_2, X_1 and X_2 are the allowed nuclear resonance transitions for the A and X nuclei; B: stick spectrum and signal assignments for positive or negative J_{AX}.

A_1 and A_2 are the allowed transitions for the A nuclei, and X_1 and X_2 are those for the X nuclei (selection rule: $\Delta m = \pm 1$; see Section 1.4.1). These transitions give the four peaks of the AX spectrum. Figure 4-3 B shows the spectrum in "stick" form with assignments. As can be deduced from the energy level scheme, if the sign of J_{AX} is changed, it is only necessary to interchange the assignments in the stick spectrum. Thus, for the two-spin system the sign has no effect on the appearance of the spectrum.

In connection with double resonance experiments, reference is sometimes made to transitions being connected either *progressively* or *regressively*. The definition of these terms will therefore be given here, although we shall not need to use them in this book.

Definition: Two transitions which have a common energy level are progressively connected if the energy of the common level lies between the other two levels. They are regressively connected if the common level lies outside the other two. If we consider the energy level scheme for the two-spin system AX – drawn in Figure 4-4 in a form slightly different from that in Figure 4-3 – the transitions A_2 and X_1 are progressively connected, as also are X_2 and A_1, whereas the transitions A_1 and X_1, and similarly A_2 and X_2, are regressively connected.

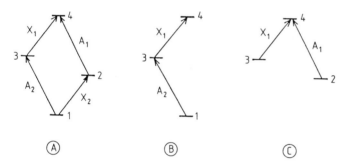

Figure 4-4.
Definition of progressively and regressively interconnected transitions. A: Energy level scheme for a two-spin AX system. B: Example of a progressively connected pair of transitions. C: Example of a regressively connected pair of transitions.

4.3.2 The AB Spin System

In the AB spin system the chemical shift difference $\Delta\nu = \nu_A - \nu_B$ is of the same order of magnitude as the coupling constant J_{AB}. As an example the 60 MHz ^{1}H NMR spectrum of 3-chloro-6-ethoxypyridazine (**5**) is shown in Figure 4-5. It can be seen that the four signals for the two ring protons in the region $\delta = 6.7$ to 7.4 do not have equal intensities in this case ("roof effect"). This makes analysis, e.g. to determine the chemical shifts and coupling constant, more difficult.

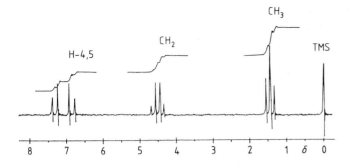

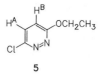

Figure 4-5.
60 MHz ^{1}H NMR spectrum of 3-chloro-6-ethoxypyridazine (**5**) in CCl$_4$ with integrated curve.

Analysis of the AB spectrum: We start by numbering the resonance frequencies of the four lines of the AB spectrum from left to right as f_1, f_2, f_3 and f_4 (Fig. 4-6). As in the AX spectrum the coupling constant J_{AB} is equal to the frequency interval between lines 1 and 2 or between 3 and 4:

$$|J_{AB}| = |f_1 - f_2| = |f_3 - f_4| \text{ [Hz]} \qquad (4\text{-}5)$$

The chemical shifts are given by the centers of gravity of the line pairs 1 and 2 and of 3 and 4. Although it would be possible from an experimental standpoint to calculate the centers of gravity from the signal intensities, this would be too inaccurate. It is easier to determine the chemical shift difference $\Delta v = v_A - v_B$ from Equation (4-6):

$$\Delta v = \sqrt{|(f_1 - f_4)(f_2 - f_3)|} \text{ [Hz]} \qquad (4\text{-}6)$$

Δv is in fact the geometrical mean of the distances between the two outer and the two inner signals. From Δv and the center frequency v_Z we obtain:

$$v_A = v_Z + \frac{\Delta v}{2} \quad \text{and} \quad v_B = v_Z - \frac{\Delta v}{2} \qquad (4\text{-}7)$$

The intensities I_1 to I_4 of the signals are in the ratio:

$$\frac{I_2}{I_1} = \frac{I_3}{I_4} = \frac{|f_1 - f_4|}{|f_2 - f_3|} \qquad (4\text{-}8)$$

The inner lines are always stronger than the outer lines, hence the name "roof effect". The smaller the chemical shift difference Δv in relation to the coupling constant J_{AB}, the stronger the inner lines become at the expense of the outer lines. In the extreme case where Δv is zero, only a singlet appears in the spectrum, the AB spectrum having changed to an A$_2$ spectrum.

A further example of an AB spectrum is that of citric acid at 250 MHz, which is shown in Figure 2-18.

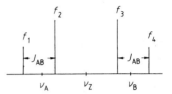

Figure 4-6.
Sketch for analyzing a two-spin AB system.

4.4 Three-Spin Systems

4.4.1 The AX_2, AK_2, AB_2 and A_3 Spin Systems

Figure 4-7 shows the spectra in "stick" form of the four three-spin systems AX_2, AK_2, AB_2 and A_3; the AK_2 and AB_2 spectra were calculated for $J_{AX} = J_{AK} = J_{AB}$, with $\Delta v = 5\,J_{AK}$ for the AK_2 spectrum and $\Delta v = J_{AB}$ for the AB_2 spectrum.

Whereas we can, by using the rules in Chapter 1, quite easily interpret the AX_2 and A_3 spectra and evaluate the parameters, this is no longer possible for the AK_2 and AB_2 spectra, as neither the numbers of lines nor the signal intensities correspond to these rules. Evidently here too, as we have already seen in the case of the AB spectrum, the appearance of the spectra depends critically on the ratio $\Delta v : J$.

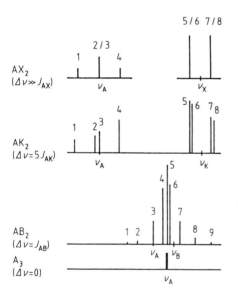

Figure 4-7.
Stick spectra for three-spin systems of the AX_2, AK_2, AB_2 and A_3 types; $J_{AX} = J_{AK} = J_{AB}$. The AK_2 spectrum was calculated with $\Delta v = 5\,J_{AK}$ and the AB_2 spectrum with $\Delta v = J_{AB}$.

The quantum mechanical calculation for the three-spin system [1] gives the result that there are eight allowed transitions with frequencies f_1 to f_8, all of which can be seen in the AB_2 spectrum. In the AX_2 spectrum, however, several ot the lines coincide: the central signal of the triplet in the A portion corresponds to two transitions (2 and 3). The same applies to the X portion, where each line of the doublet arises from the superimposition of two signals. This degeneracy is lifted in the AK_2 spectrum, and even more obviously in the AB_2 spectrum. Line 9 in the AB_2 spectrum is in most cases not detectable, as the transition to which it corresponds is forbidden by the selection rules.

The coupling between the two magnetically equivalent K or B nuclei has no effect on the spectrum.

Analysis of the AB_2 spectrum: We start by numbering the signals in the A portion (Fig. 4-7). v_A is then given by the frequency of line 3, and v_B is obtained as the average of f_5 and f_7:

$$v_A = f_3 \qquad v_B = \frac{f_5 + f_7}{2} \qquad (4\text{-}9)$$

The magnitude of the coupling constant J_{AB} is calculated from (4-10):

$$\left| J_{AB} \right| = \frac{1}{3} \left| (f_1 - f_4 + f_6 - f_8) \right| \qquad (4\text{-}10)$$

The intensities of the lines again depend on J_{AB} and Δv, but there are no simple rules for this as there were for the AB spectrum.

Examples of the types of spectra discussed here are that of benzyl alcohol (Fig. 1-24, Section 1.6.2.3), and the portions of the spectrum of 2,6-dimethylaniline (**1**) shown in Figure 4-1.

4.4.2 The AMX and ABX Spin Systems

Figure 4-8 A shows the 500 MHz ^{1}H NMR spectrum of the ring protons of thiophene-3-carboxylic acid (**6**). Here the three ring protons give an AMX spectrum, so that for each proton we obtain a doublet of doublets. As in the example of styrene discussed in Chapter 1 (Section 1.6.2.6), this spectrum can be analyzed simply as a first-order case. However, the signal intensities within each doublet of doublets deviate slightly from the ideal distribution $1:1:1:1$, indicating a small tendency towards a transition from a first-order spectrum to one of higher order. Despite this, the error involved in using the first-order analysis is quite negligible in this case.

Figure 4-8 B shows the 60 MHz ^{1}H NMR spectrum of the same compound (**6**). Here the deviation from the ideal AMX type is so great that it is no longer acceptable to analyze it as a first-order spectrum; in other words, the AMX spectrum has now been replaced by an ABX spectrum. Of the six spectral parameters $v_A, v_B, v_X, J_{AB}, J_{AX}$ and J_{BX}, only two can be read off directly from the spectrum, namely v_X and J_{AB}. v_X corresponds to the center of the X portion at $\delta = 8.24$, and J_{AB} is found by analyzing the AB resonances in the region $\delta \approx 7.7$ to 7.3.

This eight-line portion of the spectrum is made up of two AB sub-spectra with equal intensities, which we can assign on the

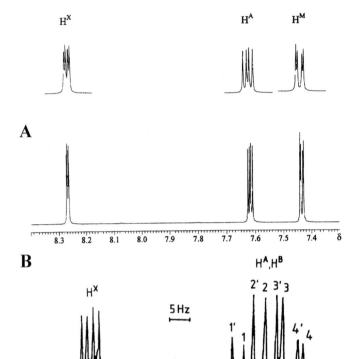

6

Figure 4-8.
^{1}H NMR spectra of thiophene-3-carboxylic acid (**6**) in DMSO-d$_6$ at two different frequencies.
A: 500 MHz spectrum (AMX type).
B: 60 MHz spectrum (ABX type).
Lines 1, 2, 3 and 4 form an AB sub-spectrum, as also do lines 1′, 2′, 3′ and 4′.

basis of the positions and intensities of the signals. In this example lines 1, 2, 3 and 4 correspond to one AB sub-spectrum and lines 1′, 2′, 3′ and 4′ to the other. We can obtain the magnitude of the coupling constant J_{AB} from the splitting of each of the line pairs 1 and 2, 3 and 4, 1′ and 2′, 3′ and 4′. The remaining parameters ν_A, ν_B, J_{AX} and J_{BX} can also be determined by a direct manual calculation, but the procedure is slow and laborious. Nowadays, therefore, it is more sensible to analyze the spectrum by the simulation method (Section 4.6).

The X part of the spectrum consists in theory of six lines. Very often, however, only four are detectable, as is the case in the spectrum of **6** (Figure 4-8 B). Despite the apparently simple appearance of this part of the spectrum, it would be wrong to assume that we can analyze it as a first-order pattern, in the same way as we did for the AMX spectra (Figure 4-8 A above and Figure 1-27 in Section 1.6.2.6). In fact this X part allows us at best only to determine the sum of the coupling constants $(J_{AX} + J_{BX})$!

Figure 4-9 shows two calculated ABX spectra. The coupling constants used in the two cases differ only in the sign of J_{BX}. In spectrum B J_{AX} and J_{BX} have the same sign (whether it is positive or negative makes no difference to the spectrum), whereas

122

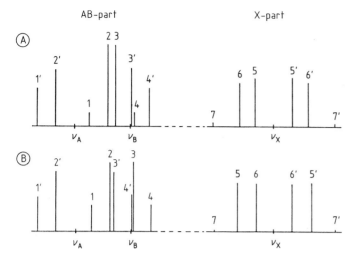

AB-part   X-part

Figure 4-9.
Two calculated ABX spectra
whose parameters differ only in
the sign of J_{BX}.

	Spectrum A:	Spectrum B:
$v_A - v_B$	6 Hz	6 Hz
J_{AB}	2 Hz	2 Hz
J_{AX}	6 Hz	6 Hz
J_{BX}	−2 Hz	2 Hz

In the AB part the two AB sub-
spectra are lines 1, 2, 3, 4 and
lines 1', 2', 3', 4'. The X part
consists of the three pairs 5/5',
6/6' and 7/7'.

in spectrum A the signs are different. This is the first example in which we can see an effect caused by the signs of coupling constants; for all the spectral types described previously only their absolute magnitudes mattered.

From a complete analysis of an ABX spectrum, taking into account both the positions and the intensities of the lines, it is thus possible to determine the relative signs of the coupling constants.

Comparing the 500 MHz and 60 MHz spectra of thiophene-3-carboxylic acid (Figure 4-8 A and B) and taking into account the above discussion, it can be seen that by recording the spectrum with a modern 500 MHz spectrometer one certainly makes the analysis simpler, but at the same time loses information about the signs of the coupling constants.

4.5 Four-Spin Systems

4.5.1 A_2X_2 and A_2B_2 Spin Systems

The spectra of four-spin systems consisting of two pairs of chemically and magnetically equivalent nuclei are relatively easy to recognize. They are symmetrical with respect to the center of the spectrum, i.e. the frequency corresponding to the overall center of gravity. The only other spectra in which this symmetry is found are those of the AX and AB types, and the AA'XX' and AA'BB' spectra treated in the next section.

Each half on the A_2X_2 spectrum consists of a triplet, which can be analyzed by first-order methods. The A_2B_2 spectrum has up to seven lines in each half of the spectrum. The values of ν_A and ν_B are usually obtained from the frequency of the strongest line. The appearance of the spectrum in each individual case depends on the ratio $\Delta \nu : J_{AB}$; J_{AA} and J_{BB} have no effect on the spectrum.

Examples of A_2X_2 spectra are those of cyclopropene (**7**) and difluoromethane (**8**). Spectra of this type only occur rarely in practice.

7

8

4.5.2 The AA′XX′ and AA′BB′ Spin Systems

Spectra of the AA′XX′ (or $[AX]_2$) and AA′BB′ (or $[AB]_2$) types are always found when a molecule contains two pairs of protons such that the protons within each pair can be imagined to exchange positions by rotation about an axis of symmetry or by reflection at a plane of symmetry. The most important examples are disubstituted benzene derivatives with two different substituents *para* to each other or with two identical substituents *ortho* to each other, as in *o*-dichlorobenzene (**9**), (Fig. 4-10).

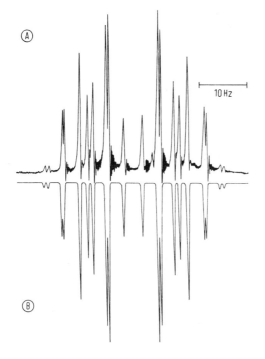

10 Hz

9

Figure 4-10.
90 MHz ^{1}H NMR spectrum of *o*-dichlorobenzene (**9**).
A: observed spectrum; B: calculated spectrum. The spectrum is of the four-spin AA′BB′ type ($[AB]_2$).

As a result of the symmetry we have:

$\nu_A = \nu_{A'}$; $\nu_B = \nu_{B'}$ (chemical equivalence)

$J_{AB} = J_{A'B'}$; $J_{AB'} = J_{A'B}$.

The spectrum also depends on the chemical shift difference $\Delta\nu = \nu_A - \nu_B$ and the coupling constants $J_{AA'}$ and $J_{BB'}$ and also on the relative signs of the four coupling constants.

Each half of the AA'XX' spectrum can have up to 10 lines, and of the AA'BB' spectrum up to 12 lines. The transition from an AA'BB' to an AA'XX' spectrum is clearly seen by comparing the 90 MHz and 250 MHz spectra of o-dichlorobenzene shown in Figures 4-10 and 2-13 C.

It is still possible with much labor to analyze spectra of this type manually, but the calculations are usually performed by simulation or iteration methods. The basic features of these methods will now be described.

4.6 Spectrum Simulation and Iteration [4]

For analyzing the spectra of asymmetrical molecules, simple rules of calculation cannot usually be applied. Even to analyze the spectrum of a three-spin ABC system requires the use of a computer. The usual procedure is as follows:

As a first approximation one "simulates" a theoretical spectrum using estimated ν- and J-values, and compares it with the observed spectrum. The input parameters are then changed in a stepwise manner until agreement is obtained between the theoretical and experimental spectra. Figure 4-10 B shows the result of such a spectrum simulation for o-dichlorobenzene (**9**). The spectrum is, as already mentioned above, of the AA'BB' (or [AB]$_2$) type.

The success of the spectrum simulation depends critically on how the initial parameters are chosen, since it is necessary not only to insert good approximate values for the frequencies and coupling constants, but also to correctly choose the relative signs of the coupling constants and the line widths. All these parameters must be estimated on the basis of experience or by comparison with other related compounds.

Computer programs are now available which compare a theoretically calculated spectrum with an experimental spectrum taken directly from the spectrometer, and automatically continue altering the parameters until an optimum fit is obtained (iteration).

Both simulation and iteration can also be carried out on normal work-station computers (PCs). Nevertheless, there is no hiding the fact that calculating and comparing spectra demands a certain amount of theoretical knowledge and experience in the use of the PC and the software, to ensure that one arrives at correct conclusions. Moreover, serious problems always arise with spectra that contain too few lines.

4.7 Analysis of ^{13}C NMR Spectra

The simplicity of ^{13}C NMR spectra with ^{1}H broad-band decoupling is in striking contrast to those without decoupling, which are very complex. The mojority of these are higher-order spectra. They also often appear deceptively simple, so that there is a temptation to analyze them by first-order methods. Whether or not this is in fact permissible can easily be tested experimentally by recording the ^{13}C NMR spectrum at different field strengths, i.e. resonance frequencies. If the appearance of the spectrum does not alter it is first-order, and the δ- and J-values can be measured directly from the spectrum. If it alters, however, a full spectrum analysis is needed. In many cases it is not sufficient to analyze only the ^{13}C NMR spectrum – the ^{1}H NMR spectrum, i.e. the ^{13}C satellites, must also be analyzed. The simple examples which follow illustrate this.

Examples:
○ In the structure (**10**) the two protons and the ^{13}C nucleus of the fragment $\mathbf{H}^A - C = {}^{13}\mathbf{C} - \mathbf{H}^B$ give an ABX type spectrum. In the ^{13}C spectrum we see only the X portion, which consists of four (possibly six) peaks. We saw in Section 4.4.2 that neither J_{AX} nor J_{BX} can be determined from the X part of the spectrum; for this one must analyze the AB part. This part can only be found by examining the ^{13}C satellites of the ^{1}H NMR spectrum!
○ In maleic ahydride (**11**) the two protons and the single ^{13}C nucleus of the carboxy group form a three-spin AA'X system, the X part of which is in the ^{13}C spectrum. For an exact analysis it is essential to have the data from the AA' part, i.e. the ^{13}C satellites of the ^{1}H spectrum.
○ For styrene (**12**) we find in the ^{13}C NMR spectrum the X part of an AKMX spectrum. An analysis is not possible without data from the AKM part. In a case such as this one does not analyze the ^{13}C satellites spectrum but instead the ^{1}H NMR spectrum of the three olefinic protons of the vinyl group, which yields J_{AK}, J_{AM} and J_{KM}. (Compare also Section 1.6.2.6 and Fig. 1-27, noting, however, the different labeling of the protons). Using these values and the chemical shifts v_A, v_K, v_M, together with estimated values for the couplings to the X nucleus, the X part or the entire spectrum can now be analyzed.

10

11

12

126

4.8 Bibliography for Chapter 4

[1] J. D. Roberts: *An Introduction to the Analysis of Spin-Spin Splitting in High-Resolution Nuclear Magnetic Resonance Spectra.* New York: W. A. Benjamin, Inc., 1962.

[2] R. A. Hoffmann, S. Forsén and B. Gestblom: Analysis of NMR Spectra. In: *NMR: Basic Principles and Progress.* Vol. 5, P. Diehl, E. Fluck and R. Kosfeld (Eds.), Berlin: Springer-Verlag, 1971.

[3] C. W. Haigh, *J. Chem. Soc. A 1970*, 1682.

[4] P. Diehl, H. Kellerhals and E. Lustig: Computer Assistance in the Analysis of High-Resolution NMR Spectra. In: *NMR: Basic Principles and Progress*, Vol. 6, P. Diehl, E. Fluck and R. Kosfeld (Eds.), Berlin: Springer-Verlag, 1972.

Additional and More Advanced Reading

G. Haegele, W. Boenigk and M. Engelhardt: *Simulation und automatisierte Analyse von Kernresonanzspektren.* Weinheim: VCH Verlagsgesellschaft, 1987.

U. Weber and H. Thiele: *NMR Spectroscopy: Modern Spectral Analysis.* Weinheim: Wiley-VCH, 1998.

5 Double Resonance Experiments

5.1 Introduction

For the majority of users of NMR spectroscopy double resonance experiments are synonymous with spin decoupling. The main purpose of such experiments in the past was to simplify spectra. However, the availability of high-field spectrometers and the consequently increased chemical shifts (when expressed in Hz) means that spectra are now in general simpler than before, and many that previously had to be treated as higher-order can now be analyzed by first-order methods. A typical example can be seen in the comparison of the 60 MHz and 500 MHz spectra of thiophene-3-carboxylic acid shown in Figure 4–8 (Section 4.4.2). In addition we now have access to a variety of special techniques using complex pulse sequences and two-dimensional methods, which often provide much more elegant ways of solving problems in spectrum analysis. Nevertheless, this does not mean that decoupling experiments in one-dimensional NMR spectroscopy have become redundant, as they can often be performed more quickly, yielding an immediate result when applied to suitable problems.

Although spin decoupling for simplifying spectra is the most familiar type of double resonance experiment, there are many others, as the following list shows.

- ^{13}C spectra are normally recorded with broad-band decoupling of the ^1H resonances (Section 5.3.1). This gives a ^{13}C spectrum consisting of single peaks, and at the same time yields an increase of up to 200 % in the signal intensities through the *nuclear Overhauser effect* (NOE). An enhancement through the NOE effect can also occur in other types of double resonance experiments, and these phenomena will be treated in detail in Chapter 10.
- Double resonance experiments can be used as an aid to determining *energy level schemes* and *relative signs of coupling constants*.
- In Chapter 8 we shall meet the methods of *selective polarization transfer* and the one-dimensional selective TOCSY experiment, both of which involve a double resonance experiment as a key step.

- In dynamic NMR spectroscopy (DNMR), special double resonance experiments involving *saturation transfer* can be used to measure rate constants in exchange processes.

In this chapter, however, we shall be concerned only with routine spin decoupling measurements, which are of two types:
- *homonuclear decoupling*, which is mainly used in ^{1}H NMR spectroscopy, and is a means of decoupling individual protons or groups of protons within a molecule from each other; thus it is a form of *selective decoupling*.
- *heteronuclear decoupling*, here applied to ^{13}C NMR spectroscopy; most commonly this is ^{13}C{^{1}H} decoupling (using the notation explained below), which can be performed either selectively or non-selectively.

It has become customary to describe the different types of experiments by a notation in which the symbol for the decoupled (irradiated) nuclide is enclosed in curly brackets, preceded by the symbol for the nuclide being observed. Thus, observing ^{13}C resonances with simultaneous decoupling of the protons is denoted by the abbreviation ^{13}C{^{1}H}.

5.2 Spin Decoupling in ^{1}H NMR Spectroscopy

5.2.1 Simplification of Spectra by Selective Spin Decoupling

When discussing the coupling mechanism (Section 3.6.3) we learned that the interaction with neighboring nuclear dipoles only causes a splitting of the signal if the nuclei maintain their spin orientation in the external magnetic field B_0 for a time longer than $\tau_1 = 1/J$. For protons and ^{13}C nuclei this condition is normally satisfied.

In the decoupling experiment one shortens the lifetime τ_1 by selectively irradiating with a second frequency ν_2 while recording the spectrum in the normal way. The frequency ν_2 is adjusted to the exact resonance frequency of the nucleus to be decoupled. This causes the signal splitting associated with the indirect spin-spin coupling to disappear.

Examples:
○ In ethyl acetate (**1**, CH$_3$COOCH$_2$CH$_3$) the methylene and methyl protons of the ethyl group are coupled, giving a quartet and a triplet

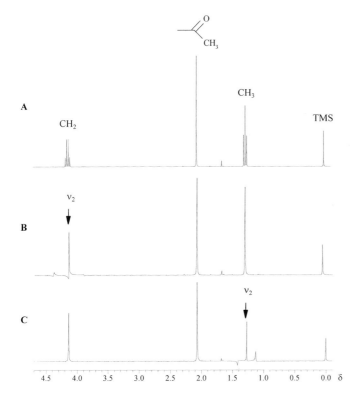

A

CH₂

CH₃

CH₃

TMS

B

ν₂

C

ν₂

4.5   4.0   3.5   3.0   2.5   2.0   1.5   1.0   0.5   0.0 δ

$$CH_3-\overset{\overset{\textstyle O}{\|}}{C}-O-CH_2-CH_3$$

1

Figure 5-1.
A: 300 MHz ¹H NMR spectra of ethyl acetate (**1**).
B: CH₂ protons decoupled.
C: CH₃ protons decoupled.

(A₂X₃ type spectrum, Fig. 5-1 A); the acetyl protons give a singlet. If we wish to decouple the methylene protons we irradiate at the decoupling frequency ν_2 during the measurement, so that all four lines of the quartet are excited and become saturated. To achieve this the decoupling irradiation must have a high power level, as it must be broad enough to excite the whole of the quartet.

By saturating the resonances of the methylene protons we induce rapid transitions between the corresponding energy levels, and the lifetime τ_1 of the methylene protons in any given spin state is shortened. As a result the additional field contribution at the position of the methyl protons is averaged to zero, and in the spectrum one observes only a singlet for the methyl protons (Fig. 5-1 B). If instead the resonances of the methyl protons are saturated, a singlet is obtained for the methylene protons (Fig. 5-1 C).

○ A second, slightly more complicated example, is shown in Figure 5-2. When H-6 in 2-chloropyridine (**2**) is decoupled the signals of protons H-3, 4 and 5 become simplified, as their couplings to H-6 disappear. The spectrum can now be analyzed simply by first-order methods (Figure 5-2 B).

Irradiation at a second auxiliary frequency can also be applied if the spectrometer includes a provision for this, allowing a further simplification of the spectrum (*triple resonance experiment*). However, all the types of decoupling experiments described here can only be successful and meaningful if the

131

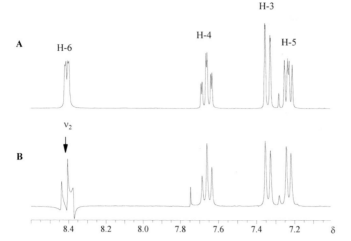

Figure 5-2.
A: 300 MHz ^{1}H NMR spectrum of 2-chloropyridine (**2**).
B: H-6 decoupled.

frequencies and amplitudes of the decoupling irradiations are precisely adjusted.

There exist variants of these spin-decoupling experiments in which not all the lines of a multiplet are saturated. In the *spin-tickling experiment*, for example, only one line of a multiplet is saturated, and this causes a doubling of same lines in other parts of the spectrum. Such experiments can be used to determine energy level schemes and the relative signs of coupling constants. Further details of this technique, along with some others that are now almost forgotten, can be found in a 1971 review article [1]. All the experiments described in that article were performed by the continuous wave (CW) method, but they are equally possible using the pulsed method.

There exist various modern procedures based on complex pulse sequences, in which selective pulses are used to decouple or saturate specific resonances, and these will be described in detail in Chapter 8.

5.2.2 Suppression of a Solvent Signal

Since deuterated solvents are generally used for ^{1}H NMR measurements, it is only in exceptional cases that the residual signals of the non-deuteriated fraction of the solvent cause difficulties. However, if it is necessary to work with aqueous media, e. g. in biochemical investigations, the water signal in the neighborhood of $\delta \approx 4$ to 5 interferes with the measurements. Whenever possible it is better to use D_2O, but even then the HDO signal is usually still larger than the sample signal. Sometimes it is possible to take advantage of the fact that the position of the water signal is strongly temperature-dependent: at 24° C the signal of H_2O and HDO is around $\delta \approx 4.8$, whereas at 80° C it is around $\delta \approx 4.4$. It is sometimes possible by raising the temperature to reveal signals hidden by the solvent. However, it is better to suppress the solvent signal. To do this an irradiating field of greater amplitude is applied at the frequency of the solvent signal during the recording of the spectrum. The resulting saturation greatly reduces the signal intensity. Unfortunately signals of the sample under examination which lie in this region may also be saturated, with the result that some or all of them escape detection. Also the saturation may be partially transferred to other exchangeable protons such as those of OH or NH groups, reducing the intensity of their signals.

This disadvantage can be at least partly avoided by switching off the decoupler a few tenths of a second before recording the data. Because of their different relaxation times the protons of the sample have already relaxed within this time, whereas the HDO signal still remains essentially saturated (Section 7.2.4). The suppression of the H_2O and HDO signal is even more effective if one uses a reduced decoupler power, and compensates for this by a cyclic repetition (about 50 times) of the irradiating and waiting phases before each observing pulse. The spectra of Figures 3-3 and 11-9 were recorded using this technique (Chapter 11, Ref. [32]). Suitable pulse programs for suppressing solvent signals are routinely provided by the instrument manufacturers.

5.3 Spin Decoupling in ^{13}C NMR Spectroscopy

5.3.1 ^{1}H Broad-band Decoupling

^{13}C NMR spectra are complex and have many lines due to coupling with the hydrogen nuclei, and usually this causes difficulties in analyzing them. Furthermore, it has the effect of dividing the already low intensities of the ^{13}C signals into multiplets. It is usual to avoid these undesirable effects of the C,H couplings by decoupling the protons during the recording of the spectrum. In the proton-decoupled spectrum all the carbon nuclei give single peaks, provided that the molecule does not contain other coupled nuclei such as fluorine, phosphorus or deuterium.

It can be seen from Figure 5-3 C that all the C,H couplings are removed in this procedure, which is called *^{1}H broad-band decoupling*, abbreviated to ^{1}H BB decoupling.

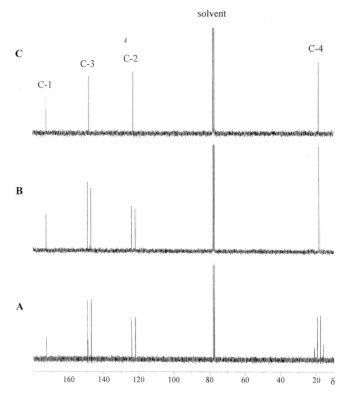

solvent

C

C-1   C-3   C-2   C-4

B

A

160   140   120   100   80   60   40   20   δ

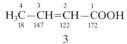

$$\underset{18}{\overset{4}{H_3C}}-\underset{147}{\overset{3}{CH}}=\underset{122}{\overset{2}{CH}}-\underset{172}{\overset{1}{COOH}}$$

3

Figure 5-3.
75.47 MHz ^{13}C NMR spectra of crotonic acid (**3**) in CDCl$_3$.
A: With C,H couplings, recorded by the gated decoupling technique.
B: With selective decoupling of the methyl protons.
C: With ^{1}H broad-band (BB) decoupling.

In practice ^{13}C{^{1}H} decoupling is performed by irradiating with a continuous sequence of "composite pulses". To explain in detail the mechanism of this *composite pulse decoupling* (CPG) method would be outside the scope of the level of treatment in this book; instead the reader is referred to more advanced works on the subject [2,3]. With this technique all the protons, even though they are not equivalent, undergo rapid reversal of their spin orientation relative to the magnetic field, causing the C,H coupling to disappear from the spectrum. The principle of ^{1}H BB decoupling is shown schematically in Figure 5-4.

Spectra recorded by this method contain only one parameter, the chemical shift. This disadvantage is, however, outweighed by two advantages. The total intensity is concentrated into one line, and in addition it is increased by up to 200 % through the nuclear Overhauser effect mentioned earlier (see also Chapter 10.2.2). This results in a considerable reduction in the recording time needed.

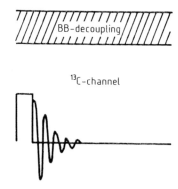

^{1}H-channel

BB-decoupling

^{13}C-channel

Figure 5-4.
Principle of broad-band (BB) ^{1}H decoupling. The BB decoupler is switched on during the entire ^{13}C data acquisition period.

5.3.2 The Gated Decoupling Experiment

When ^{13}C NMR spectra are recorded with simultaneous ^{1}H BB decoupling, the shortening of the recording time is achieved at the cost of a sacrifice of information, as the C,H couplings disappear. If one wants to avoid losing this information, it is necessary to record the coupled (i. e. non-decoupled) ^{13}C NMR spectrum. As explained in Section 5.3.1, this requires long recording times, firstly because the intensities are divided into multiplets, and secondly because there is no longer a gain in intensity from the NOE. However, it is possible, by means of a specially designed pulse experiment, to record the ^{13}C spectrum with the C,H couplings while retaining some, though not all, of the NOE enhancement.

Figure 5-5 shows schematically the pulse sequence used to achieve this. To begin with, the BB decoupling in the ^{1}H channel is switched on (this phase is indicated in the diagram by hatching). During this time the NOE acts on the system to establish the population ratios which give the increase in the intensities of the signals (Chapter 10). Before the observing pulse and the recording of the data, the BB decoupler is switched off. Whereas the indirect spin-spin coupling reappears immediately, the level populations only return slowly to their equilibrium values. This return to equilibrium is controlled by the relaxation times T_1 (Chapter 7), which are in general greater than the time needed to record the data. In this way it is possible to observe C,H couplings in the ^{13}C NMR spectrum while retaining the amplification of the signals by the NOE.

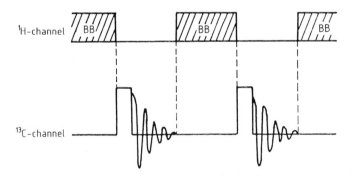

Figure 5-5.
Principle of the gated decoupling experiment (two cycles are shown). The ^{1}H BB decoupler is switched off during the ^{13}C excitation pulse and data acquisition.

This *gated decoupling* method was used to record the spectrum of crotonic acid (**3**) which is shown in Figure 5-3 A. Another spectrum of this type can be found in Figure 8-18 B.

In Section 1.6.3.2 we have already met the *inverse gated decoupling* experiment. In this the ^{1}H BB decoupler is switched on only for the duration of the observing pulse and data recording in the ^{13}C channel (Fig. 5-6). This eliminates the C,H

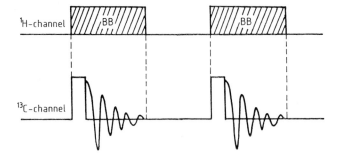

coupling, but no NOE is allowed to build up, and the intensities of the ^{13}C signals are not distorted.

This technique can also be useful where the magnetogyric ratio γ is negative (as is the case for ^{15}N, for example), since the NOE might otherwise suppress the signal.

5.3.3 ^{1}H Off-Resonance Decoupling

Assigning the singlets in ^{13}C NMR spectra recorded with ^{1}H BB decoupling often presents difficulties, since the only data available for making the assignments are the chemical shifts. In some cases it would be very useful to know the number of directly bonded hydrogen atoms, which one would normally expect to obtain from the multiplicities of the signals. A now rather outdated method of achieving this is by ^{1}H off-resonance decoupling. In this one records the ^{13}C NMR spectrum by the usual pulse method, while simultaneously irradiating all the ^{1}H resonances with a radiofrequency power level that is sufficient to significantly reduce the C,H couplings in the spectrum without removing them completely. By suitably adjusting the instrumental variables one can arrive at a situation where only residual couplings through one bond are seen in the spectrum. Thus, in accordance with the number of directly attached hydrogen atoms, ^{13}C nuclei in CH_3, CH_2 and CH groups give quartets, triplets and doublets respectively, while quaternary carbons, those without any directly attached hydrogen atoms, give singlets. Thus we obtain just the information that is required, and even manage to retain most of the NOE enhancement.

However, the multiplets in the off-resonance spectrum do not always have the appearance of doublets, triplets or quartets; instead they are often even more complex than in the normal, non-decoupled spectrum, as the ^{13}C NMR spectrum is then the X part of a higher-order spectrum.

In Chapter 8 we shall meet the DEPT and APT method, which have now entirely replaced the off-resonance technique.

5.3.4 Selective Decoupling in ^{13}C NMR Spectroscopy

In a selective decoupling experiment the resonances of particular nuclei in the molecule are saturated specifically, so that all couplings to these nuclei are eliminated. ^{1}H BB decoupling is not selective according to this definition. On the other hand, all homonuclear proton decoupling experiments are selective.

Another type of experiment which has become of practical importance is *selective heteronuclear decoupling*, especially selective ^{13}C{^{1}H} decoupling. To explain this technique we again take crotonic acid as an example.

In the ^{1}H NMR spectrum of crotonic acid the methyl proton signals appear at $\delta = 1.89$ and the olefinic proton signals at $\delta = 5.82$ and 7.04. If we irradiate selectively at the resonance frequency of the methyl protons, these protons are decoupled from the ^{13}C nucleus of the methyl group. The signals in the ^{1}H NMR spectrum which are of importance here are not the main methyl signal at $\delta = 1.89$ but the ^{13}C satellites! These come from the 1.1 % of the molecules whose resonances we observe in the ^{13}C spectrum. In order to achieve decoupling the decoupler power must be so great that both the ^{13}C satellites, which are separated by about 125 Hz, are excited. Only then will the four peaks of the methyl group in the ^{13}C spectrum become a singlet, while all the other couplings, and therefore the multiplet structures of the other signals in the ^{13}C spectrum, remain unchanged. Figure 5-3 B shows the resulting spectrum. Fortunately the ^{1}H resonances of the chemically non-equivalent hydrogen nuclei in crotonic acid are well separated. Usually the conditions are not so favorable, and complications occur due to the fact that decoupling with such a broad frequency band (> 125 Hz) also saturates nearby, or even overlapping, ^{13}C satellites belonging to other protons. For the protons thus affected the conditions which apply are similar to those in off-resonance decoupling.

In order to carry out a selective decoupling experiment it is first necessary to assign the signals in the ^{1}H NMR spectrum and accurately determine their frequencies. With modern spectrometers selective decoupling is a fairly straightforward routine procedure. There are advantages in combining the two techniques of selective ^{1}H decoupling and gated decoupling in one experiment, so as to increase the signal intensities by means of the NOE.

In Section 9.4.1 we shall learn about a two-dimensional procedure, known as C,H-COSY or HETCOR, whereby one can determine which protons are coupled to which ^{13}C nuclei.

5.4 Bibliography for Chapter 5

[1]   W. v. Philipsborn, *Angew. Chem. Int. Ed. Engl. 10* (1971) 472.

[2]   A. E. Derome: *Modern NMR Techniques for Chemistry Research.* Oxford: Pergamon Press, 1987, p. 159.

[3]   R. Freeman: *A Handbook of Nuclear Magnetic Resonance.* New York: Longman Scientific & Technical, 1987, p. 43 ff.

Additional and More Advanced Reading

R. A. Hoffman and S. Forsén: High Resolution Nuclear Magnetic Double and Multiple Resonance. In: *Prog. Nucl. Magn. Reson. Spectrosc. 1* (1966) 15.

J. K. M. Sanders and J. D. Mersh: Nuclear Magnetic Double Resonance; The Use of Difference Spectroscopy. In: *Prog. Nucl. Magn. Reson. Spectrosc. 15* (1982) 353.

6 Assignment of ^{1}H and ^{13}C Signals

6.1 Introduction

NMR studies are most commonly undertaken with structure determination as the aim. The required information is contained in the spectral parameters, of which the three most important are the chemical shifts (δ), the coupling constants (J) and the intensities (I). The methods whereby these parameters are obtained from the analysis of the spectra have been described in Chapters 1, 4 and 5. In this procedure one does not in principle need to have in mind a molecule with a certain structure; usually, however, there is a proposed structure to be considered, and the questions then posed are: does the ^{1}H or ^{13}C NMR spectrum contain the expected signals for each of the hydrogen or carbon nuclei in the molecule, and does the spectrum confirm the proposed structure?

If, on the other hand, the structure of the molecule is completely unknown, one searches in the spectrum for characteristic signals which provide evidence for the presence of important structural features or functional groups. In practice the two stages – analysis and assignment – are not rigidly separated. Often a partial assignment may lead to further analysis of the spectrum.

Even the experienced spectroscopist makes frequent use of data collections. Up to fairly recent times these generally consisted of monographs, review articles and spectra catalogs in which spectra were reproduced. Such a catalog of complete spectra can certainly be very illuminating to the user, but a collection containing thousands of spectra is awkward to handle. Furthermore, it is difficult to make comparisons when the spectra have been recorded at different frequencies. This is especially true of ^{1}H NMR spectra.

In the computer era catalogs can be produced in the form of magnetic tapes, CD-ROM, and disks. This opens up entirely new possibilities for building up data files and for data processing (Section 6.4). Data files on tape or disk do not contain the complete spectral curve with all the signals, their intensities and the line shapes, but just the spectral parameters such as chemical shifts, coupling constants, multiplicities and intensities. Storing these data for a large number of compounds at present still requires the memory capacity of a main-frame computer. Through the spectrometer's computer one can have access to a central data bank, e. g. via a telephone data link. In addition it is already possible to have data collections of limited size stored in situ by the spectrometer's computer.

Developments are at a more advanced stage for ^{13}C NMR spectroscopy than for ^1H. The reasons for this become apparent when one compares ^1H and ^{13}C spectra.

In the following sections the use of standard spectra catalogs as aids to assignment will be deliberately avoided. Instead we will indicate other methods which can help to solve problems of assignment in ^1H and ^{13}C NMR spectroscopy (Sections 6.2 and 6.3). Section 6.4 describes how a computer can be used for spectral assignment.

The techniques of one- and two-dimensional NMR spectroscopy using special pulse sequences, which have led to such spectacular progress in methods of assignment in recent years, call for some theoretical background knowledge, and will therefore be described in Chapters 8 and 9.

6.2 ^1H NMR Spectroscopy

6.2.1 Defining the Problem

In many spectra it is not possible to assign all the signals. Indeed, this is often not essential for solving the problem, as for example when all that is required is to find out whether a substitution reaction, such as a methylation or acetylation, has been successful. In these examples it is sufficient to show that there is a singlet in the expected region with an intensity corresponding to three protons. If, on the other hand, one is interested in the positions of the substituents in a benzene ring, it is only necessary to analyze and assign the signals of all the aromatic protons. An example of this is the spectra of the three isomeric dichlorobenzenes shown in Figure 2-13. There it was possible from the spectral types to unambiguously determine the positions of the two chlorine substituents in each case. But which signals should be assigned to which protons?

Reference was made in Chapter 2 to empirical correlations and additivity rules which can be used to predict approximate chemical shifts for protons and ^{13}C nuclei in the various classes of compounds. As these methods can be used without any additional experimental work, without large collections of data, and without a computer, we shall deal with them first. After that we will discuss methods of assignment which require changes to the recording conditions and the samples.

6.2.2 Empirical Correlations for Predicting Chemical Shifts

All rules for predicting chemical shifts are based on the observation that within a particular class of compounds the contribution of a substituent to the chemical shift is nearly constant.

These contributions, called the *substituent increments*, have been determined by analyzing large quantities of experimental data. Using these in conjunction with simple rules enables one to estimate the chemical shifts (see also Section 6.4.2). The increments have been obtained by averaging experimental data from a more or less arbitrary selection of compounds, and the chemical shifts calculated from them cannot be any more reliable than the increments themselves. For this reason they will be referred to throughout this book as only estimates. Provided that this fact is kept in mind, δ-values in ppm obtained using these empirical correlations are a valuable aid to assignment. In the following sections the most important of these rules will be described together with examples. (The sequence of the different classes of compounds is the same as in Chapter 2.)

6.2.2.1 Alkanes (Shoolery's Rule)

The chemical shifts of protons in disubstituted, and to a limited extent trisubstituted, methanes can be predicted using *Shoolery's rule* (Equation 6-1). For a disubstituted methane $X-CH_2-Y$ this is:

$$\delta(CH_2) = 0.23 + S_x + S_y \text{ [ppm]} \qquad (6-1)$$

The reference point for δ is the chemical shift of the protons in methane. The increments S are listed in Table 6-1.

As an example the δ-value for the methylene protons in benzylmethylamine has already been calculated in Section 2.2.1.

Table 6-1.

Substituent increments S for estimating ^1H chemical shifts in disubstituted methanes X-CH$_2$-Y using *Shoolery's rule*:

$$\delta(CH_2) = 0.23 + S_x + S_y \text{ [ppm]} \tag{6-1}$$

Substituent	S	Substituent	S
$-CH_3$	0.47	$-NRR'$	1.57
$-CF_3$	1.14	$-SR$	1.64
$-CR=CR'R''$	1.32	$-I$	1.82
$-C\equiv CH$	1.44	$-Br$	2.33
$-COOR$	1.55	$-OR$	2.36
$-CONH_2$	1.59	$-Cl$	2.53
$-COR$	1.70	$-OH$	2.56
$-C\equiv N$	1.70	$-OCOR$	3.13
$-C_6H_5$	1.83	$-OC_6H_5$	3.23

6.2.2.2 Alkenes

Chemical shifts in ethylene derivatives can be estimated using the *Pascual-Meier-Simon rule* [1] (Equation 6-2):

$$\delta(H) = 5.28 + S_{gem} + S_{cis} + S_{trans} \text{ [ppm]} \tag{6-2}$$

The reference point in this case is the chemical shift of the protons in ethylene ($\delta = 5.28$). The increments are given in Table 6-2. The values given in parentheses in Examples 1 and 2 below and in Section 6.2.2.3 were calculated using the program "Spectool" (see Section 6.4.2).

Example 1: trans-crotonic acid (**1**)

$\delta(H-2) = 5.28 + S_{gem} (COOH) + S_{cis} (CH_3)$
$\qquad = 5.28 + 0.69 \qquad\quad - 0.26 \quad = 5.71 \; (5.83)$
$\qquad\qquad\qquad\text{experimental value: } 5.82$

$\delta(H-3) = 5.28 + S_{gem} (CH_3) \quad + S_{cis} (COOH)$
$\qquad = 5.28 + 0.44 \qquad\quad + 0.97 \quad = 6.69 \; (6.68)$
$\qquad\qquad\qquad\text{experimental value: } 7.04$

Also: $\delta(CH_3) = 1.89$; $\delta(COOH) \approx 12$

Example 2: cis- and *trans*-stilbene (**2** and **3**)

In this special case the position of the phenyl substitutent cannot be determined from the value of ^3J (H,H), as the two olefinic protons are equivalent in both isomers, and consequently each spectrum shows only a singlet in this region. However, an assignment can be made by using the rule (6-2).

Table 6-2.

Substituent increments $S^{1)}$ for estimating ^1H chemical shifts in alkenes using the expression:

$$\delta(H) = 5.28 + S_{gem} + S_{cis} + S_{trans} \text{ [ppm]} \qquad (6\text{-}2)$$

Substituent	S_{gem}	S_{cis}	S_{trans}
$-H$	0	0	0
$-CH_3$ (alkyl)	0.44	-0.26	-0.29
$-F$	1.51	-0.43	-1.05
$-Cl$	1.00	0.19	0.03
$-Br$	1.04	0.40	0.55
$-I$	1.11	0.78	0.85
$-NR_2$ (aliph.)	0.69	-1.19	-1.31
$-OAlkyl$	1.18	-1.06	-1.28
$-OCOCH_3$	2.09	-0.40	-0.67
$-C_6H_5$	1.35	0.37	-0.10
$-CH=CH_2$ (conj.)	1.26	0.08	-0.01
$-COOH$ (conj.)	0.69	0.97	0.39
$-NO_2$	1.84	1.29	0.59

$^{1)}$ Data from [1] and [2].

cis-stilbene (**2**):
$$\delta(H) = 5.28 + S_{gem} \text{ (Ph)} + S_{trans} \text{ (Ph)}$$
$$= 5.28 + 1.35 \quad - 0.1 \quad = 6.53 \ (6.56)$$
$$\text{experimental value: } 6.55$$

trans-stilbene (**3**):
$$\delta(H) = 5.28 + S_{gem} \text{ (Ph)} + S_{cis} \text{ (Ph)}$$
$$= 5.28 + 1.35 \quad + 0.37 \quad = 7.00 \ (6.99)$$
$$\text{experimental value: } 7.1$$

6.2.2.3 Benzene Derivatives

For benzene derivatives we have the following empirical equation:

$$\delta(H) = 7.27 + \Sigma S \text{ [ppm]} \qquad (6\text{-}3)$$

The reference point here is the chemical shift of the protons in benzene. The substituent increments S are given in Table 6-3. For multiple substitution the appropriate increments corresponding to the positions of the substituents must be added together.

143

Table 6-3.

Substituent increments $S^{1)}$ for estimating ^1H chemical shifts in arenes using the expression:

$$\delta(H) = 7.27 + \Sigma S \text{ [ppm]} \tag{6-3}$$

Substituent	S_o	S_m	S_p
$-CH_3$	-0.17	-0.09	-0.18
$-CH_2CH_3$	-0.15	-0.06	-0.18
$-F$	-0.30	-0.02	-0.22
$-Cl$	$+0.02$	-0.06	-0.04
$-Br$	$+0.22$	-0.13	-0.03
$-I$	$+0.40$	-0.26	-0.03
$-OH$	-0.50	-0.14	-0.4
$-OCH_3$	-0.43	-0.09	-0.37
$-OCOCH_3$	-0.21	-0.02	0.0
$-NH_2$	-0.75	-0.24	-0.63
$-N(CH_3)_2$	-0.60	-0.10	-0.62
$-C_6H_5$	$+0.18$	0.0	$+0.08$
$-CHO$	$+0.58$	$+0.21$	$+0.27$
$-COCH_3$	$+0.64$	$+0.09$	$+0.3$
$-COOCH_3$	$+0.74$	$+0.07$	$+0.20$
$-NO_2$	$+0.95$	$+0.17$	$+0.33$

1) Data from [3].

Example: p-nitroanisole (**4**)

$\delta(H\text{-}2,6) = 7.27 + S_o (OCH_3) + S_m (NO_2)$
$\quad\quad\quad = 7.27 - 0.43 \quad\quad + 0.17 \quad\quad = 7.01 \ (7.04)$
$\quad\quad\quad\quad\quad$ experimental value: 6.88

$\delta(H\text{-}3,5) = 7.27 + S_o (NO_2) \ + S_m (OCH_3)$
$\quad\quad\quad = 7.27 + 0.95 \quad\quad - 0.09 \quad\quad = 8.13 \ (8.12)$
$\quad\quad\quad\quad\quad$ experimental value: 8.15

Also: $\delta(OCH_3) = 3.90$

The agreement between the calculated and observed values is good. Experience shows that the deviations become greater with increasing numbers of substituents, with *ortho* substitution, and with bulky substituents.

6.2.3 Decoupling Experiments

Spin decoupling makes spectra simpler, i. e. it reduces the number of lines (Section 5.2), and it is therefore one of the routine experiments of every NMR laboratory. In many cases, despite decoupling, the spectra are still not easy to analyze, but from the change in the spectrum one does at least find the positions of the signals of nuclei that are mutually coupled. Figure 5-2 shows an example. In principle one can start from the resonance of one proton which has been assigned and, by a series of successive decoupling experiments, assign the whole spectrum. In practice, however, difficulties always arise when the resonance frequencies of the coupled nuclei are too close together (i. e. when $\Delta\nu$ is less than about 100 Hz).

For example, if one wished to assign the complex spectrum of the mixture of α- and β-glucose shown in Figure 3-3 by using H,H decoupling, one would start with the resonances of the proton H-1, as these are well separated from the others. In this way one would easily locate the H-2 signals of both anomers. However, to then assign further signals by saturating the H-2 resonances is virtually impossible, since decoupling these will always simultaneously saturate other signals.

A much more elegant way of elucidating such connectivity relationships is by modern one- and two-dimensional NMR experiments (Chapters 8 and 9).

6.2.4 Effects of Solvent and Temperature

Often there are interactions between dissolved molecules and the solvent. This can cause the signals to shift, especially if the solvent molecules arrange themselves in preferred orientations around the solute molecule, or if hydrogen bonding can occur. Effects of this sort are mainly found with polar compounds. The most extreme solvent effects are obtained when the spectrum is measured first in a non-polar solvent (CCl_4, $CDCl_3$, hydrocarbons), then in one which causes large anisotropy effects (benzene or other aromatic compounds).

When OH groups are present as substituents, dimethylsulfoxide is a suitable solvent. The formation of strong hydrogen bonds to the solvent molecules then slows down the proton exchange to such an extent that even vicinal couplings between the OH protons and protons at the α-position can be observed. The multiplet pattern of the OH signal – triplet, doublet or singlet – gives a direct indication of whether a primary, secondary or tertiary alcohol is present. Analogous effects are also observed for NH_2 groups.

Changing the sample temperature sometimes causes certain signals to move relative to others. This is nearly always an indication of molecular association. Such effects are most familiar for OH signals. Raising the temperature can cause coupling effects to disappear, for example through inducing faster proton exchange. Other temperature effects and the uses of paramagnetic shift reagents and of chiral compounds as aids to assignment will be described in Chapters 11 and 12.

6.2.5 Altering the Chemical Structure of the Sample

Introducing a chemical change into the molecule means that a new spectrum is obtained. If one wishes to alter a compound by chemical means with the aim of analyzing the spectrum of the original compound, the molecule must not be altered too radically! Two possible methods are:

- *replacement of hydrogen by deuterium*, and
- *derivatization* – in cases where this is chemically possible.

Replacing hydrogen by deuterium causes the corresponding signal to disappear from the spectrum. It also simplifies the signals of protons that were previously coupled to the proton replaced (Section 3.6.2). OH, NH and SH protons can readily undergo exchange with deuterium; it is often sufficient to simply shake the CCl_4 or $CDCl_3$ solution in the sample tube with a few drops of D_2O. The signals of the exchangeable protons are eliminated or greatly reduced, and can thus be identified.

Selevtive derivatization usually only affects the resonance positions of the nearest neighbors. A routine method which is important in carbohydrate chemistry is the acetylation of OH groups. This causes a reduction of about 1 ppm in the shielding of the CH or CH_2 protons in the a-position (i. e. a shift to higher δ-values). Another advantage of acetylation is that the resulting carbohydrate derivatives are more readily soluble in organic solvents, which usually give better resolution than in H_2O or D_2O. Both acetylation and methylation are commonly used. Alcohols and phenols undergo conversion by trichloroacetyl isocyanate to give the corresponding carbamates.

6.3 ¹³C NMR Spectroscopy

6.3.1 Defining the Problem

Let us suppose that we need to assign the signals in the ¹³C NMR spectrum of the complex neuraminic acid derivative (**5**, Fig. 6-1). Since the molecule contains 13 carbon atoms and has no symmetry elements, we find 13 singlets in the spectrum recorded with ¹H broad-band decoupling. The chemical shifts can easily be read off from the spectrum.

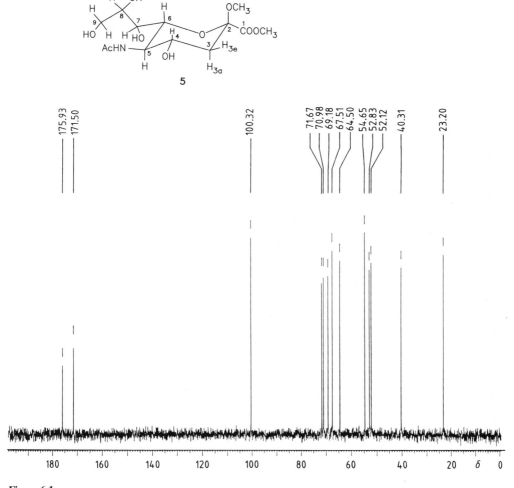

Figure 6-1.
100.617 MHz ¹³C NMR spectrum of the methylketoside of *N*-acetyl-*β*-D-neuraminic acid methyl ester (**5**) in D₂O; 256 FIDs, 16 *K* data points, 2.441 Hz/point, 1 s delay.

What aids are available to help us in assigning the signals to the individual carbon nuclei? The first possibility is to compare the observed spectrum with those of known compounds of similar structure. The degree of success with this method depends on one's personal experience and on what collections of data and spectra are available.

As in ^{1}H NMR spectroscopy (Section 6.2.2), so too in ^{13}C NMR spectroscopy there are various empirical rules which allow one to estimate ^{13}C chemical shifts without a great deal of labor (see also Section 6.4.2). All other assignment procedures call for additional experiments, which either use special equipment and techniques or involve altering the chemical structure.

We will deal first with empirical correlations, then with some special techniques. We shall return to our example of the neuraminic acid derivative (**5**) in the final section (6.4.2) of this chapter, and we shall also consider the ^{1}H and ^{13}C NMR spectra of this compound in much more detail in Chapters 8 and 9.

6.3.2 Empirical Correlations for Predicting Approximate Chemical Shifts

No universally valid correlation exists between calculated σ- and π-electron densities and chemical shifts. Similarly, attempts to correlate chemical shifts with electronegativities and with Hammett or Taft constants have resulted in simple linear relationships in only a few cases. None of these are suitable as aids to assignment for the NMR spectroscopist. The only relationships which have become important in practice are those whereby chemical shifts are estimated using empirically determined substituent increments. These are based on the assumption that the substituent effects are additive. A selection of these are described in the following sections. In each of the examples the experimentally determined δ-values, taken from Ref. [4], are given below the structural formula.

6.3.2.1 Alkanes

Linear and branched alkanes: The spectra of a large number of pure hydrocarbons have been analyzed by Grant and Paul [5], and from these they developed an incremental system which can be used to predict the chemical shifts of the carbon nuclei in alkanes:

$$\delta_i = -2.3 + 9.1\,n_\alpha + 9.4\,n_\beta - 2.5\,n_\gamma + 0.3\,n_\delta + 0.1\,n_\varepsilon + \Sigma\,S_{ij}\ [\text{ppm}]$$

$$(6\text{-}4)$$

where:

δ_i = chemical shift of the carbon nucleus of interest

n = numbers of carbon atoms in the α-, β-, γ-, δ- and ε-positions relative to this nucleus

S_{ij} = steric correction terms taking account of branching.

The increments S_{ij} are given in Table 6-4. The values given in parentheses in the example below and in Sections 6.3.2.2 and 6.3.2.3 were calculated using the program "Spectool" (see Section 6.4.2).

Table 6-4.
Steric correction factors S_{ij}[1]) for estimating ^{13}C chemical shifts in branched alkanes by the method of Ref. [5].

i \ j	primary	secondary	tertiary	quaternary
primary	0	0	− 1.1	− 3.4
secondary	0	0	− 2.5	− 7.5
tertiary	0	−3.7	− 9.5	−15.0
quaternary	−1.5	−8.4	−15.0	−25.0

i = observed nucleus; J = neighbor nucleus
[1]) from [6].

Example: 2-methylbutane (**6**)

C-1: $n_\alpha = 1$, $n_\beta = 2$, $n_\gamma = 1$
Steric corrections: primary with adjacent tertiary → − 1.1
S_{ij} = − 1.1
δ(C-1) = − 2.3 + (9.1 x 1) + (9.4 x 2) − (2.5 x 1) − 1.1 =   22.0
(22.3)

C-2: $n_\alpha = 3$, $n_\beta = 1$
Steric corrections: 1. tertiary with adjacent primary   →   0
2. tertiary with adjacent secondary → − 3.7
S_{ij} = − 3.7
δ(C-2) = − 2.3 + (9.1 x 3) + (9.4 x 1) − 3.7 =   30.7
(30.1)

C-3: $n_\alpha = 2$, $n_\beta = 2$
Steric corrections: 1. secondary with adjacent tertiary → − 2.5
2. secondary with adjacent primary →   0
S_{ij} = − 2.5
δ(C-3) = − 2.3 + (9.1 x 2) + (0.4 x 2) − 2.5 =   32.2
(32.0)

C-4: $n_\alpha = 1$, $n_\beta = 1$, $n_\gamma = 3$
Steric corrections: primary with adjacent secondary →   0
S_{ij} =   0
δ(C-4) = − 2.3 + (9.1 x 1) + (9.4 x 1) − (2.5 x 2) =   11.2
(11.5)

$$\underset{21.9}{H_3\overset{1}{C}} - \underset{29.9}{\overset{2}{C}H} - \underset{31.6}{\overset{3}{C}H_2} - \underset{11.5}{\overset{4}{C}H_3}$$
with CH_3 attached to C-2

6

On the basis of this calculation the signals found in the spectrum at $\delta = 21.9$ and 11.5 can be assigned to C-1 and C-4 respectively. The assignment of the C-2 and C-3 signals is not quite so unambiguous, as the difference between the δ-values is too small. Here, however, one can easily decide between the two alternative assignments with the aid of an off-resonance decoupled spectrum (Section 5.3.3), as this gives a doublet for C-2 and a triplet for C-3, or with the aid of an APT or DEPT spectrum (Section 8.6).

Substituted alkanes (R ≠ alkyl): For a substituted alkane one first calculates the δ-values for the corresponding unsubstituted hydrocarbon from Equation (6-4), then corrects these using the increments listed in Table 6-5.

Table 6-5.
Substituent increments $S^{1)}$ for estimating ^{13}C chemical shifts in substituted alkanes $X - C_\alpha - C_\beta - C_\gamma - C_\delta$.

Substituent	S_α	S_β	S_γ	S_δ
−D	−0.4	−0.12	−0.02	−
−CH$_3$	9.1	9.4	−2.5	0.3
−CH=CH$_2$	22.3	6.9	−2.2	0.2
−C≡CH	4.5	5.5	−3.5	−
−C$_6$H$_5$	22.3	8.6	−2.3	0.2
−CHO	31.9	0.7	−2.3	−
−COCH$_3$	30.9	2.3	−0.9	2.7
−COOH	20.8	2.7	−2.3	1.0
−CN	3.6	2.0	−3.1	−0.5
−MH$_2$	28.6	11.5	−4.9	0.3
−NO$_2$	64.5	3.1	−4.7	−1.0
−OH (prim.)	48.3	10.2	−5.8	0.3
−OH (sec.)	44.5	9.7	−3.3	0.2
−OH (tert.)	39.7	7.3	−1.8	0.3
−OR	58.0	8.1	−4.7	1.4
−OCOCH$_3$	51.1	7.1	−4.8	1.1
−SH	11.1	11.8	−2.9	0.7
−F	70.1	7.8	−6.8	−
−Cl	31.2	10.5	−4.6	0.1
−Br	20.0	10.6	−3.1	0.1
−I	−6.0	11.3	−1.0	0.2

$^{1)}$ Adapted from [4] and [6].

Example: isobutanol (**7**)

$$\overset{3}{(CH_3)_2}\overset{2}{CH} - \overset{1}{CH_2} - OH \quad \mathbf{7}$$
20.4   32.0   70.2

We treat this as being formally derived from isobutane (**8**) by substitution of a hydroxyl group.

$$(CH_3)_2CH - CH_3 \quad \mathbf{8}$$
24.6   23.3   24.6

For **8**, by the method used in the example of 2-methylbutane (**6**), we calculate the following δ-values:

δ (CH$_3$) = 24.5
δ (CH) = 25.0

Using the values $S_\alpha = 48.3$, $S_\beta = 10.2$ and $S_\gamma = -5.8$, as given in Table 6-5 for a (primary) OH group, we obtain

δ (C-1) = 24.5 + 48.3 = 72.8
δ (C-2) = 25.0 + 10.2 = 35.2
δ (C-3) = 24.5 − 5.8 = 18.7

The deviations between the calculated and observed values give an indication of the accuracy of the method; in our example the assignment of the signals is unambiguous.

For cycloalkanes different additivity rules and substituent increments apply.

6.3.2.2 Alkenes

Unsaturated hydrocarbons: The chemical shifts of alkenes can be calculated from Equation (6-5) [7].

$$C_\gamma - C_\beta - C_\alpha - \overset{1}{C} = \overset{2}{C} - C_{\alpha'} - C_{\beta'} - C_{\gamma'}$$
$$\uparrow$$
$$^{13}C \text{ nucleus}$$
$$\text{being observed}$$

$$\delta\,(\text{C-1}) = 123.3 + 10.6n_\alpha + 7.2\,n_\beta - 1.5n_\gamma \qquad (6\text{-}5)$$
$$-7.9n_{\alpha'} - 1.8n_{\beta'} + 1.5n\gamma' + \Sigma\,S \text{ [ppm]}$$

where n is the number of neighbor carbon atoms of each type and S is a steric term given by:

$S = \quad 0 \quad$ if C_α and $C_{\alpha'}$ are in the E-configuration (aa', *trans*)
$S = -1.1 \quad$ if C_α and $C_{\alpha'}$ are in the Z-configuration (aa', *cis*)
$S = -4.8 \quad$ for two alkyl substituents at C-1 (aa)
$S = +2.5 \quad$ for two alkyl substituents at C-2 (aa')
$S = +2.3 \quad$ for two or three alkyl substituents at C_β.

Example: 2-methylbut-1-ene (**9**)

C-1: $n_{\alpha'} = 2$, $n_{\beta'} = 1$, $S = +2.5$

$$\delta(\text{C-1}) = 123.3 - (7.9 \times 2) - (1.8 \times 1) + 2.5 = 108.2$$
$$(108.2)$$

C-2: $n_{\alpha} = 2$, $n_{\beta} = 1$, $S = -4.8$

$$\delta(\text{C-2}) = 123.3 + (10.6 \times 2) + (7.2 \times 1) - 4.8 = 146.9$$
$$(146.9)$$

$$\begin{array}{c} 22.5 \\ \overset{}{CH_3} \\ | \\ H_2\overset{1}{C} = \overset{2}{C} - \overset{3}{CH_2} - \overset{4}{CH_3} \\ 109.1 \quad 147.0 \quad 31.1 \quad 12.5 \end{array}$$

9

Substituted alkenes are treated as ethylene derivatives, to which Equation (6-6) can be applied [6].

$$\delta = 123.3 + \Sigma\, S_i \; [\text{ppm}] \qquad (6\text{-}6)$$

The substituent increments S_i are listed in Table 6-6.

Table 6-6.
Substituent increments S_i[1] for estimating ^{13}C chemical shifts for the double-bonded carbon nuclei in alkenes using the expression:
$\delta = 123.3 + \Sigma\, S_i \; [\text{ppm}]$

$$X - \overset{1}{CH} = \overset{2}{CH_2}$$

Substituent	S_1	S_2	Substituent	S_1	S_2
$-H$	0	0	$-OCH_3$	29.4	-38.9
$-CH_3$	10.6	-7.9	$-OCOCH_3$	18.4	-26.7
$-CH_2CH_3$	15.5	-9.7			
			$-C_6H_5$	12.5	-11.0
$-F$	24.9	-34.3	$-CH=CH_2$	13.6	-7.0
$-Cl$	2.6	-6.1	$-COOH$	4.2	8.9
$-Br$	-7.9	-1.4			
$-I$	-38.1	7.0	$-NO_2$	22.3	-0.9

[1] Data from [6].

Example: crotonic acid (**10**)

$$\delta(\text{C-2}) = 123.3 + S_1(\text{COOH}) + S_2(\text{CH}_3)$$
$$= 123.3 + 4.2 \qquad - 7.9 \qquad = 119.6 \; (120.5)$$

$$\delta(\text{C-3}) = 123.3 \; S_1(\text{CH}_3) \qquad + S_2(\text{COOH})$$
$$= 123.3 + 10.6 \qquad + 8.9 \qquad = 142.8 \; (142.5)$$

$$\begin{array}{c} \overset{4}{H_3C} - \overset{3}{CH} = \overset{2}{CH} - \overset{1}{COOH} \\ 18 \qquad 147 \qquad 122 \qquad 172 \\ (16.4) \qquad\qquad\qquad (170.0) \end{array}$$

10

6.3.2.3 Benzene Derivatives

The chemical shifts of the carbon nuclei in substituted benzenes are calculated according to Equation (6-7)

$$\delta = 128.5 + \Sigma S \ [\text{ppm}] \qquad (6\text{-}7)$$

The reference point is the ^{13}C resonance of benzene. The substituent increments S are given in Table 6-7.

Table 6-7.
Substituent increments $S^{1)}$ for estimating ^{13}C chemical shifts in substituted benzenes using the expression:
$\delta = 128.5 + \Sigma S \ [\text{ppm}]$ $\qquad (6\text{-}7)$

Substituent	S_1	S_o	S_m	S_p
$-CH_3$	9.2	0.7	−0.1	− 3.1
$-CH_2CH_3$	15.6	− 0.5	0.0	− 2.7
$-F$	34.8	−13.0	1.6	− 4.4
$-Cl$	6.3	0.4	1.4	− 1.9
$-Br$	− 5.8	3.2	1.6	− 1.6
$-I$	−34.1	8.9	1.6	− 1.1
$-OH$	26.9	−12.8	1.4	− 7.4
$-OCH_3$	31.4	−14.4	1.0	− 7.7
$-OCOCH_3$	22.4	− 7.1	0.4	− 3.2
$-NH_2$	18.2	−13.4	0.8	−10.0
$-N\,(CH_3)_2$	22.5	−15.4	0.9	−11.5
$-C_6H_5$	13.1	− 1.1	0.4	− 1.1
$-CHO$	8.4	1.2	0.5	5.7
$-COCH_3$	8.9	0.1	−0.1	4.4
$-COOCH_3$	2.0	1.2	−0.1	4.3
$-NO_2$	19.9	− 4.9	0.9	6.1

$^{1)}$ Data from [4].

*Example: p-nitrophenol (**11**)*

$\delta(C\text{-}1) = 128.5 + S_1(OH) \quad + S_p(NO_2)$
$\qquad = 128.5 + 26.9 \qquad + 6.1 \qquad = 161.5 \ \ \text{Exp.: } 161.5 \ (163.4)$
$\delta(C\text{-}2) = 128.5 + S_o(OH) \quad + S_m(NO_2)$
$\qquad = 128.5 - 12.8 \qquad + 0.9 \qquad = 116.6 \qquad 115.9 \ (116.6)$
$\delta(C\text{-}3) = 128.5 + S_m(OH) \quad + S_o(NO_2)$
$\qquad = 128.5 + 1.4 \qquad - 4.9 \qquad = 125.0 \qquad 126.4 \ (125.0)$
$\delta(C\text{-}4) = 128.5 + S_p(OH) \quad + S_1(NO_2)$
$\qquad = 128.5 - 7.4 \qquad + 19.9 \qquad = 141.0 \qquad 141.7 \ (141.0)$

11

153

6.3.3 Decoupling Experiments

The ^{1}H off-resonance decoupling and selective ^{1}H decoupling methods described in Chapter 5 are nowadays only used in rare cases. The off-resonance method has been largely replaced by the more elegant APT and DEPT methods, which we shall meet in Sections 8.3 and 8.6 respectively, and the selective ^{1}H decoupling method has essentially been superseded by two-dimensional methods (Chapter 9).

6.3.4 T_1 Measurements

The spin-lattice relaxation time T_1 depends on the molecular motions, and the more rapid these motions are the greater is T_1 (Chapter 7). On the basis of this dependence it has been possible, for example, to assign the NMR signals of carbon nuclei in mobile side-chains of large molecules. Similarly the signals of carbon nuclei which lie on the longitudinal axis (backbone) of the molecule can be distinguished from those lying off this axis, as the latter are more mobile. We shall meet examples of this sort in Chapter 7.

6.3.5 Solvent and Temperature Effects and Shift Reagents

The effects of solvent and temperature on ^{13}C resonances are complicated, and unlike the situation in ^{1}H NMR spectroscopy they are rarely used as an aid to assignment.

On the other hand shift reagents are more commonly used. This topic is so large that a separate chapter (Chapter 12) will be devoted to it.

6.3.6 Chemical Changes to the Sample

All the assignment techniques described so far are of an instrument-orientated kind. What contributions can the chemist make to solving assignment problems?

Of the many possibilities that exist for chemically altering a molecule, the three listed below are of special importance to the NMR spectroscopist; none of these involves interfering too drastically with the molecular structure:

- *¹³C enrichment* at specific sites in the molecule by appropriate synthetic methods using labeled starting materials;
- *specific deuteration*, either by synthesis or by H,D exchange;
- *derivatization*.

In a ¹³C enrichment experiment it is usually sufficient to achieve an enrichment of only a few percent (typically 5 %); this increases the intensities of the corresponding signals to such an extent as to provide an unambiguous assignment.

Besides its use as an aid to assignment, ¹³C-labeling plays an important role in elucidating reaction mechanisms and biochemical pathways (Section 14.1) and protein structures.

Specific deuteration enables one to identify the signal of the carbon nucleus bonded to the deuterium by its multiplet structure in the ¹H broad-band decoupled ¹³C NMR spectrum, as the C,D coupling is not eliminated. For example, the signals of the deuterated solvent can be quickly recognized in the ¹³C NMR spectrum by their multiplet structure. For $CDCl_3$ a triplet with three peaks of equal intensity at $\delta = 77.0$ is always found; the spectrum of C_6D_6 is shown in Figure 2-6.

In addition to the splitting or broadening of the signal a shift is also observed (Section 2.1.2.5 and Fig. 2-6). This *isotope effect* can be used as an aid to assignment [8].

Isotopic labeling is expensive and time-consuming. It is often easier to prepare derivatives, and here one must particularly mention the methylation of OH, NH and SH groups. This reduces the shielding of the carbon nucleus directly bonded to the substituent, i.e. increases its δ-value, by about 10 ppm. Acetylation, on the other hand, reduces the shielding, though only by about 2–3 ppm. It will be recalled that in ¹H NMR spectroscopy the shift caused by acetylation was greater than that caused by methylation.

6.4 Computer-aided Assignment of ^{13}C NMR Spectra

6.4.1 Searching for Identical or Related Compounds

It is only in rare cases that nothing whatever is known about the compound whose spectrum has been recorded. Here we consider the much more usual case where a proposed structure exists. How does the spectroscopist then go about the analysis?

First he makes use of his experience, then he refers to spectra catalogs or tables, and in some cases he calculates chemical shifts "by hand" using incremental relationships. Much of this work can be carried out quicker and more easily by a computer – provided that it has access to a data bank [9–11] and is correctly programmed. A file of data is therefore essential. This should contain as many reference spectra as possible, in the form of all the experimental data such as numbers of lines, line positions, coupling constants, multiplicities etc.

Just as when using conventional catalogs, one can ask the computer to search in the data file for the required compound and its spectrum by name, molecular formula, molecular mass or chemical shifts.

Another approach is to start from the observed ^{13}C NMR spectrum and search the data bank for the compound concerned, by feeding either the line positions ("peak-picking" method), or the complete spectrum obtained by Fourier transformation of the FID, directly into a search program (retrieval software). The program then searches for an *identical* spectrum. The program must be such that it tolerates some deviations from the experimental values, since even two spectra from the same compound can show differences depending on the instrument and the recording conditions.

If the computer fails to find an identical spectrum in the data bank, one can then instruct the program to search for similar spectra, and to print out the names and structures of all the compounds thus found, arranged in order of diminishing agreement with the experimental spectrum (automatic ranking). The spectroscopist then has to decide which of the similar compounds found he will consider further, and which he will reject.

If at the outset one does not specify all the lines but only a selection, the computer will probably find several reference compounds even in the initial search for identical spectra. If the data file includes the correct compound, it should be in this list.

In the rarer cases where no proposed structure exists, the list of compounds with similar spectra printed out by the computer may enable one to identify common structural features, and thus to obtain important information about the structure of the unknown compound.

6.4.2 Spectrum Prediction

When using NMR spectroscopy for determining structures the problem very often takes the form: is the observed spectrum consistent with the proposed structure? To answer this question one must try to predict the spectrum that would be expected. For predicting ^{1}H or ^{13}C chemical shifts one can choose between two different strategies, both of which are successfully used in practice.

The first method is based on incremental relationships of the kind that we have already met in Sections 6.2.2 and 6.3.2. However, it was not until the advent of computers and the development of suitable programs that it became possible to quickly predict entire spectra of complicated molecules by this method. To predict a spectrum for a given compound one first draws the structural formula on the monitor screen using a "structure editor", then instructs the computer to calculate the ^{1}H and/or ^{13}C chemical shifts. An approximate idea of the accuracy of the predictions thus obtained can be gained from the examples given in Sections 6.2.2 and 6.3.2. There the observed chemical shifts and those calculated "by hand" using the incremental relationships are compared with those generated by the commercially available program "Spectool" [12]. However, the more complicated the molecular structures to which one applies such programs, the less accurate are the predictions.

In the second method one similarly begins by drawing the structural formula on the monitor screen. Each carbon nucleus in the molecule must then be coded by a predefined system according to its environment in the proposed structure, a procedure that is performed automatically by the computer. Next the computer searches in the file to find the expected chemical shift ranges for all the carbon and hydrogen nuclei according to their environments, and prints out the resulting δ-values or the spectrum. This procedure is based on values originally determined from the spectra of molecules containing carbon or hydrogen atoms with the same, or at least very similar, environments. The accuracy achieved with such spectrum predictions depends on the size of the data file, and especially on how frequently each of the structural elements in question occurs in the file.

Table 6-8.

Observed δ-values [ppm] for the neuraminic acid derivative **5** compared with predicted values obtained using the programs ACD/CNMR Predictor 3.0 [14] and Specedit [11].

Carbon nucleus	δ (observed)	δ (predicted) ACD	δ (predicted) Specedit[c]
C-1	171.50	169.3	167
C-2	100.32	102.5	101
C-3	40.31	39.7	40
C-4	67.51	69.4	68
C-5	52.83	51.2	52
C-6	71.67	73.9	72
C-7	69.18	67.6	69
C-8	70.98	72.6	71
C-9	64.50	63.7	64
OCH_3[a]	54.65	53.3	53
OCH_3[b]	52.12	53.1	51
NC=O	175.93	175.3	175
CH_3 (Ac)	23.20	22.6	23

[a] ester

[b] ketoside

[c] prediction based on 22 000 ^{13}C NMR spectra

As an example Table 6-8 lists the measured ^{13}C chemical shifts for the methylketoside of N-acetyl-β-D-neuraminic acid methyl ester (**5**) taken from the spectrum in Figure 6–1, and alongside these the predicted values obtained by using the programs ACD/CNMR Predictor 3.0 [13] and SpecEdit [11]. The agreement between the predicted and experimental spectra is very good.

There are also programs for estimating ^1H chemical shifts [12, 14], but the accuracy of the predictions for complicated molecules is much poorer than that achieved for ^{13}C chemical shifts. This is mainly because it is more difficult and troublesome to build up a data file for ^1H, as the information in normal ^1H NMR spectra is present in a more complex form than in ^{13}C spectra. One need only consider, for example, how higher-order ^1H NMR spectra change their appearance according to the measuring frequency. Developments in this area are continuing.

Lastly it should be mentioned that programs are being developed which, on the basis of partial structural information, can propose a complete structure for the molecule. Some of these can also use information obtained from other spectroscopic methods such as infrared spectroscopy or mass spectrometry.

6.5 Bibliography for Chapter 6

[1]   C. Pascual, J. Meier and W. Simon, *Helv. Chim. Acta 48* (1969) 164.

[2]   M. Hesse, H. Meier and B. Zeeh: *Spektroskopische Methoden in der organischen Chemie*. Stuttgart: Georg Thieme Verlag, 1984.

[3]   L. M. Jackman, S. Sternhell: *Applications of Nuclear Magnetic Resonance Spectroscopy in Organic Chemistry*. Oxford: Pergamon Press, 1969.

[4]   H.-O. Kalinowski, S. Berger and S. Braun: *Carbon-13 NMR Spectroscopy*. Chichester: John Wiley & Sons, 1988.

[5]   D. M. Grant and E. G. Paul, *J. Amer. Chem. Soc. 86* (1964) 2984.

[6]   E. Pretsch, T. Clerc, J. Seibl and W. Simon: *Tabellen zur Strukturaufklärung organischer Verbindungen mit spektroskopischen Methoden*. 2nd Edition, Berlin: Springer, 1981.

[7]   D. E. Dorman, M. Jautelat and J. D. Roberts, *J. Org. Chem. 36* (1971) 2757.

[8]   F. W. Wehrli and T. Wirthlin: *Interpretation of Carbon-13 NMR Spectra*. London: Heyden, 1976, p. 107.

[9]   ACD/CNMR & HNMR Databases, Advanced Chemistry Development, Toronto, Canada.

[10] SpecInfo Online on STN International, Karlsruhe, Germany, and SpecInfo Inhouse, Workstation based spectroscopic Database for NMR, IR, MS and UV Spectra. Database Producer: Chemical Concepts GmbH, Weinheim, Germany.

[11] WinSpecEdit, Bruker/Chemical Concepts, Germany.

[12] Spectool, Chemical Concepts GmbH, Weinheim, Germany.

[13] E. Pretsch and J. T. Clerc: *Spectra Interpretation of Organic Compounds*. Weinheim: Wiley-VCH, 1997.

[14] ACD/CNMR Predictor 3.0, Advanced Chemistry Development, Toronto, Canada.

[15] ACD/HNMR Predictor 3.0, Advanced Chemistry Development, Toronto, Canada.

Additional and More Advanced Reading

N. A. B. Gray: Computer Assisted Analysis of Carbon-13 NMR Spectral Data. In: *Progr. Nucl. Magn. Reson. Spectrosc. 15* (1982) 201.

U. Weber and H. Thiele: *NMR Spectroscopy: Modern Spectral Analysis*. Weinheim: Wiley-VCH, 1998.

L. Griffiths: Automatic Analysis of NMR Spectra. In: *Annual Reports on NMR Spectroscopy*, G. A. Webb (Ed.), Vol. 50. Kidlington, Oxford, Elsevier Science Ltd., 2003, p. 217.

7 Relaxation

7.1 Introduction

In the NMR experiment the thermal equilibrium of the spin system is disturbed by irradiating at the resonance frequency (Section 1.5). This
- alters the population ratios, and
- causes transverse magnetic field components (M_x and M_y) to appear.

When the perturbation ceases the system relaxes until it once again reaches equilibrium.

We need to distinguish between two different relaxation processes:
- the relaxation in the applied field direction, which is characterized by the *spin-lattice* or *longitudinal relaxation time* T_1, and
- the relaxation perpendicular to the field direction, which is characterized by the *spin-spin* or *transverse relaxation time* T_2.

In contrast to electronic, vibrational and rotational excited states, the relaxation of nuclear systems is very slow, especially in cases where the nuclear spin I is 1/2. The time needed for complete relaxation may be seconds, minutes, or even hours.

For protons under high-resolution NMR conditions the spin-lattice relaxation times T_1 are of the order of a second, and they do not vary greatly for protons in different bonding situations. This is one reason why T_1-values for protons are not often measured. A second reason is the complexity of ^{1}H NMR spectra.

For ^{13}C nuclei the situation is quite different. Here the T_1-values vary from milliseconds in large molecules to several hundred seconds in small molecules. Because of these large differences the spin-lattice relaxation time T_1 for ^{13}C nuclei has become an additional spectral parameter of importance to the chemist.

Although the spin-spin relaxation time T_2 is less important from a chemical point of view than T_1, it will nevertheless be treated in detail in this chapter, because the corresponding relaxation processes and methods for measuring T_2 form the basis for many pulse experiments, some of them quite complicated, which will be described in Chapters 8 and 9.

Initially we shall confine our attention to the relaxation of ^{13}C nuclei. In this chapter we shall return to proton relaxation only in connection with line-widths and with the suppression of solvent signals. The relaxation times T_1 and T_2 of protons also play an important role in NMR tomography (Section 14.4).

7.2 Spin-Lattice Relaxation of ^{13}C Nuclei (T_1)

7.2.1 Relaxation Mechanisms

As already mentioned, the equilibrium of the spin system is disturbed in the NMR experiment: the macroscopic magnetization vector M_0 ($= M_z$ initially) is rotated by a $90^\circ_{x'}$ pulse into the y'-axis direction, or by a $180^\circ_{x'}$ pulse into the ($-z$) direction. The new value of M_z after a $90^\circ_{x'}$ pulse is zero, and after a $180^\circ_{x'}$ pulse it is $-M_0$ (see Fig. 7-1a).

In both cases the population ratio changes. A $90^\circ_{x'}$ pulse equalizes the populations of the two energy levels, whereas a $180^\circ_{x'}$ pulse inverts the population ratio. After the perturbation the equilibrium condition $M_z = M_0$ reasserts itself (Fig. 7-1b). The rate at which this occurs is determined by the *spin-lattice relaxation time* T_1. Felix Bloch described this process by the differential equation (7-1) (which is the same as Eq. (1-15) of Section 1.5.3).

$$\frac{dM_z}{dt} = -\frac{M_z - M_0}{T_1} \tag{7-1}$$

From the standpoint of chemical kinetics T_1^{-1} is the rate constant of the relaxation, which is a *first-order* process.

Spin-lattice relaxation is always associated with a change in the energy of the spin system, as the energy absorbed from the pulse must be given up again. It is transferred to the surroundings, the *lattice*, whose thermal energy therefore increases. Here the "lattice" means the neighboring molecules in the solution, and even the wall of the vessel.

Various intra- and intermolecular interactions are recognized as contributing to the ^{13}C spin-lattice relaxation. Thus one distinguishes between different *relaxation mechanisms*, such as:

- dipole-dipole (DD), or simply "dipolar" relaxation
- spin-rotation (SR) relaxation
- relaxation due to chemical shift anisotropy (CSA)
- relaxation due to scalar coupling (SC)

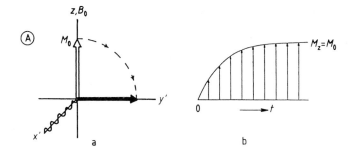

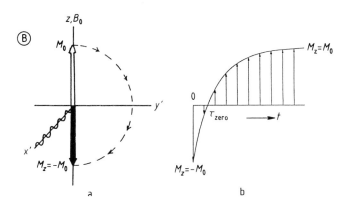

Figure 7-1.
Evolution with time of the
longitudinal component M_z of the
macroscopic magnetization, in the
rotating coordinate system x', y', z.
A: after a $90^\circ_{x'}$ pulse;
B: after a $180^\circ_{x'}$ pulse. The wavy line
along the x'-axis of the rotating
coordinate system indicates the
direction of the effective $\boldsymbol{B}_1$ field.

- electric quadrupolar (EQ) relaxation, and
- relaxation due to interactions with unpaired electrons in
  paramagnetic compounds.

The main contribution to the spin-lattice relaxation of ^{13}C
nuclei is the dipole-dipole (DD) coupling. This is also the most
intensively studied of the various mechanisms, and – importantly in this context – it can be directly measured through the
nuclear Overhauser effect (NOE, Chapter 10). This interaction
between the nuclear dipoles arises from the fact that each
nucleus is surrounded by other magnetic nuclei, in the same
or in adjacent molecules, which are in motion. Their motion
causes fluctuating magnetic fields at the position of the nucleus
being observed. The frequency band is comparatively broad,
and is largely influenced by the viscosity of the solution. If
these fluctuating fields have components of the appropriate frequency, they can induce nuclear spin transitions.

The theoretical description of dipolar relaxation for an
assembly of molecules leads to the proportionality relationship
$T_1^{-1} \propto \tau_c$, which can be expressed in the form of a useful rule:
the faster a molecule moves, the greater is T_1. Here τ_c is the *correlation time*, and corresponds roughly to the interval between
two successive reorientations or positional changes of the molecule (by vibration, rotation or translation).

The dipole-dipole relaxation mechanism is especially effective when the observed carbon nucleus has directly bonded hydrogen atoms, as in CH, CH_2 or CH_3 groups.

Of the remaining relaxation mechanisms, the dipole-dipole interaction of the nuclei with unpaired electrons in paramagnetic molecules is of special practical importance. Owing to the large magnetic moment of the electron this interaction is very strong. Thus, for example, a high concentration of paramagnetic impurities in the sample causes the nuclear Overhauser effect (Chapter 10) to be completely lost, and also results in broad NMR signals (Section 7.3.3).

7.2.2 Experimental Determination of T_1; the Inversion Recovery Experiment

Of the various methods for determining the spin-lattice relaxation time T_1, the only one we shall discuss here is the basic *inversion recovery* experiment. In this method one records a series of ^{13}C NMR spectra using the pulse sequence $180^\circ_{x'} - \tau - 90^\circ_{x'} - $ FID, which is shown diagrammatically in Figure 7-2, while simultaneously eliminating the C,H couplings by ^1H BB decoupling.

As an example Figure 7-3 shows seven spectra recorded by this method for ethylbenzene (**1**); the delay time τ between the $180^\circ_{x'}$ and $90^\circ_{x'}$ pulses was set at a different fixed value for each spectrum, as shown at the right-hand end of the figure. The assignments of the six signals to the six chemically different carbon nuclei in the molecule are shown in the top spectrum. As τ is varied the amplitudes of each of the signals change in different ways.

So that we can better understand the action of this pulse sequence on the spin system and the changes in the amplitudes, we first consider a simpler case, that of a sample with only one sort of chemically equivalent ^{13}C nuclei, e. g. the ^{13}C nuclei in chloroform ($^{13}CHCl_3$). After the first stage, namely the $180^\circ_{x'}$ pulse, M_0 lies along the $(-z)$ direction. During the time τ the system relaxes with the rate constant $k = T_1^{-1}$. Figure 7-2 B shows the stages of evolution reached through relaxation by the magnetization vector M $(= M_z)$ for five different delay times τ. Diagram 'a' corresponds to $\tau = 0$ and $M_z = -M_0$; 'b' shows the stage reached after a short time τ in which M_z still remains negative. The situation shown in diagram 'c', in which $M_z = 0$, is important for the quantitative measurement of T_1. For greater values of τ, M_z again becomes positive, and in 'e' the system has finally returned to the equilibrium value $M_z = M_0$.

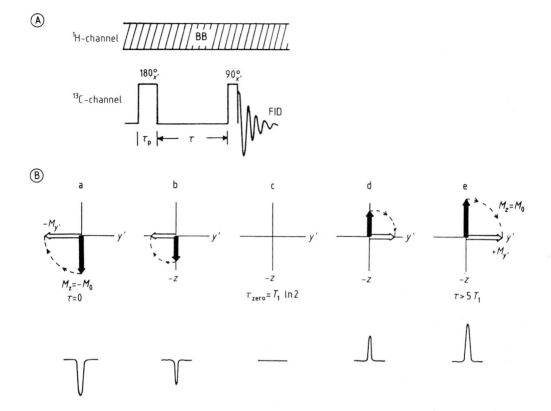

Figure 7-2.
A: Pulse sequence for determining the ^{13}C spin-lattice relaxation time T_1 by the *inversion recovery method* with continuous ^1H BB decoupling: $180°_{x'} - \tau - 90°_{x'} -$ FID. (The pulse durations along the time axis are not to scale: for a 180° pulse τ_P is several μs, whereas τ is of the order of seconds.)
B: The vector diagrams a to e and the signals below them show, for five different values of τ, the effects of spin-lattice relaxation on M_z and on the amplitudes of the signals obtained after applying a $90°_{x'}$ pulse and performing the Fourier transformation of the FID.

M_0 and M_z are not directly measurable quantities, since a signal is only induced in the receiver when the magnetization vector has a transverse component ($M_{x'}$, $M_{y'}$). Therefore, in the second phase of the pulse sequence after the delay τ the magnetization component M_z is rotated into the direction of the y'-axis by a $90°_{x'}$ pulse, thus giving a transverse magnetization component $M_{y'}$ which can be observed. The intensities I of the NMR signals obtained after the Fourier transformation are proportional to the components $M_{y'}$. Since M_z is still negative initially, the transverse component after the $90°_{x'}$ pulse is in the negative y'-direction, and signals with negative amplitudes appear in the spectrum. For $M_z = 0$ ($\tau = \tau_{zero}$) no signal can be detected. At greater τ-values the signal amplitude is again positive, with a magnitude determined by the component M_z in each case.

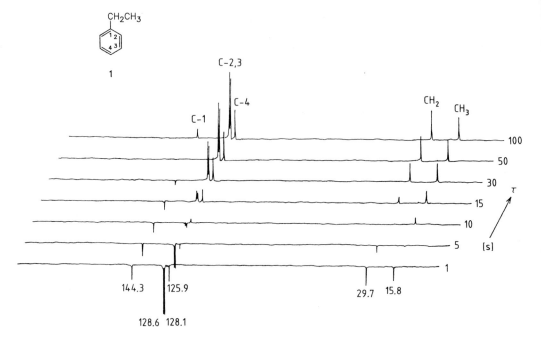

Figure 7-3.
22.63 MHz ^{13}C NMR spectra of ethylbenzene (**1**), recorded by the inversion recovery method (Fig. 7-2) with $\tau = 1, 5, 10, 15, 30, 50$ and 100 s [1].

The starting point for the *quantitative treatment* of the spectra is Equation (7-1). By integrating the equation we obtain:

$$M_0 - M_z = A\, e^{-t/T_1} \qquad (7\text{-}2)$$

M_z is the magnetization in the z-direction at the time $t = \tau$; A is a constant whose value depends on the initial conditions. In our inversion recovery experiment we have, at the instant $\tau = 0$ immediately after the $180^\circ_{x'}$ pulse, $M_z = -M_0$, and therefore $A = 2\,M_0$. Equation (7-2) becomes:

$$M_0 - M_z = 2M_0\, e^{-t/T_1} \qquad (7\text{-}3)$$

and by taking logarithms of both sides of the equation we obtain:

$$\ln(M_0 - M_z) = \ln 2M_0 - \frac{t}{T_1} \qquad (7\text{-}4)$$

Replacing the magnetizations M in Equation (7-4) by the signal intensities I gives:

$$\ln(I_0 - I_z) = \ln 2I_0 - \frac{t}{T_1} \qquad (7\text{-}5)$$

Here I_0 is the maximum measurable signal intensity, and I_z is the intensity at $t = \tau$.

166

If we plot $\ln(I_0 - I_z)$ against t, for discrete intervals of t corresponding to the delay times τ, we obtain a straight line whose gradient is $-1/T_1$. At the time τ_{zero} when the signal intensity I_z is just zero (the zero-crossing point), Equation (7-5) simplifies to

$$\tau_{zero} = T_1 \ln 2 \qquad (7\text{-}6)$$

Equation (7-6) can be used to determine or estimate T_1 for each signal in a ^{13}C NMR spectrum, if one wishes to avoid making measurements over the entire period of relaxation.

The differences in Figure 7-3 between the zero-crossing points for the different carbon nuclei of ethyl benzene show how greatly the T_1 values can vary from one nucleus to another. For the signal of the CH$_3$ group τ_{zero} is about 5 s, and from Equation (7-6) the gives $T_1 \approx 7.2$ s; for the signal of the quaternary nucleus C-1 τ_{zero} is about 50 s, giving $T_1 \approx 72$ s.

Usually it is necessary to accumulate many FIDs to obtain a good quality ^{13}C NMR spectrum. Before repeating a pulse sequence (Fig. 7-2) a delay time of at least 5 T_1 must be inserted in each cycle, and the value used here for T_1 must take account of the most slowly relaxing ^{13}C nucleus in the molecule. The system needs this length of time to return to equilibrium, so that the intensities and relaxation times can be correctly measured. The complete pulse sequence is therefore:

$$(5\ T_1 - 180^\circ_{x'} - \tau - 90^\circ_{x'} - FID)_n$$

As T_1 can be 50 s or more for quaternary carbon nuclei, this implies delay times of 4 to 5 min. This results in very long total measuring times. Modern NMR spectrometers allow the measurements to be carried out automatically, so that experiments of this kind can be run overnight.

The accuracy of the calculated T_1-values depends on how precisely the intensity I_0 at the instant $t = \tau = 0$ is measured. It is usual, therefore, to determine I_0 at the beginning of the experiment, during it, and at the end, and to take the average value.

A further source of error is the presence in the sample of paramagnetic impurities, the commonest being dissolved oxygen. Therefore each sample must be carefully degassed before the T_1 measurement. If this is not done, or if the sample contains other paramagnetic impurities, all the relaxation times will be shortened and the measured values will be very difficult to interpret. These effects are especially serious in cases where the true T_1 is more than 50 s.

7.2.3 Relationships between T_1 and Chemical Structure

The T_1-values for the ^{13}C nuclei in organic molecules are in the approximate range from 0.1 to 300 s. The smaller values, 0.1 to 10 s, belong to carbon nuclei with directly bonded protons, while the larger values (more than 10 s) are for those without protons (quaternary carbons) and for small symmetrical molecules. Despite the fact that the amount of available experimental T_1 data has always been less than for chemical shifts and coupling constants, various relationships between T_1-values and molecular structure which are of interest to chemists have been discovered and discussed.

7.2.3.1 Influence of Protons in CH, CH$_2$ and CH$_3$ Groups

Where the main contribution to spin-lattice relaxation comes from the dipole-dipole (DD) interactions, directly bonded protons are expected to have a large effect on the T_1-values of the corresponding ^{13}C nuclei. It is in fact found that the more hydrogen atoms are attached to a carbon the shorter is its T_1. In the ideal case T_1 is inversely proportional to the number of attached protons. This effect is clearly apparent in the example of iso-octane (**2**) [2]; although the ratio $T_1(CH)$: $T_1(CH_2)$ does not quite have the ideal value of 2, the observed ratio of 23:13 is not greatly different from this. Possible reasons for the deviation from the ideal ratio are that the DD relaxation is additionally affected by more distant protons, and that mechanisms other than the DD interaction also contribute to the relaxation. In the case of the CH$_3$ groups, $T_1(CH_3)$ is appreciably greater than $\tau_1(CH)/3$, showing that here the methyl rotation slows down the DD relaxation by shortening the effective correlation time τ_C (Section 7.2.1).

The large T_1-values of quaternary carbon nuclei are responsible for the fact that, under the usual recording conditions of ^1H BB decoupling and short pulse repetition times, their signals are very weak, and sometimes cannot even be found (Section 1.6.3.2).

The effects on T_1 caused by neighboring nuclei with magnetic moments increase with the magnitudes of their magnetic moments and with their proximity to the observed nucleus. This can be used as an aid to assignment. For example, one or more of the protons in the molecule can be selectively replaced by deuterons, which have a smaller magnetic moment. This reduces the DD interaction effects, and so one expects an increase in T_1. This is in fact borne out by experimental results. The effects are seen especially clearly when the relaxation times are long, as is the case for quaternary carbon nuclei.

Example:

○ In phenanthrene (**3**) the ^{13}C resonances of the two carbon nuclei 11 and 12 lie close together at $\delta = 130.1$ and 131.9, and they cannot be unambiguously assigned. C-11 is expected to have the shorter T_1 of the two, as two nearest neighbor protons contribute to its relaxation compared with only one for C-12; however, in the undeuterated compound **3 A** the T_1-values also are not greatly different (51 and 59 s). Replacing H-4 and H-5 by deuterium (see **3 B**) increases the two T_1-values to 59 and 80 s respectively. This confirms the assignment shown in the formula, since the T_1 showing the largest increase must belong to C-12. For the more distant C-11 nucleus the weaker DD interaction affects the T_1-value to a smaller extent [3].

3A

3B

7.2.3.2 Influence of Molecular Size

The T_1-values found for cycloalkanes [4] decrease as the ring becomes larger. For example, a value of 37 s was found for cyclopropane, whereas for cyclohexane it was only 20 s. Eventually, at a ring size of $n = 20$, a limiting value of about 1–2 s is reached. These results reflect the influence of the mobilities of the molecules, since the mobility of the cyclopropane molecule as a whole is much greater than that of the larger rings. For the small rings one cannot rule out the possibility that, in addition to the dominant DD interaction mechanism, spin-rotation too contributes to relaxation. For large rings, on the other hand, the relaxation times can also be affected by the intramolecular mobility of individual CH_2 segments, even though these effects have not been quantified.

For macromolecules T_1-values of less than 1 s are usually found, and they can fall to as little as 10^{-1} to 10^{-2} s, as in the case of polystyrene (**4**)

In biopolymers the ^{13}C NMR signals are generally poorly resolved, and consequently it is in most cases not possible to obtain relaxation times for the individual carbon nuclei. Often only group relaxation times are measured; from these it is possible to reach conclusions about the mobilities of sub-units of the molecule.

4

7.2.3.3 Segmental Mobilities

Often the individual parts of a molecule move at different rates. For example, the side-chains of steroids or the ester side-groups in acrylate polymers are more mobile than the ring system or carbon main chain. The latter more rigid parts of the molecules act as anchors for the side-chains. For flexible alkyl groups a bromine atom or an amide group is sufficient to provide anchoring. Hydroxyl or carboxy groups have a similar effect, as they form hydrogen bonds and can thereby limit the mobilities of entire molecular segments. In all compounds of these types the T_1-values increase with distance from the anchor point. There are, of course, additional effects arising from the structural peculiarities of each molecule.

These effects can be illustrated by three *examples* (all T_1-values are in seconds):

$$CH_3 - CH_2 - CH_2 - CH_2 - CH_2 - C_5H_{11} \qquad \textbf{5}$$
$$\;\;8.7 \quad\;\; 6.6 \quad\;\; 5.7 \quad\;\; 5.0 \quad\;\; 4.4$$

$$BrCH_2 - CH_2 - (CH_2)_5 - CH_2 - CH_2 - CH_3 \qquad \textbf{6}$$
$$\;2.8 \quad\;\;\; 2.7 \quad\; 2.0 \quad\;\;\; 3.1 \quad\;\;\; 3.9 \quad\;\; 5.3$$

$$HOCH_2 - (CH_2)_5 - CH_2 - CH_2 - CH_2 - CH_3 \qquad \textbf{7}$$
$$\;0.7 \quad\quad\;\; 0.8 \quad\;\; 1.1 \quad\;\; 1.6 \quad\;\; 2.2 \quad\;\; 3.1$$

Even in unsubstituted *n*-alkanes the CH_2 segments are found to have different T_1-values, as the mobility in the middle of the chain is less than near the ends [2].

7.2.3.4 Anisotropy of the Molecular Mobility

In monosubstituted benzenes it is found that ^{13}C nuclei on the long axis of the molecule have noticeably smaller T_1-values than do the other ring carbons. In most cases the difference is approximately a factor of two, as can be seen in the examples of toluene (**8**), phenylacetylene (**9**), biphenyl (**10**) and diphenyldiacetylene (**11**). This observation indicates that the molecular mobility is anisotropic, i. e. that the molecule rotates preferentially about its long axis. Owing to this preferred rotation the carbon nuclei lying off the axis move faster than those on the axis [2].

170

7.2.4 Suppression of the Water Signal

Often one cannot avoid using water as the solvent. This is especially so in biochemical investigations. Even when D_2O is used as the solvent, the residual HDO signal is likely to be much larger than the signals of the substances in which one is interested (Section 5.2.2). This leads to various difficulties:

- some signals may be hidden
- integration is not possible near to the water signal
- the computer finds it difficult to process weak signals together with very strong signals.

These problems can be avoided by an experiment which is based on an observation previously referred to in connection with measuring T_1 for ethylbenzene (Fig. 7-3). The ^{13}C NMR signal of C-1 in ethylbenzene disappeared when the delay time was set at $\tau_{zero} = 50$ s (zero-crossing condition). A comparison of the three spectra recorded after delay times of 30, 50 and 100 s shows that all the other ^{13}C nuclei have already almost completely relaxed at this stage and their signal intensities have reached their equilibrium values. The water signal (H_2O or HDO) can be suppressed in the same way as is seen here for C-1. The T_1-value for water is about 3 s, and is thus greater than the values for the protons in organic molecules. One therefore records the ^1H NMR spectrum using the same pulse sequence as for T_1 measurements i. e. $(180^\circ_{x'} - \tau - 90^\circ_{x'} - FID)_n$. The value of τ is then chosen so that the water signal disappears. During this time interval the protons of the sample can undergo complete relaxation, and one obtains only the ^1H NMR spectrum of the sample. Also the signals that were previously hidden under the water signal can now be seen. (The optimal conditions must in each case be determined by preliminary experiments).

Unfortunately this method can only be used for solvents whose relaxation times are considerably greater than those of the samples to be examined.

7.3 Spin-Spin Relaxation (T_2)

7.3.1 Relaxation Mechanisms

We have already learned in Chapter 1 (Section 1.5) that immediately after a $90^\circ_{x'}$ pulse the z-component of the macroscopic magnetization vector is zero, i. e. $M_z = 0$. This is only possible if the populations N_α and N_β have become equal.

Instead of M_z there is now a transverse magnetization component $M_{y'}$. According to the classical picture a small proportion of the nuclear dipoles are now bunched together (i. e. in phase) and precessing around the surface of the double cone (Fig. 1-12). This condition is called *phase coherence*. The magnetization $M_{y'}$ is the sum of the y'-components of all the individual spins. The evolution of this transverse magnetizsation after the pulse is determined by the Bloch equations which were given in Section 1.5.3; here we need only the differential equation for $M_{y'}$, which is:

$$\frac{dM_{y'}}{dt} = -\frac{M_{y'}}{T_2} \qquad (7\text{-}7)$$

The time constant T_2 is the *spin-spin* or *transverse relaxation time*. It determines how rapidly the transverse magnetization components $M_{x'}$ and $M_{y'}$ decay. According to the classical picture this means that the precessing nuclear spins which are bunched together gradually lose their phase coherence, i. e. the bunch fans out (see Fig. 1-14). As the fanning-out process continues $M_{y'}$ becomes smaller, and with it the induced signal in the receiver coil.

The energy of the spin system is not altered by *spin-spin relaxation*, as the level populations are not affected. Only the phase coherence between the bunched precessing nuclear spins is lost. This type of relaxation is therefore sometimes described as an *entropy process*.

What causes the spin-spin relaxation? From a classical viewpoint one can visualize energy being transferred from one nucleus to another via fluctuating magnetic fields. In this process one of the nuclei changes from a higher to a lower energy state, while another is simultaneously raised from a lower to a higher level. Equal amounts of energy are released and absorbed, and only the phase coherence is lost.

However, the main contribution to the spin-spin relaxation is of quite a different nature. Since the sample being examined has finite dimensions and the magnetic field B_0 is not homogeneous throughout this volume, not all the nuclei experience the same magnetic field. The effective magnetic field at the position of the nucleus varies from one nucleus to another by an amount ΔB_0. These magnetic field inhomogeneities cause even nuclei that are chemically equivalent to precess with slightly different Larmor frequencies, some faster and some slower than the resultant transverse magnetization $M_{y'}$. This leads to the fanning-out with time which is shown in Figure 1-14.

The main practical significance of T_2-values lies in their relationship to the *line-width* of the observed NMR signals (Section 7.3.3).

To end this section we will consider how the magnitudes of the spin-spin and spin-lattice relaxation times compare. According to all that we have learned up to now about relaxation processes, T_2 can never be greater than T_1. This is because it is possible for the transverse magnetization $M_{y'}$ to have already decayed away completely by relaxation before the longitudinal magnetization M_z has reached the equilibrium value M_0. On the other hand, M_z cannot grow to its equilibrium value M_0 until the transverse magnetization $M_{y'}$ has completely disappeared. Therefore we have:

$$T_1 \geq T_2$$

7.3.2 Experimental Determination of T_2; the Spin-Echo Experiment

As explained in Section 7.3.1, T_2 is mainly determined by the field inhomogeneities ΔB_0. However, this contribution to the relaxation is of no interest to the chemist, being merely an instrumental parameter. As long ago as 1950 E. L. Hahn [5] suggested an elegant method for determining T_2, whereby the inhomogeneity contribution is eliminated; this is the *spin-echo* method. It can be understood more easily and clearly in the form of the variant developed by Carr and Purcell [6], in which the $90°_{x'}$ excitation pulse is followed not by another $90°_{x'}$ pulse, as in Hahn's version, but by a $180°_{x'}$ pulse. The complete pulse sequence is:

$90°_{x'} - \tau - 180°_{x'} - \tau \,(\text{1st echo}) - \tau - 180°_{x'} - \tau \,(\text{2nd echo}) \ldots .$

Figure 7-4 C a-h shows how each step in this pulse sequence affects the spin system.

Diagram 'a' shows the initial equilibrium condition, with the macroscopic magnetization along the z-direction ($M_z = M_0$). The $90°_{x'}$ pulse turns M_0 into the direction of the y'-axis ($M_{y'} = M_0$). This transverse magnetization $M_{y'}$ now rotates with an average Larmor frequency ν_L; the x', y', z coordinate system also rotates with this frequency, and consequently the transverse magnetization $M_{y'}$ remains along the direction of the y'-axis.

Let us single out two individual nuclei A and B from all those which are precessing in phase. Due to the field inhomogeneities nucleus A experiences a slightly higher field than the average for all the nuclei, and nucleus B a slightly lower field. In accordance with Equation (1-6) ($\nu_L = \gamma B_0$) the Larmor frequency for A is higher than that for B. The precession of nucleus A is therefore faster than the average, whereas that of B is slower. Thus nucleus A gets ahead of the coordinate system which is rotating with the average frequency, while nucleus B falls behind. In dia-

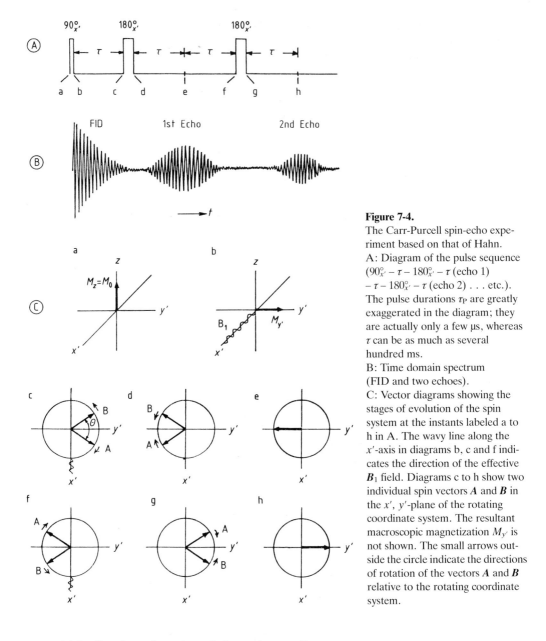

Figure 7-4.
The Carr-Purcell spin-echo experiment based on that of Hahn.
A: Diagram of the pulse sequence
($90^\circ_{x'} - \tau - 180^\circ_{x'} - \tau$ (echo 1)
$- \tau - 180^\circ_{x'} - \tau$ (echo 2) . . . etc.).
The pulse durations τ_P are greatly exaggerated in the diagram; they are actually only a few µs, whereas τ can be as much as several hundred ms.
B: Time domain spectrum (FID and two echoes).
C: Vector diagrams showing the stages of evolution of the spin system at the instants labeled a to h in A. The wavy line along the x'-axis in diagrams b, c and f indicates the direction of the effective B_1 field. Diagrams c to h show two individual spin vectors A and B in the x', y'-plane of the rotating coordinate system. The resultant macroscopic magnetization $M_{y'}$ is not shown. The small arrows outside the circle indicate the directions of rotation of the vectors A and B relative to the rotating coordinate system.

gram 'c' the directions of rotation relative to the coordinate system are indicated by the small arrows outside the circle. After a time $t = \tau$, typically up to a few hundred ms, the vectors A and B are separated by an angle Θ as in diagram 'c', i. e. they are out of phase. This fanning-out of the vectors reduces the magnitude of $M_{y'}$.

To understand the effect of the $180^\circ_{x'}$ pulse which follows, we need to anticipate to some extent the discussion in Section 8.2 by including here a short explanation. We will first confine our

attention to the vector A after a time $t = \tau$, i.e. immediately
before the $180^\circ_{x'}$ pulse (Fig. 7-4 C, diagram 'c'). In Figure 7-5c
the vector diagram for nucleus A is shown again on an enlarged
scale. The projection of A on the y'-axis is $A_{y'}$, and that on the
x'-axis is $A_{x'}$. A 180° pulse applied in the x'-direction turns the
component $A_{y'}$, into the $(-y')$ direction, but has no effect on
$A_{x'}$. By vector addition of $A_{x'}$ and $-A_{y'}$ we obtain the new direction of A after the $180^\circ_{x'}$ pulse. The angle between A and the
y'-axis is the same as before the pulse (if we disregard sign),
and the direction or rotation relative to the coordinate system
remains unchanged (Fig. 7-5 'd'). In other words, the effect of
the $180^\circ_{x'}$ pulse is that the vector A has undergone reflection in
the x', z plane, without changing its direction of rotation. The
same argument also applies to the vector B.

We return now to Figure 7-4 C, vector diagram 'd'. This
shows the state of the spin system for the vectors A and B after
the $180^\circ_{x'}$ pulse. As before, A is moving faster than B, but now
A is lagging behind B! After a further time interval of exactly
τ, A has caught up with B and the two are once more in phase.
This condition is reached at a time of exactly $2\,\tau$ after the first
$90^\circ_{x'}$ pulse (Fig. 7-4 C, diagram 'e'). The arguments here set
out in detail for nuclei A and B apply equally to all the other
nuclei in the macroscopic sample which contribute to the fanning-out process caused by the field inhomogeneities. After
the interval $2\,\tau$ all the transverse components of the spin vectors
become refocussed. The resultant transverse magnetization now
points in the $(-y')$ direction, and its amplitude again reaches a
maximum. However, the absolute magnitude of $M_{y'}$ is now not
quite as great as it was immediately after the first $90^\circ_{x'}$ pulse, as

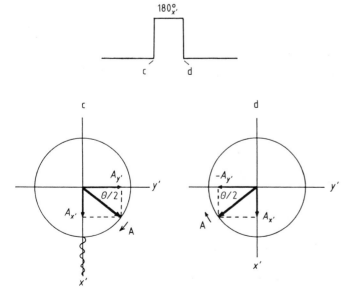

Figure 7-5.
Effect of a $180^\circ_{x'}$ pulse on the vector
A at the instant c (Fig. 7-4 Cc)
when, after the evolution time τ, it
is at an angle $\Theta/2$ to the y'-axis of
the rotating coordinate system. $A_{y'}$
undergoes a 180° rotation, whereas
$A_{x'}$ is unaffected. The direction of
rotation of A relative to the rotating
coordinate system is not altered by
the pulse.

175

the system has undergone *true* spin-spin relaxation with the relaxation time T_2 during the time 2τ.

During the following interval of length τ, the spins again fan out (Fig. 7-4 C, 'f'), and a further $180^\circ_{x'}$ pulse at this instant (diagram 'g') produces, after a further interval τ, a second echo (diagram 'h'), whose phase differs from the first by 180°. Thus, by applying $180^\circ_{x'}$ pulses at the instants τ, 3τ, 5τ, . . . etc., we obtain echoes with alternating phases at intervals of 2τ, i. e. at the times 2τ, 4τ, 6τ, . . . etc. The decay in the intensities of these echoes, i. e. of the succession of signals which follow the FID, is determined solely by T_2 (Fig. 7-4 B).

To determine T_2 we again start from the differential equation (7-7), which can be solved for $M_{y'}$ to give:

$$M_{y'} = A \, e^{-t/T_2} \tag{7-8}$$

For $t = 0$ we have $A = M_0$. Taking logarithms of (7-8) gives:

$$\ln M_{y'} = \ln M_0 - \frac{t}{T_2} \tag{7-9}$$

and since I is proportional to $M_{y'}$,

$$\ln I(t) = \ln I_0 - \frac{t}{T_2} \tag{7-10}$$

One therefore measures the intensities of the echoes or signals and plots $\ln I(t)$ against t; here $I(t)$ is the intensity of the echo at time t ($t = 2\tau$, 4τ, . . . etc.). This yields a straight line whose gradient is $-1/T_2$. The chief sources of error in the measurement are inaccurate setting of the 90° and 180° pulse angles.

Figure 14-18 (Chapter 14) shows, in the context of NMR tomography, the exponential decay of the intensity $I(t)$ for a series of eight echoes. (In this case $I(t)$ is plotted directly, not as its logarithm.)

During the spin-echo experiment there is, of course, a gradual recovery of the magnetization along the field direction due to the spin-lattice relaxation (time constant T_1). The M_z component which has thereby built up then becomes inverted by the $180^\circ_{x'}$ pulses. However, this does not contribute to the transverse magnetization nor, therefore, to the observed signal.

The system described here contained only nuclei that were chemically equivalent and precessed at different Larmor frequencies as a result of field inhomogeneities. However, the arguments are equally valid if the spin-echo experiment is carried out on a system with several chemically non-equivalent sorts of nuclei, or one in which the nuclei are coupled. For example, if two different sorts of nuclei A and B are present, their magnetization vectors M_A and M_B are similarly refocussed after a time 2τ, and an echo is induced in the receiver coil. To understand this one only needs to replace the vectors A and B

in the vector diagrams of Figure 7-4 C by M_A and M_B. Fourier transformation of the echo then gives two signals at the frequencies ν_A and ν_B.

^{1}H NMR spectra are usually too complex for T_2 measurements because of the many H,H couplings present, and such measurements are therefore almost entirely restricted to ^{13}C nuclei, as the couplings to protons can in this case be eliminated by ^{1}H BB decoupling, leaving a spectrum made up entirely of single peaks.

We shall return in Section 8.3 to the two-spin AX system with coupled ^{1}H and ^{13}C nuclei.

7.3.3 Line-widths of NMR Signals

Signals which lie very close together can only be seen separately (i. e. resolved) if the resonance lines are sharp. A measure of the resolution is the line-width at half height (*half-height width $\Delta\nu_{1/2}$*). Problems with inadequate resolution occur mainly in ^{1}H NMR spectroscopy, but less often in ^{13}C NMR spectroscopy.

Both the spin-lattice and spin-spin relaxation processes contribute to the line-width. Each of these shortens the lifetime of a nucleus in a particular energy state. According to *Heisenberg's uncertainty principle* (Eq. 7-11), the shorter the lifetime τ_1 of a particle in a given stationary state, the greater is the degree of uncertainty in the energy of that state. In NMR spectroscopy the effect of this is to introduce an uncertainty δE into the energy of a transition, and therefore into the frequency of the transition and of the resonance signal. As a result the lines are broadened, by an amount which increases as T_1 and T_2 become shorter.

$$\delta E \, \tau_1 \geq \frac{h}{2\pi} \qquad (7\text{-}11)$$

For nuclei with spin 1/2 in low-viscosity liquids the relaxation times T_1 and T_2 are approximately equal, and are quite long, resulting in very small line-widths. For protons these "natural" line-widths are in general less than 0.1 Hz.

In solids and viscous liquids T_1 can be very long, especially at low temperatures, being sometimes of the order of minutes or even hours; T_2, on the other hand, is very short, of the order of 10^{-5} s, since the magnetic coupling to neighboring spins is very large. The line-width is then determined essentially by T_2.

Usually the shape of the resonance line can be described by a *Lorentzian function*; the half-height width is then given by:

$$\Delta v_{1/2} = \frac{1}{\pi T_2{}^*} \qquad (7\text{-}12)$$

From this equation and the observed width of the signal one can obtain a transverse relaxation time $T_2{}^*$, but this is of little interest in itself, as it is determined mainly by magnetic field inhomogeneities ΔB_0 (Section 7.3.1). Instead the contribution of the field inhomogeneities to the half-height width must be subtracted so as to obtain the "true" or natural spin-spin relaxation time T_2. This is possible by using Equation (7-13).

$$\frac{1}{T_2{}^*} = \frac{\gamma \Delta B_0}{2} + \frac{1}{T_2} \qquad (7\text{-}13)$$

In addition to the contribution from field inhomogeneities, there is often a further line broadening caused by interactions with neighboring nuclei in cases where these have a spin $I \geq 1$, and consequently also an electric quadrupole moment Q. Here we must particularly mention interactions with ^{14}N nuclei, which have quite a large electric quadrupole moment Q. Deuterium too is a quadrupolar nuclide ($I = 1$), but its quadrupole moment is smaller than that of ^{14}N, and accordingly the broadening of proton resonance lines through coupling to deuterium is small. Nevertheless, if broad signals are found in such a case the probable cause is unresolved couplings.

As has already been mentioned, paramagnetic impurities shorten the relaxation times and give severely broadened lines. This applies especially to dissolved oxygen. Consequently, the best resolution can only be obtained by careful degassing of the samples.

The effects of exchange processes on line shapes will be considered in detail in Chapter 11.

7.4 Bibliography for Chapter 7

[1] H. Schneider, Dissertation, Ruhr-Universität Bochum 1975.

[2] R. J. Abraham and P. Loftus: *Proton and Carbon-13 NMR Spectroscopy*. London: Heyden, 1978, p. 131.

[3] F. W. Wehrli: Organic Structure Assignments using ^{13}C Spin-Relaxation Data. In: *Topics in Carbon-13 NMR Spectroscopy*, Vol. 2. New York: John Wiley & Sons, 1976.

[4] H.-O. Kalinowski, S. Berger, S. Braun: *Carbon-13 NMR Spectroscopy*. Chichester: John Wiley & Sons, 1988, p. 640.

[5] E. L. Hahn, *Phys. Rev. 80* (1950) 580.

[6] H. Y. Carr and E. M. Purcell, *Phys. Rev. 94* (1954) 630.

Additional and More Advanced Reading

J. R. Lyerla, Jr. and G. C. Levy: Carbon-13 Nuclear Spin Relaxation. In: *Topics in Carbon-13 NMR Spectroscopy*, Vol. 1, Ch. 2. New York: John Wiley & Sons, 1974.

R. Kitamaru: Carbon-13 Nuclear Spin Relaxation Study as an Aid to Analysis of Chain Dynamics and Conformation of Macromolecules. In: *Applications of NMR Spectroscopy to Problems in Stereochemistry and Conformational Analysis*, Ch. 3, Y. Takeuchi and A. P. Marchand (eds.). New York: VCH Publishers, 1986.

W. S. Price: Water Signal Suppression in NMR Spectroscopy. In: *Annual Reports on NMR Spectroscopy*, G. A. Webb (Ed.), Vol. 50. London, Academic Press, 1999, p. 290.

8 One-Dimensional NMR Experiments using Complex Pulse Sequences

8.1 Introduction [1]

Two problems are in the foreground in the practice of NMR spectroscopy. The first of these is the low sensitivity. This applies especially to the insensitive nuclides with low natural abundances, of which ^{13}C and ^{15}N are examples. The second problem area is that of assignments. Difficulties with this are by no means confined to higher order spectra or spectra of large molecules; problems with assignments occur also, or perhaps even especially, in spectra containing only singlets, as in ^{13}C NMR spectra with ^{1}H BB decoupling.

Modern pulse spectroscopy opens up new possibilities for overcoming these problems, through the use of selective pulses, complex pulse sequences and pulsed field gradients. The most important of these experiments, and the advantages that they offer, will be described in this and the following chapter.

To make the most effective use of these new techniques it is essential to have, in addition to the appropriate instrumentation, a good knowledge of the theoretical fundamentals of pulse experiments and of the effects of pulsed field gradients. The next two sections aim to provide this knowledge. These are followed by detailed descriptions of particular experiments.

We begin with the *J-modulated spin-echo* experiment. Using this as an example, we will consider how the amplitudes and phases of the signals in coupled systems are affected by the coupling constants. The discussion of this particular technique serves two purposes: firstly it will prepare us for when we come to consider two-dimensional *J*-resolved NMR spectroscopy (Section 9.3), and secondly it has grown in practical importance as an aid to assignment in ^{13}C NMR spectroscopy (*attached proton test*, APT). Next we learn about the *pulsed gradient spin-echo experiment*, a simple example of the use of field gradients. This will be followed by the SPI (*selective population inversion*) experiment, which can be used to increase the sensitivity for insensitive nuclei such as ^{13}C and ^{15}N by polarization transfer. Next we will deal with another technique that is based on the same principle and has been given the name INEPT. It is also possible to perform a *reverse* INEPT experiment, in which polarization is transferred from an insensitive nuclide to a sensitive nuclide, and the advantages offered by this and other types of reverse procedures will be explained. This is followed by a description of the DEPT method, which is of great practical importance. Lastly we come to the *selective TOCSY* experiment and the INADEQUATE method for determining C,C coupling constants. The latter will involve us in considering the phenomenon of double quantum coherence. We will be returning to these topics later in Sections 9.5 and 9.4.3 with discussions of the two-dimensional variant of the INADEQUATE experiment and the reverse two-dimensional H,C correlation method.

At this point we will anticipate one of the findings of this chapter. These newer types of experiments provide, in addition to the already familiar spectral parameters of chemical shifts and coupling constants, new information such as correlations between chemical shifts and couplings, between ^{1}H and ^{13}C chemical shifts, or between other pairs of nuclides. By giving information about the neighbor relationships (connectivities) between nuclei whose spins are coupled via the scalar or dipole–dipole (through-space) mechanisms, they provide new insights into the structures of molecules. Sometimes, too, these experiments allow one to measure chemical shifts and coupling constants better and faster as a result of the gain in intensity through polarization transfer.

8.2 Basic Techniques Using Pulse Sequences and Pulsed Field Gradients

The one- and two-dimensional NMR experiments described in this and the following chapter are based on the use of complex pulse sequences. By this we mean a series of pulses applied one after another with fixed or variable time delays between them. Most sequences for observing ^{13}C resonances involve applying such a pulse sequence in the ^{13}C channel while simultaneously another sequence is applied in the ^{1}H channel. Changes can be made to the pulse angles and to the directions (in the rotating frame) of the pulsed $\boldsymbol{B}_1$ fields which act on the individual spins and on the macroscopic magnetization. The most frequently used pulse angles are 90° and 180°.

The techniques will be explained wherever possible by means of vector diagrams, which show in a clear way what happens as a result of applying the different pulses. Nevertheless, such vector diagrams are sometimes difficult to understand without previous knowledge. Therefore the aim in the following sections is to learn how some of the most frequently recurring types of pulses affect simple magnetization vectors, and to practice using this knowledge. This should help in quickly attaining a good understanding of the techniques.

The methods whereby the different pulse sequences are generated, and their effective field directions are controlled by altering the r.f. phase, are matters of electronic engineering which will not be considered here.

The latest generation of NMR experiments that have been developed includes some that employ not only complex (often two-dimensional) pulse sequences, but also pulsed field gradients. These represent a considerable advance in that they can greatly reduce the time needed to complete a measurement. Consequently many types of NMR measurements that were previously resorted to only in exceptional cases can now be performed routinely. Procedures using pulsed field gradients have thus become very important, and the principles of the technique will be explained in Section 8.2.3 and illustrated by a simple example.

8.2.1 The Effect of the Pulse on the Longitudinal Magnetization (M_z)

Figure 8-1 shows how the longitudinal magnetization M_z, whose magnitude corresponds to the equilibrium magnetization, rotates under the influence of 90° and 180° pulses with different effective field directions in the rotating frame, namely with B_1 along the x', y', and $(-x')$ directions (see Section 1.5.2), as determined by the phase of the applied r.f. field. After a $90°_{x'}$ pulse the magnetization vector lies along the direction of the y'-axis, whereas after a $90°_{y'}$ pulse it is along the $(-x')$ direction. A $90°_{-x'}$ pulse rotates M_z into the $(-y')$ direction. The same condition would also be reached after a $270°_{x'}$ pulse (Fig. 8-1 a–c). (The rotation always takes place in the anticlockwise sense as seen by an observer looking from the origin of the coordinate system along the direction of B_1.)

A 180° pulse tips M_z into the $(-z)$ direction. In this case the direction of the B_1 field does not matter, provided that it lies in the x', y' plane. The end result is the same in every case, the only difference being that for a $180°_{x'}$ pulse the vector rotates in the y', z plane, whereas for a $180°_{y'}$ pulse it rotates in the x', z plane. This difference is of no practical importance for our purposes.

In the spin conditions existing after a 180° pulse the level populations are inverted. We shall return to this point in Section 8.5.

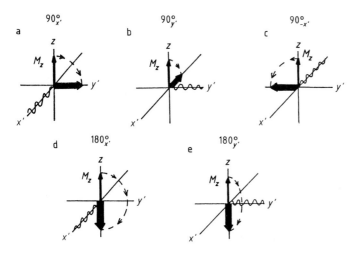

Figure 8-1.
Effects of 90° and 180° pulses on a longitudinal magnetization component M_z. The wavy line indicates the effective direction of the r. f. field B_1 in the rotating coordinate system x', y', z. The thick arrow represents the magnetization vector after applying the pulse.

8.2.2 The Effect of the Pulse on the Transverse Magnetization Components ($M_{x'}$, $M_{y'}$)

Figure 8-2 shows four vector diagrams which illustrate the effects of 90° and 180° pulses on a magnetization component $M_{y'}$ lying along the direction of the y'-axis. As in the experiments described above, the rotation of the vector takes place in the plane perpendicular to the B_1 field. Figure 8-2b is intended to show that a $90°_{y'}$ pulse has no effect on the magnetization $M_{y'}$, as this has no component at right angles to the B_1 field.

This is made clearer in the examples shown in Figure 8-3. Let us assume that the initial magnetization vector A has the direc-

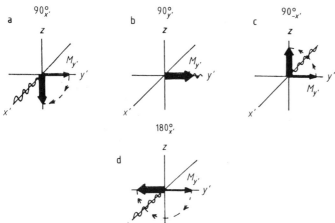

Figure 8-2.
Effects of 90° and 180° pulses on a transverse magnetization component $M_{y'}$. The wavy line indicates the effective direction of the r. f. field B_1 in the rotating coordinate system x', y', z. The thick arrow represents the magnetization vector after applying the pulse.

184

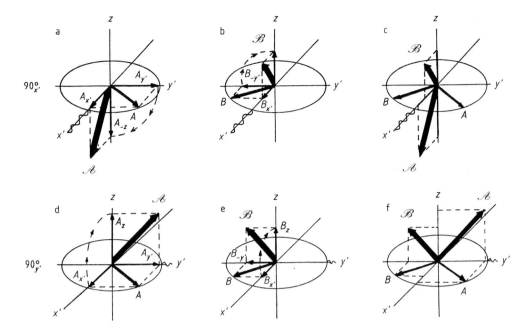

Figure 8-3.
Effects of a $90^\circ_{x'}$ pulse (upper diagrams, a to c) and a $90^\circ_{y'}$ pulse (lower diagrams, d to f) on two arbitrary magnetization vectors **A** and **B**. In diagrams a, b, d and e the vectors **A** and **B** are shown resolved into their x'- and y'-components. $\mathscr{A}$ and $\mathscr{B}$ (thick arrows) are the new vectors after applying the pulses. Diagrams c and f show the effects on both vectors **A** and **B** together.

tion shown in the front right quadrant of the diagram. We can resolve **A** into two components $A_{x'}$ and $A_{y'}$, which are its projections onto the x'- and y'-axes. The $90^\circ_{x'}$ pulse affects only the component $A_{y'}$, not $A_{x'}$. The resulting vector $\mathscr{A}$ after the $90^\circ_{x'}$ pulse has the components A_{-z} and $A_{x'}$. Vector addition thus gives the direction of $\mathscr{A}$, which is in the x', z-plane, downwards and to the front.

A $90^\circ_{y'}$ pulse, on the other hand, affects only the x'-component $A_{x'}$ (Fig. 8-3 d), rotating it into the $(+z)$ direction. Vector addition of $A_{y'}$ *and* A_z shows that the new vector $\mathscr{A}$ is in the y', z-plane, pointing upwards and to the rigth.

Figure 8-3 also includes the diagrams for the vector $\mathscr{B}$ which results from a different initial orientation **B** (Fig. 8-3 b and e), and for a system with both vectors $\mathscr{A}$ and $\mathscr{B}$ (Fig. 8-3 c and f). The case illustrated in Figure 8-3 c will turn out to be important later when polarization transfer experiments are discussed (Section 8.5).

In the next examples we consider the effects of 180° pulses on a magnetization vector **A**, and also on a pair of vectors **A** and **B**. This time only the x', y'-plane is shown, as the vectors lie in this plane both before and after the pulses. We also assume that the x'- and y'-axes are rotating with the *average* Larmor frequency

$(v_A + v_B)/2$; the vector A rotates faster than the coordinate system, and the vector B slower. Such a situation might be caused by field inhomogeneities, or by interactions with neighboring nuclei, for example a spin-spin coupling. The directions of rotation of the vectors relative to the coordinate system are shown by arrows beside the circular path.

Both vectors have components in the x'- and y'- directions. First we consider only the vector A with its components $A_{x'}$ and $A_{y'}$ (Fig. 8-4, top center). A $180^\circ_{x'}$ pulse, which acts in the direction of the x'-axis, rotates $A_{y'}$ from the $(+y')$- to the $(-y')$-direction; $A_{x'}$ lies along the direction of the B_1 field, and is therefore unaffected. Vector addition of $A_{x'}$ and $A_{-y'}$ gives the new vector $\mathscr{A}$. However, as before, $\mathscr{A}$ rotates faster than the coordinate system, since neither the field inhomogeneities nor the spin-spin coupling are affected by the pulse. Therefore the direction of rotation relative to the coordinate system, indicated by the small arrow, remains unchanged. The new direction of the vector $\mathscr{A}$ after the $180^\circ_{x'}$ pulse corresponds to a reflection through the x'-axis (Fig. 8-4 a).

The effect of a $180^\circ_{y'}$ pulse is to reflect the vector A through the y'-axis (Fig. 8-4 b), or, expressing this in another way, $A_{x'}$ is turned through 180° from the $(+x')$- to the $(-x')$-direction. Vector addition of $A_{-x'}$ and $A_{y'}$ gives the new vector $\mathscr{A}$.

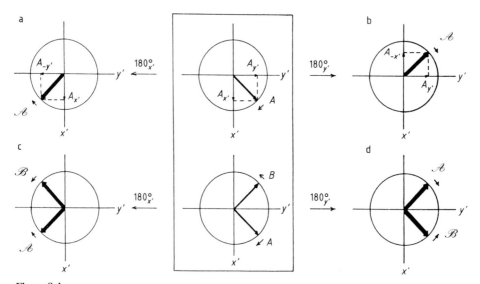

Figure 8-4.
Effects of $180^\circ_{x'}$ and $180^\circ_{y'}$ pulses on the transverse magnetization vector A (upper diagrams) and on the two vectors A and B (lower diagrams). The coordinate system x', y', z rotates with the average Larmor frequency $(v_A + v_B)/2$. $\mathscr{A}$ and $\mathscr{B}$ (thick arrows) are the new vectors after applying the pulses.
The initial situation is shown in the central box, with A rotating faster than the coordinate system and B slower, as indicated by the small arrows outside the circles. Diagrams a and c show the situation after a $180^\circ_{x'}$ pulse, b and d after a $180^\circ_{y'}$ pulse. The directions of rotation relative to the coordinate system are not altered by the pulses.

186

Again, the direction of rotation relative to the rotating coordinate system remains unchanged; the small arrow beside the circle points in the same direction as before the pulse.

To conclude this example we now consider the effects of 180° pulses on a pair of magnetic vectors A and B. As before, we assume that A rotates slightly faster than the coordinate system and B slightly slower. The situations after a $180°_{x'}$ or a $180°_{y'}$ pulse are shown in Figure 8-4 c and d respectively. The $180°_{x'}$ pulse reflects both vectors through the x'-axis, their original directions of rotation being preserved. If the two vectors are moving apart before the $180°_{x'}$ pulse, they will be moving towards each other after it (Fig. 8-4 c). After a certain time they are again parallel, and an echo is recorded by the receiver coil. The $180°_{x'}$ pulse has great practical importance; we have already met it in connection with the determination of the spin-spin relaxation time T_2 (spin-echo experiment, Section 7.3.2 and Fig. 7-6).

Fig. 8-4 d shows the situation in the spin system after a $180°_{y'}$ pulse. The reflection through the y'-axis causes A and B to exchange places. The only change in the appearance of the diagram is that the directions of rotation of the vectors are reversed, so that after the $180°_{y'}$ pulse the faster vector $\mathcal{A}$ is behind the slower vector $\mathcal{B}$. This leads to a refocussing of $\mathcal{A}$ and $\mathcal{B}$ after a certain time, and again one observes an echo in the receiver coil. Therefore we can perform a spin-echo experiment by using either a $180°_{x'}$ or a $180°_{y'}$ pulse. The only difference between the two experiments is that the phase of the echo is shifted by 180°. It is also found in many other experiments that one can use different methods with different pulse sequences to achieve the same result.

8.2.3 The Effect of Pulsed Field Gradients on the Transverse Magnetization

It was mentioned in Section 1.5.5 that when one is accumulating successive FIDs to obtain a spectrum the spin system should ideally be fully relaxed, i.e. in equilibrium, before each new pulse. However, in order to achieve this one would typically have to allow a delay time (Δ) of many seconds, or even several minutes, since the relaxation times can sometimes be quite long (see Chapter 7). In practice this would usually mean that the total duration of the measurement would be unacceptably long. In most cases, therefore, one reaches a compromise by setting the value of Δ at no more than a few seconds; a typical delay time is, say, 2 s. This means that the system will not be fully relaxed at the beginning of the next pulse: M_z has not yet reached its equilibrium value M_0, and there are still residual transverse magnetization components $M_{x'}$ and $M_{y'}$, which can cause artefacts when using two-dimensional methods (see Chapter 9). This problem can be solved by using methods based on *pulsed field gradients* (PFG). What are these field gradients? How do they affect the spin system, and how can they be used to eliminate the residual transverse magnetization?

In high-resolution NMR spectroscopy one normally wishes the magnetic field throughout the observed region of the sample to be as homogeneous as possible, since field variations ΔB_0 cause undesirable broadening of the resonances (see Section 7.3.3). However, deliberately introduced inhomogeneities can be used to advantage, as they are in the pulsed field gradient technique. Using special gradient coils one applies a *linear* magnetic field gradient along the direction of B_0 (the *z*-axis), so that the nuclei within the observed volume experience different field strengths depending on their positions. In principle one could also apply gradients in the *x*- and *y*-directions (as one does in magnetic resonance tomography – see Chapter 14), but that is not relevant for our present purpose.

The situation is shown in Figure 8-5. The applied field gradient is so chosen that the additional field contribution at the center of the observed sample volume is zero; in other words, the nuclei in a thin slice at the middle experience simply the field B_0 produced by the cryomagnet. Above the middle of the sample the field is greater (by an amount g_n), whereas below the middle it is smaller. This is represented in Figure 8-5 by showing five equally spaced thin slices (1–5), in which the magnetic flux densities (from the bottom to the top) are $B_0 - g_1$, $B_0 - g_2$, B_0

$$v_n = \frac{\gamma}{2\pi}(B_0 + g_n)$$

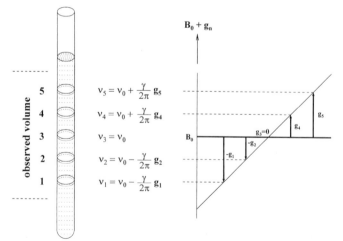

Figure 8-5.
Effect of applying a linear field gradient along the direction of the field B_0 (the *z*-direction). The quantities g_n are the contributions of the field gradient to the field B_0. v_1 to v_5 are the resonance frequencies for the five arbitrarily chosen slices 1 to 5. In the center of the observed sample (slice 3 in the example shown) g_n is zero, while it is positive above the center and negative below it.

(since $g_3 = 0$), $B_0 + g_4$ and $B_0 + g_5$. The quantities g_n are the additional field contributions at the different positions. In accordance with the resonance condition [Eq. (1-12)], the precession frequencies of the nuclei at the different positions are given by $v_n = (\gamma/2\pi)(B_0 + g_n)$, resulting in the values v_1 to v_5 given in Figure 8-5. However, in our experiment the field gradient is only applied for a certain time τ. After it is switched off all the nuclei in the sample experience only the field B_0, and precess with the same frequency v_0.

For simplicity we will consider an ideal rectangular gradient pulse (shaped like the r.f. pulse shown in Figure 1-7), so that the field gradient is established instantaneously at the start and disappears instantaneously at the end. In practice the transition is not quite so abrupt. In principle one can also use other pulse shapes for special experiments.

We will now consider how a field gradient affects the transverse magnetization M_n following a $90°_{x'}$ r.f. pulse (Figure 8-1 a). This is shown schematically in Figure 8-6. Before the $90°_{x'}$ pulse all the nuclei are in the same homogeneous field B_0, and the macroscopic magnetization M_0 lies along the field direction (z-axis). The $90°_{x'}$ pulse turns the magnetization into the y'-direction in the rotating frame, which rotates at the precession frequency of the nuclei, v_0 (as explained in Section 1.5.2). At this instant we apply the linear field gradient. We will focus our attention on five slices of the sample, as shown earlier in Figure 8-5. The slices are assumed to be sufficiently thin that all the nuclei within each one experience the same magnetic field, differing from slice to slice due to the field gradient contributions g_n, as explained above. Immediately after the $90°_{x'}$ pulse the nuclei within a given slice are combined to give a total transverse magnetization vector M_n directed along the y'-axis. The sum of the projections of all such vectors M_n onto the x',y' plane is the macroscopic transverse magnetization $M_{y'}$ (see Figure 8-6 c). While the field gradient is applied, the vectors M_n rotate with different frequencies v_n (see also Figure 8-5). The coordinate system x', y' rotates at the frequency v_0, which is also the precession frequency of the nuclei in the middle slice (since $g_0 = 0$). The vectors of the two lower slices (1 and 2) rotate more slowly, and therefore fall behind in the rotating frame, whereas those of the two upper slices (4 and 5) rotate faster, and move ahead in the rotating frame. These motions are indicated by the small arrows outside the ellipses in Figure 8-6 b. Evidently the angle θ through which one of these vectors moves in the rotating frame during the time τ of the gradient pulse depends on the frequency difference Δv, and thus on the position of the slice; it is given by $\theta = 2\pi\tau\Delta v$. If we join the tips of all such vectors, we obtain a helix extending through the volume of the sample. The pitch of the helix, and thus also the number of revolutions from top to bottom, depends on the

189

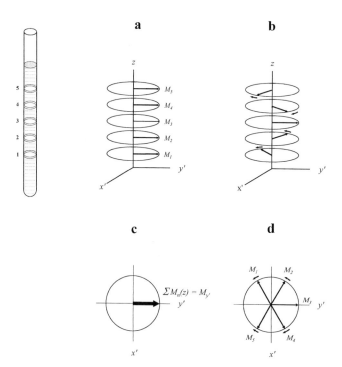

Figure 8-6.
The behavior of the transverse magnetization under the influence of a field gradient. As in Figure 8-5, five slices are shown, the gradient field contribution g_3 in slice 3 at the center of the observed volume being zero. A $90^\circ_{x'}$ pulse establishes the transverse magnetization vectors M_1 to M_5. Owing to the field gradient contributions g_1 to g_5 these precess with different frequencies v_1 to v_5. M_3 remains along the y'-direction in the rotating frame, since both the magnetization and the frame have the same frequency v_0, while the other transverse magnetization vectors rotate in the directions indicated by the small arrows (diagrams b and d). As a result of the fanning-out process the macroscopic transverse magnetization $M_{y'}$ for the sample as a whole (diagram d) eventually falls to zero.

applied field gradient, which we represent by G. A typical value of G is 10 Gs cm^{-1} (in SI units 0.1 T m^{-1}). If we consider the projections of the vectors M_n onto the x', y' plane, as shown in Figure 8-6 d, it can be seen that the phase coherence of these vectors is apparently lost during the time τ. After a sufficient length of time (typically 1 ms) the transverse magnetization component for the sample as a whole will have completely disappeared, and no signal will be induced in the receiver coil.

However, the phase coherence produced by the $90^\circ_{x'}$ pulse has not been permanently destroyed. This can be shown by the following experiment. If, after the field gradient pulse, we apply to the spin system a second field gradient pulse exactly like the first but in the opposite direction, we find that the spins become refocussed, so that all the transverse magnetization vectors M_n are once more aligned along the y'-direction, as they were after the initial $90^\circ_{x'}$ pulse. Figure 8-7 shows this sequence of events. Immediately after the $90^\circ_{x'}$ pulse, each of the partial transverse magnetization vectors M_n lies along the y'-direction (Figure 8-7 a). During the first gradient pulse G these components precess relative to the rotating frame at a frequency $v_n - v_0 = (\gamma/2\pi)g_n$, which depends on the position of the slice in the sample (cf. Figure 8-5). In the time τ during which the gradient is applied, the vector rotates through an angle $\theta = 2\pi(v_n - v_0)\tau$ (see Figure 8-7 b). During the second field gradient pulse, applied in the opposite sense $(-G)$, the

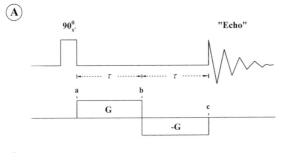

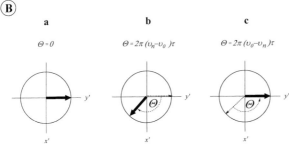

Figure 8-7.
The gradient echo experiment.
A: Pulse sequence. After the $90^\circ_{x'}$
pulse the field gradient G is applied
for a time τ. A gradient of the same
magnitude and duration is then
applied in the opposite direction
$(-G)$.
B: The vector diagrams show the
behavior of the transverse magnetization for slice n. During the first
interval τ, under the influence of the
gradient G, it precesses through an
angle Θ in the rotating frame
(diagram b). During the second
interval τ the process is reversed by
the gradient $-G$, so that the transverse magnetization vector is again
along the y'-direction, producing an
echo.

field experienced by the same slice is less than B_0 by exactly the same amount that it was greater than B_0 during the first gradient pulse. The partial magnetization vector then precesses in the rotating frame at a rate equal and opposite to the previous rate, and therefore rotates through the angle $-\theta$ during the time τ, so that then it once more lies exactly along the y'-axis (Figure 8-7 c). At that instant a signal (the "gradient echo") is induced in the receiver coil. Thus the phase coherence is not in fact destroyed during the first gradient pulse. This is in contrast to the effects of random field gradients in an insufficiently homo-geneous magnetic field, which eliminate the phase coherence and thus permanently destroy the transverse magnetization – a *macroscopic* effect.

We began this discussion by considering a practical problem: how can we get rid of the residual transverse magnetization before we apply the next radiofrequency pulse, without having to wait until the spin system has fully relaxed (which would involve a delay determined by T_2)? The pulsed field gradient method has provided a solution to this difficulty, thereby making it possible to greatly reduce the time needed to complete a measurement, especially when using two-dimensional methods.

The experiment described above is the simplest application of pulsed field gradients, but is by no means the only one. We shall meet others later.

8.3 The *J*-Modulated Spin-Echo Experiment

In Section 7.3.2 we met the spin-echo experiment and the pulse sequence on which it is based, i. e. the sequence:

$$90^\circ_{x'} - \tau - 180^\circ_{x'} - \tau \,(\text{echo})$$

The objective of that experiment was to achieve refocussing of spin vectors which had fanned out during a time τ as a result of field inhomogeneities or different Larmor frequencies (chemical shifts).

It is now necessary to explain how scalar couplings affect the spin-echo. For this we consider an AX spin system with A = ^{1}H and X = ^{13}C, and we assume that the ^{13}C resonances are to be observed. An example of such a heteronuclear two-spin system is chloroform, ^{13}CHCl$_3$.

The evolution of the spin system up to the echo will be illustrated by vector diagrams, in which we will confine our attention to the ^{13}C magnetization vector M_C.

The following preliminary remarks will help in understanding the later discussion. In Section 1.5.2 it was shown that a $90^\circ_{x'}$ pulse has the effect of bunching together a small fraction of the nuclear spins (Fig. 1-12). The transverse magnetizsation $M_{y'}$ is the vector sum of the transverse components of all the precessing bunched individual spins. $M_{y'}$, like the individual spins, rotates about the z-axis with the Larmor frequency.

In the two-spin system there are two different Larmor frequencies as a result of the C,H coupling: one for ^{13}C nuclei in molecules whose protons are in the α-state, which we denote by $\nu(^{13}\text{CH}_\alpha\text{Cl}_3)$, and one for those whose protons are in the β-state, i. e. $\nu(^{13}\text{CH}_\beta\text{Cl}_3)$. These two frequencies are given by:

$$
\begin{aligned}
\nu(^{13}\text{CH}_\alpha\text{Cl}_3) &= \nu_c - \tfrac{1}{2}\,J(\text{C,H}) \\
\nu(^{13}\text{CH}_\beta\text{Cl}_3) &= \nu_c + \tfrac{1}{2}\,J(\text{C,H})
\end{aligned}
\tag{8-1}
$$

where ν_c is the Larmor frequency of the ^{13}C nuclei in the absence of C,H coupling (e. g. as measured with ^{1}H BB decoupling). Thus the ^{13}C NMR spectrum of chloroform consists of two signals whose separation is $^1J(\text{C,H}) = 209$ Hz, with a chemical shift $\delta = 77.7$ (center of the doublet).

A macroscopic sample contains nearly equal numbers of ^{13}CH$_\alpha$Cl$_3$ and ^{13}CH$_\beta$Cl$_3$ molecules (see Sections 1.3.3 and 4.3.1). Therefore there are two magnetization vectors $M_C^{H\alpha}$ and $M_C^{H\beta}$ whose magnitudes are nearly equal. $M_C^{H\alpha}$ belongs to the 50 % of the chloroform molecules ^{13}CH$_\alpha$Cl$_3$ in which the proton is in the α-state, and $M_C^{H\beta}$ to the other 50 %, ^{13}CH$_\beta$Cl$_3$, in which it is in the β-state (Fig. 8-8).

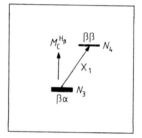

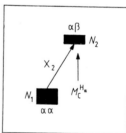

Figure 8-8.
Energy level scheme for a two-spin AX system with A = ^{1}H and X = ^{13}C; example: ^{13}CHCl$_3$. N_1 to N_4 are the populations, with $N_1 > N_2 > N_3 > N_4$. $M_C^{H\alpha}$ is the macroscopic ^{13}C magnetization vector for the $N_1 + N_2$ chloroform molecules whose protons are in the α-state (^{13}CH$_\alpha$Cl$_3$), while $M_C^{H\beta}$ is that for the $N_3 + N_4$ molecules whose protons are in the β-state (^{13}CH$_\beta$Cl$_3$).

We now return to the spin-echo experiment. The $90^\circ_{x'}$ pulse turns both magnetization vectors $M_C^{H\alpha}$ and $M_C^{H\beta}$ into the direction of the $(+y')$-axis, and they begin to rotate about the z-axis. Since the coupling constant $^1J(C,H)$ is always positive, $M_C^{H\alpha}$ rotates slower than $M_C^{H\beta}$, because:

$$\nu(^{13}CH_\alpha Cl_3) < \nu(^{13}CH_\beta Cl_3).$$

The difference between the two frequencies is exactly equal to the coupling constant $^1J(C,H)$. Therefore, after a time τ the phase angle Θ between the two vectors is:

$$\Theta = 2\pi J(C,H)\tau \qquad (8-2)$$

The continuation of this analysis follows exactly that for the vector diagrams c, d and e in Figure 7-5 of Section 7.3.2; all that is necessary to replace the vectors A and B by $M_C^{H\beta}$ and $M_C^{H\alpha}$ respectively. The $180^\circ_{x'}$ pulse reflects the vectors through the x'-axis, without changing the directions of rotation of the vectors relative to the x', y', z coordinate system. After an interval τ, $M_C^{H\alpha}$ and $M_C^{H\beta}$ are again parallel and along the $(-y')$ direction. This shows that in the spin-echo experiment refocussing after a time 2τ also occurs for vectors which have fanned out as a result of a spin-spin coupling. In this discussion we have neglected field inhomogeneities in the interest of simplicity. This additional complication will be treated later (Section 9.3.1). All we need to know at this point is that after the interval 2τ all the spins are once more in phase, regardless of whether the fanning out has been caused by field inhomogeneities, chemical shifts or scalar couplings.

A completely new situation arises if, after the $180^\circ_{x'}$ pulse and during data acquisition, we switch on the ^1H BB decoupler. Figure 8-9 A shows the pulse sequence for such a modified spin-echo experiment. The vector diagrams (Fig. 8-9 B) show again the effects of the pulse sequence on the ^{13}C magnetization vectors $M_C^{H\alpha}$ and $M_C^{H\beta}$. As in Figure 7-5 we use a coordinate system x', y', z which rotates at the frequency ν_c and we look at the behavior of the vectors in the x', y'-plane. After the $90^\circ_{x'}$ pulse the two vectors $M_C^{H\alpha}$ and $M_C^{H\beta}$ rotate in this plane with the frequencies $\nu_c - J(C,H)/2$ and $\nu_c + J(C,H)/2$ respectively (Equation (8-1)), i.e. $M_C^{H\beta}$ rotates faster than the coordinate system by an amount $J(C,H)/2$, and $M_C^{H\alpha}$ slower than it by the same amount. The small arrows in the vector diagrams show the directions of rotation of the vectors relative to the coordinate system. After an arbitrary time τ the vectors are separated by a phase angle Θ as given by Equation (8-2). Table 8-1 lists these phase angles for five special values of τ. The vector diagrams for these five values of the delay time τ, at each of the instants marked 'a' to 'e' in the pulse sequence, are shown in Figure 8-9 B.

Table 8-1.
J-modulated spin-echo experiment. Phase angles Θ as given by Equation (8-2) for five special values of τ.

τ	Θ
0	$0°$
$[4J(C,H)]^{-1}$	$90°$
$[2J(C,H)]^{-1}$	$180°$
$3[4J(C,H)]^{-1}$	$270°$
$[J(C,H)]^{-1}$	$360°$

193

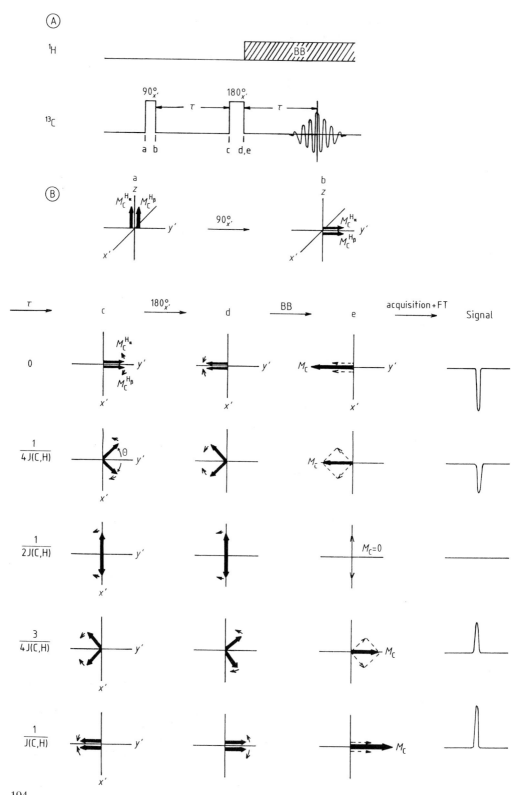

Vector diagram 'a' shows the initial situation, and 'b' that after the $90°_{x'}$ pulse. During the subsequent delay time τ, $M_C^{H\alpha}$ and $M_C^{H\beta}$ evolve so that they are separated by a phase angle Θ (at 'c'). The $180°_{x'}$ pulse then reflects the vectors through the x'-axis ('d'). If one now switches on the ^{1}H BB decoupler the spin-spin coupling, which is the cause of the different rates of rotation of the two vectors, is removed, so that they both rotate with the frequency v_c. Vector addition gives us the vector M_C ('e'); its direction is along either the positive or the negative y'-axis, depending on the delay time τ, unless it happens to be exactly zero. Data acquisition begins after the time 2τ, i. e. one records the second half of the echo.

Since the signal induced in the receiver coil is proportional to the length of the vector along the y'-axis (in either the positive or the negative direction), after Fourier transformation one obtains an absorption signal with either a positive or a negative amplitude; for $\tau = [2J(C,H)]^{-1}$, however, the amplitude is just zero. The amplitude of the absorption signal obtained, shown schematically in the right-hand column of Figure 8-9 B, thus depends on $J(C,H)$, and therefore on the value chosen for the delay time τ.

There remains the question as to why one waits for the echo, rather than recording the signal immediately after a time τ with simultaneous ^{1}H BB decoupling. For the example of chloroform chosen here, in which all the X-nuclei are identical, it would in fact be possible to dispense with the $180°_{x'}$ pulse and the second delay time τ. However, if we have several different sorts of ^{13}C nuclei with different Larmor frequencies and C,H coupling constants, their magnetization vectors will evolve with different frequencies during the time τ. As an example, Figure 8-10 B shows the vector diagrams for three different CH groups (I, II and III). Altogether there are six vectors, I_+, I_-, II_+, II_-, III_+ and III_-, which fan out during the time τ following the $90°_{x'}$ pulse, since their rotation frequencies differ as a consequence of the different Larmor frequencies v and C,H coupling constants ^1J:

$$v(I_+, I_-) \quad = v_I \pm {}^1J_I$$
$$v(II_+, II_-) \quad = v_{II} \pm {}^1J_{II}$$
$$v(III_+, III_-) \quad = v_{III} \pm {}^1J_{III}$$

◀

Figure 8-9.
J-modulated spin-echo experiment.
A: Pulse sequence (the pulse durations are shown greatly exaggerated); only BB decoupling is applied (intermittently) in the ^{1}H channel.
B: Vector diagrams in the rotating coordinate system x', y', z for an AX spin system with A = ^{1}H and X = ^{13}C. Diagrams a to e show the evolution of the ^{13}C magnetization vectors $M_C^{H\alpha}$ and $M_C^{H\beta}$ up to the correspondingly labeled instants in the pulse sequence A. Diagrams c to e, which show only the x', y'-plane, are for five different special values of τ as indicated. The ^{13}C NMR signals after data acquisition and Fourier transformation are shown schematically on the right.

195

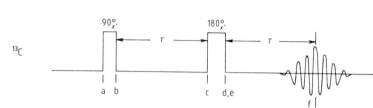

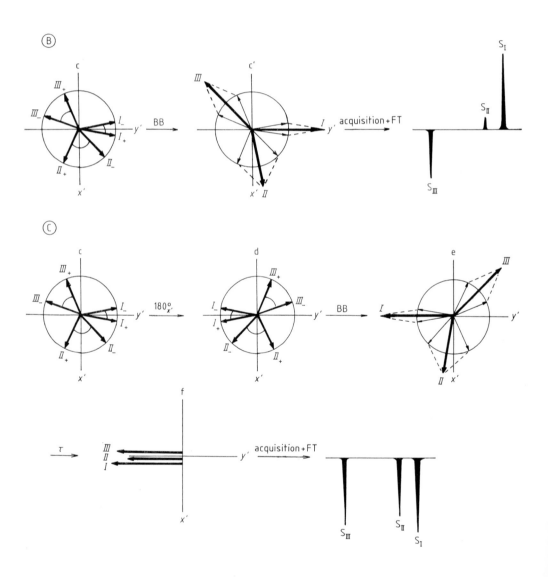

196

The two diagrams c of Figure 8-10 (corresponding to the instant 'c' marked in the pulse sequence A) have been drawn with the assumption

$$\nu_I \quad < \nu_{II} \quad < \nu_{III}$$

$$\text{and } {}^1J_I \quad < {}^1J_{III} \quad < {}^1J_{II}.$$

Also the coordinate system is defined as rotating with the frequency ν_I. Consequently the vector I_+ rotates faster than the coordinate system and the vector I_- slower than it by an equal amount: II_+ and II_- are both faster than the coordinate system, while III_+ and III_-, having the highest Larmor frequencies, are the fastest of all. The three phase angles Θ, given by Equation (8-2), depend on the C,H coupling constants and on τ.

If one switches on the BB decoupler immediately at the end of the interval τ, thereby shortening the pulse sequence, only the three vectors I, II and III are obtained (diagram c'), as the C,H coupling is eliminated. The magnitudes (i. e. the lengths) of the vectors are functions of the C,H coupling constants, and the directions are determined by the respective Larmor frequencies.

After data acquisition and Fourier transformation the spectrum contains three signals, S_I, S_{II} and S_{III}, whose amplitudes are proportional to the y'-components of the vectors I, II and III (Fig. 8-10 B). Thus the amplitudes and signs of the absorption signals depend not only on the C,H coupling constants, which is what we want, but also on the Larmor frequencies (chemical shifts).

This unwanted effect can be avoided by using the complete spin-echo pulse sequence (Fig. 8-10 A). In Figure 8-10 C, vector diagram c again shows, as in Figure 8-10 B, the situation at the instant 'c'. The $180^\circ_{x'}$ pulse which follows reflects all the vectors through the x'-axis (diagram d). After switching on the BB decoupler there are only three vectors I, II and III (diagram e), and these become refocussed at the end of the second delay

◀

Figure 8-10.
J-modulated spin-echo experiment for three different CH groups (I, II and III) whose Larmor frequencies ν and C,H coupling constants ${}^1J$ increase in the order: $\nu_I < \nu_{II} < \nu_{III}$ and ${}^1J_I < {}^1J_{III} < {}^1J_{II}$.
A: Pulse sequence (as in Fig. 8-9).
B: Vector diagrams and spectrum (schematic) for a truncated experiment without the second 180°_x ^{13}C pulse and without the second delay time τ (i. e. $90^\circ_x - \tau$ (BB) – acquisition). Diagram c shows, in the x', y'-plane of the rotating coordinate system, the precessing ^{13}C magnetization vectors $I_+, I_-, II_+, II_-, III_+$ and III_-, all rotating at different frequencies, at the instant c following the evolution time τ. Applying BB decoupling reduces these to the three vectors I, II and III (diagram c'). Acquisition and Fourier transformation of the y'-components of these vectors yields the signals S_I, S_{II} and S_{III}.
C: Vector diagrams and spectrum (schematic) for the full pulse sequence shown in A. Diagrams c to f again show the evolution of the six ^{13}C magnetization vectors. Acquisition of the second half of the echo after the time 2τ (diagram f) followed by Fourier transformation yields three signals with negative amplitudes.

time τ (diagram f). In this way the effect of the different Larmor frequencies is eliminated, while retaining the effects of the C,H couplings on the evolution of the spin system during the time between the $90^\circ_{x'}$ and $180^\circ_{x'}$ pulses. The observed absorption signals S_I, S_{II} and S_{III} all have the same sign, but they have different amplitudes which depend on $J(C,H)$.

So far we have considered only CH groups, i.e. two-spin systems. But what sort of ^{13}C NMR signals does one get under these conditions for a quaternary carbon (C_q) or for a CH$_2$ or CH$_3$ group? In contrast to the CH group, a quaternary carbon nucleus has only one magnetization vector, while for a CH$_2$ group there are three and for a CH$_3$ group four (Fig. 8-11, column a). The evolution of these vectors on applying the pulse sequence of Figure 8-9 A is shown in the vector diagrams b–e in Figure 8-11, for the special case where $\tau = [J(C,H)]^{-1}$. First the $90^\circ_{x'}$ pulse tips the vectors into the direction of the y'-axis (Fig. 8-11, column b).

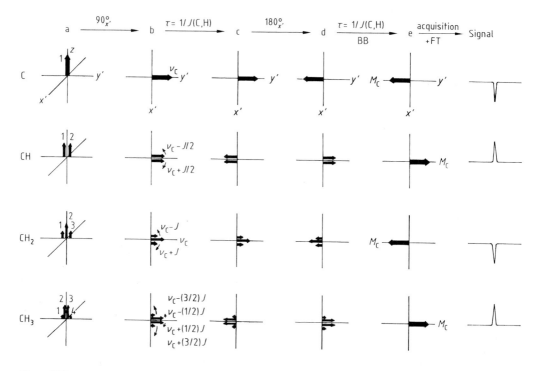

Figure 8-11.
J-modulated spin-echo experiment for ^{13}C nuclei with 0, 1, 2 or 3 directly attached protons. The pulse sequence is the same as in Figures 8-9 and 8-10, but here only one special case, $\tau = [J(C,H)]^{-1}$, is considered. The vector diagrams a to e show the starting situation and the evolution of the magnetization vectors for quaternary, CH, CH$_2$ and CH$_3$ carbon nuclei. Acquisition of the second half of the echo followed by Fourier transformation gives the signals shown (schematically) on the right.

The precession frequencies of the vectors are then as follows:

for C_q : ν_C

for CH : $\nu_C \pm J(C,H)/2$

for CH_2: ν_C, $\nu_C \pm J(C,H)$

for CH_3: $\nu_C \pm 3J(C,H)/2$, $\nu_C \pm J(C,H)/2$

After the interval $\tau = [J(C,H)]^{-1}$ the several vectors are again exactly parallel in each of these cases as a result of their different frequencies; those for CH and CH_3 lie along the $(-y')$ direction, whereas those for CH_2 and C_q lie along the $(+y')$ direction, in the x', y', z coordinate system which rotates with the frequency ν_C (column c). The $180^\circ_{x'}$ pulse which follows reflects all these

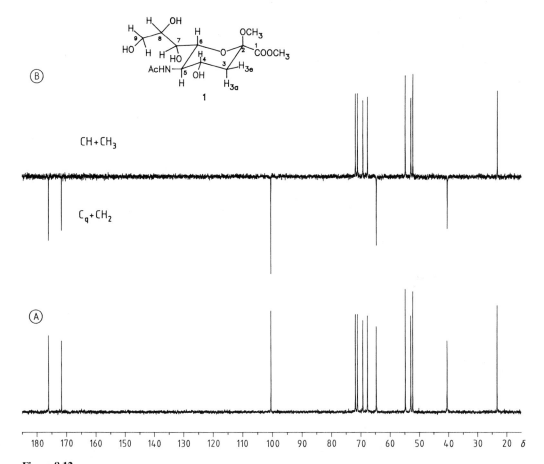

Figure 8-12.

Example of a *J*-modulated spin-echo experiment.

A: 50.3 MHz ^{13}C NMR spectrum of the neuraminic acid derivative **1** with ^{1}H BB decoupling.

B: 50.3 MHz ^{13}C NMR spectrum of the same sample, recorded using the *J*-modulated spin-echo pulse sequence (Fig. 8-9 A). Signals with positive amplitudes are assigned to CH or CH_3 groups, those with negative amplitudes to quaternary carbons or CH_2 groups.

(Experimental conditions for B:

20 mg of the compound in 0.5 ml D_2O; 5 mm sample tube; 224 echoes recorded; 32 K data points; $\tau = 7.14$ ms; duration of experiment: approx. 5 min.)

vectors through the x'- axis (column d). The couplings are then eliminated by BB decoupling, so that in all four cases we are left with just one vector M_C precessing with the frequency v_C; for C_q and CH_2 it lies along the $(-y')$ direction, while for CH and CH_3 it lies along the $(+y')$ direction (column e). After data acquisition and Fourier transformation we therefore obtain negative signals for C_q and CH_2 and positive signals for CH and CH_3.

Figure 8-12 B shows such a spin-echo spectrum for the methylketoside of N-acetyl-D-neuraminic acid methyl ester (**1**), which we met earlier in Chapter 6, together with the BB-decoupled ^{13}C NMR spectrum (Fig. 8-12 A). It can be seen that there are five negative signals, which must be assigned to quaternary or CH_2 carbon nuclei, and eight positive signals from CH or CH_3 carbon nuclei.

As this example illustrates, the J-modulated spin-echo method can be used as an aid to signal assignment, as it enables one to quickly determine from a single experiment whether the number of hydrogen atoms attached to each carbon is odd (1 or 3) or even (0 or 2). The experiment is therefore sometimes called the *attached proton test* (APT) [2].

8.4 The Pulsed Gradient Spin-Echo Experiment

In Section 8.2.3 we learned how pulsed field gradients can be used to eliminate an unwanted residual transverse magnetization. In this section we will look at another experiment in which pulsed field gradients rather than radiofrequency pulses again play the key role; it is fairly simple but of fundamental importance.

Figure 8-13 A shows the pulse sequence used in the *pulsed gradient spin-echo experiment*. The r.f. pulse sequence (upper trace) is exactly the same as in the normal spin-echo experiment, which was described in detail in Section 7.3.2. Here, however, the $90^\circ_{x'}$ and $180^\circ_{y'}$ pulses are each followed immediately by a field gradient pulse (G_1 and G_2 respectively), both with the same sign, magnitude G, and duration τ. The effects of this combined pulse sequence on the spin system are shown by the vector diagrams a–f in Figure 8-13 B, which correspond to the labeled instants in the pulse sequence. We will assume that all the nuclei in the sample are in identical environments, for example the protons in H_2O. This assumption is made only to simplify the discussion here; in principle it does not have to be so, since we learned in Section 7.3.2 that the r.f. pulse sequence $t - 180^\circ_{y'} - t$ has the effect of refocussing the spins, by canceling the earlier fanning out that was caused by chemical

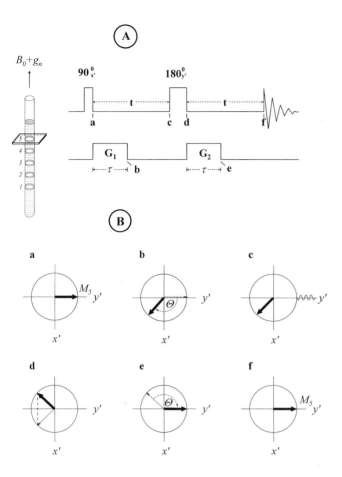

Figure 8-13.
The spin–echo experiment with pulsed gradient.
A: Spin–echo pulse sequence with the addition of a gradient field G_1 after the $90^\circ_{x'}$ pulse and a second (identical) gradient field G_2 after the $180^\circ_{y'}$ pulse.
B: The vector diagrams show the state of the transverse magnetization M_5 of slice 5 at each of the instants a to f in diagram A.

shift differences, spin–spin couplings and magnetic field inhomogeneities. Implicit in our assumption is that B_0 is taken to be constant throughout the sample volume, so that in the absence of an applied field gradient all the nuclei precess with the same frequency v_0, which is the frequency of the rotating frame. Applying the arguments of Section 7.3.2 to the H_2O sample that we take as our example, it is evident that at the instant f, immediately before recording the FID, we have just a single vector M_H for all the protons in the sample. However, we must now consider the effect of the field gradient pulses.

We will focus our attention on just one slice of the sample, arbitrarily choosing slice 5 (Figure 8-13, cf. also Figure 8-5). After the $90^\circ_{x'}$ pulse the magnetization vector M_5 lies along the y'-axis (diagram a). During the time τ for which the field gradient G (pulse G_1) is applied, the nuclei in slice 5 are in a field $B_0 + g_5$, and the precession frequency of M_5 is v_5 (see Figure 8-5). This frequency exceeds that of the rotating frame by $(\gamma/2\pi)g_5$, so that during the time τ the vector advances by the angle $\theta = \gamma g_5 \tau$ (diagram b). During the remaining time

201

$(t - \tau)$ between the end of the gradient pulse and the beginning of the $180^\circ_{y'}$ pulse, the angle of M_5 relative to the coordinate system remains unchanged, as both rotate with the same frequency ν_0, then at c the $180^\circ_{y'}$ pulse reflects M_5 through the y'-axis (d). The field gradient pulse G_2 that follows, and is exactly equivalent to the first, again gives a precession frequency ν_5, so that after the time τ the vector M_5, precessing faster than the frequency of the rotating frame, is turned into the direction of the y'-axis (e). It stays in this orientation after the gradient is removed. Similar arguments apply to all such slices in our sample, and thus at the beginning of the data acquisition (f) all the magnetization vectors M_n are refocussed along the y'-axis, producing an "echo".

Up to now we have implicitly assumed that all the nuclei within a slice remain within that slice throughout the measurement. However, that is not in general the case. In all spectrometers except those of the very latest generation, it is necessary to spin the sample to obtain the best resolution, and this causes some mixing in the sample. To avoid this, experiments using pulsed field gradients must be performed without spinning. Even then we are still left with effects due to diffusion of the molecules in solution. Diffusion within the slice has no effect, but diffusion in the axial direction, which is the direction of the field gradient, means that the value of g_n, and therefore of the precession frequency ν_n, can change between the first gradient pulse (G_1) and the second (G_2). Consequently the affected nuclear spins are not refocussed, and do not contribute to the intensity at the middle of the echo. Thus we see that diffusion reduces the magnitude of the echo. This effect can be used to measure diffusion coefficients D, or to suppress the solvent signal in cases where the solvent molecules diffuse faster than those of the solute, e. g. in polymer solutions.

8.5 Signal Enhancement by Polarization Transfer

8.5.1 The SPI Experiment [3]

It has already been mentioned that one of the main problems encountered in NMR spectroscopy is the low sensitivity for some important nuclides. In Section 1.4.1 we became familiar with the main reason for this, namely that the signal intensity is proportional to $N_\alpha - N_\beta$, the difference between the populations of the two energy levels, which is quite small.

According to Equations (1-9) and (1-10) the ratio of the populations is given by:

$$\frac{N_\beta}{N_\alpha} \approx 1 - \frac{\gamma \hbar B_0}{k_B T} \qquad (8\text{-}3)$$

Therefore $N_\alpha - N_\beta$ becomes greater as the magnetic flux density B_0 is increased. This is one of the reasons why the trend in spectrometer development is towards ever higher magnetic field strengths.

However, the population difference, and therefore the signal intensity, also depends on the magnetogyric ratio γ. Since the nuclides ^{1}H, ^{19}F and ^{31}P have large values of γ, their nuclear resonances are relatively easy to detect, much more so than those of ^{13}C and ^{15}N (see Table 1-1). A distinction can therefore be made, on the basis of the different γ-values, between the sensitive nuclides ^{1}H, ^{19}F and ^{31}P and the insensitive nuclides ^{13}C and ^{15}N. It often happens that the insensitive nuclides also have low natural abundances, and the two disadvantages are combined.

By means of special experiments using selctive pulses it is possible, in coupled systems, to increase the signal intensities for the insensitive nuclear species. What is the principle on which these experiments are based?

Let us consider a two-spin AX system with scalar coupling, in which A is a sensitive nucleus (a proton in our case) and X is an insensitive nucleus (^{13}C). An example of such a system is chloroform (^{13}CHCl$_3$).

In the ^{13}C NMR spectrum, which is the only part of the AX spectrum in which we are interested at this point, the coupling to the proton gives a doublet with a separation $^1J(C,H) = 209$ Hz. If ^{1}H BB decoupling were applied, only one signal would appear.

Figure 8-14 A shows the spectrum and the energy level scheme. The transitions X_1 and X_2 correspond to the two resonance lines observed in the ^{13}C NMR spectrum. The proton transitions A_1 and A_2 appear in the ^{1}H NMR spectrum as ^{13}C satellites of the main signal, if the sample is normal chloroform, without ^{13}C enrichment (Fig. 1-28).

From Equation (8-3) we can deduce relationships between the populations N_1 to N_4, and hence the relative intensities of the signals. All we need to know is that $\gamma(^1$H) is about four times as large as $\gamma(^{13}$C). The differences between N_1 and N_2 and between N_3 and N_4 are small, as they are determined by $\gamma(^{13}$C). On the other hand, the differences between N_1 and N_3 and between N_2 and N_4 depend on $\gamma(^1$H), which is four times larger. As a result the populations of levels 1 and 2 are significantly greater than those of levels 3 and 4. In Figure 8-14 A this is indicated by the thicknesses of the slabs.

A selective 180° pulse which excites only the A_2 transition causes an exact inversion of the populations of levels 1 and 3!

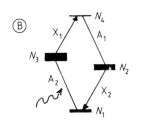

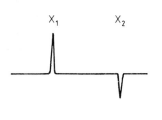

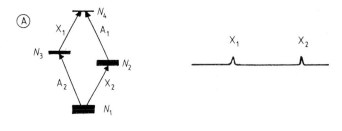

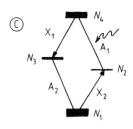

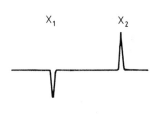

Figure 8-14.
Energy level scheme for a two-spin AX system such as $^{13}CHCl_3$ (A = ^1H, X = ^{13}C) showing schematically the resulting X (^{13}C) spectra.
A: Equilibrium state.
B: A selective 180° pulse exciting only the A_2 transition inverts the populations of levels 1 and 3, as indicated by the thicknesses of the slabs. The X_1 transition then gives an enhanced absorption signal, while the X_2 transition gives an enhanced emission signal.
C: A selective 180° pulse exciting the A_1 transition inverts the populations of levels 2 and 4. In the ^{13}C NMR spectrum we again obtain two enhanced signals, an absorption signal for the X_2 transition and an emission signal for the X_1 transition.
(The argument can also be worked through as a numerical simulation by inserting the values $N_1 = 6$, $N_2 = 5$, $N_3 = 2$ and $N_4 = 1$ for the equilibrium populations. Using these numbers leads directly to the enhancement factors of +5 and −3 for the two-spin ^1H, ^{13}C system.)

N_3 is then greater than N_1, and in the new energy level scheme (Fig. 8-14 B) the slab for level 3 is shown thicker than that for 1.

As a result of this the situation regarding the ^{13}C transitions X_1 and X_2 is completely altered. The intensity of the signal corresponding to X_1 increases, since the population difference between N_3 and N_4 has become greater due to this *polarization transfer*, which is better described as a *magnetization transfer*. For the X_2 transition we have $N_2 > N_1$. This signal too increases, but becomes an emission signal.

In the case of a 180° pulse which selectively excites the A_1 transition, an analogous argument can be applied. This inverts the populations of levels 2 and 4, and one observes an increased emission signal for X_1 and an increased absorption signal for X_2 (Fig. 8-14 C). Theory shows that the signal amplification factors, which are functions of the magnetogyric ratios, are:

$$1 + \frac{\gamma_A}{\gamma_X} \quad \text{and} \quad 1 - \frac{\gamma_A}{\gamma_X} \qquad (8\text{-}4)$$

For chloroform, since $\gamma\,(^1\text{H})/\gamma\,(^{13}\text{C}) \approx 4$, the intensity of the absorption signal is five times larger than the normal signal, and the emission signal is three times larger.

The experiment described above is called *selective population inversion* (SPI). Carrying out the experiment is not a simple matter. First it is necessary to measure the frequencies of the ^1H transitions (A_1 and A_2). As already mentioned, these are not the main signals but the ^{13}C satellites, which are often complex. Secondly one needs to generate and apply a selective $180°$ pulse – this again is by no means a trivial task [1]. Consequently new methods using special pulse sequences have been developed, and these have replaced the older SPI experiment. Nevertheless, these too are based on the polarization transfer principle. Two such methods, the INEPT and DEPT techniques, will be described in detail in the following sections.

8.5.2 The INEPT Experiment [4]

Compared with the SPI technique described in Section 8.5.1, the technique known as INEPT (*insensitive nuclei enhanced by polarization transfer*) has the great advantage that the polarization transfer is achieved using non-selective pulses. The pulse sequence for the INEPT experiment is shown in Figure 8-15 A. The pulses in the ^1H channel (which affect only the protons) and those in the ^{13}C channel (which affect only the ^{13}C nuclei) are shown separately.

The principle of the experiment can again be understood by taking as an example the two-spin AX system of $^{13}\text{CHCl}_3$ in which $A = {}^1\text{H}$ and $X = {}^{13}\text{C}$. The vector diagrams a to h in Figure 8-15 B show how the pulse sequence affects this heteronuclear two-spin system. To begin with we will consider only the magnetization vectors of the protons (M_H). In doing so we will carry over from Section 8.3 the arguments developed there for the macroscopic ^{13}C magnetization vectors M_C and apply them to M_H. Accordingly, the magnetization vector $M_\text{H}^{\text{C}\alpha}$ arises from the 50 % of the chloroform molecules in which the ^{13}C nuclei are in the α-state ($^{13}\text{C}_\alpha\text{HCl}_3$) and $M_\text{H}^{\text{C}\beta}$ from the other 50 % of the molecules with ^{13}C nuclei in the β-state ($^{13}\text{C}_\beta\text{HCl}_3$).

The first vector diagram a shows the situation at the start. Here we use a coordinate system x', y', z which rotates at the frequency $\nu_\text{H} = [\nu(^{13}\text{C}_\alpha\text{HCl}_3) + \nu(^{13}\text{C}_\beta\text{HCl}_3)]/2$, which corresponds to the Larmor frequency of the protons in chloroform, $^{12}\text{CHCl}_3$. A $90°_{x'}$ pulse then turns both magnetization vectors into the direction of the y'-axis (b).

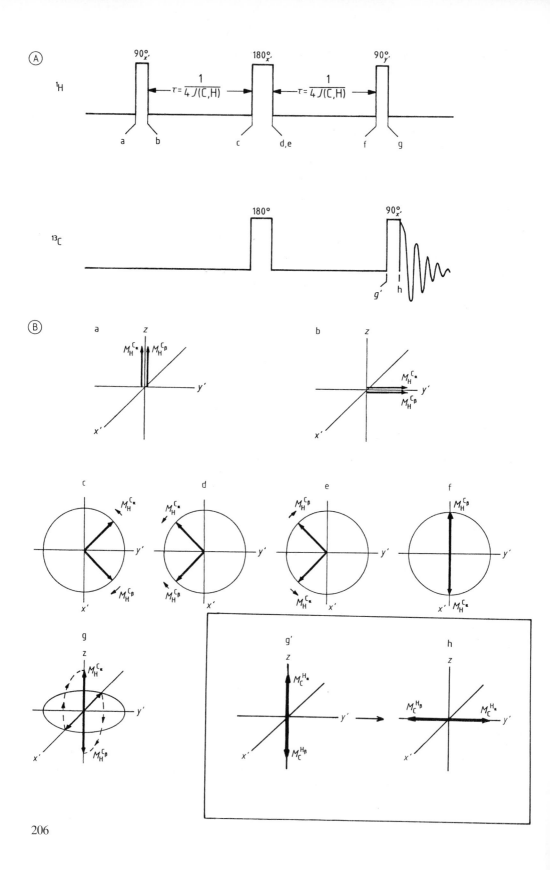

These vectors precess about the z-axis with the frequencies:

$$\nu(^{13}C_\alpha HCl_3) = \nu_H - J(C,H)/2$$
$$\nu(^{13}C_\beta HCl_3) = \nu_H + J(C,H)/2 \qquad (8\text{-}5)$$

Thus the vector $M_H^{C\beta}$ rotates faster than the rotating coordinate system by an amount $J(C,H)/2$, while $M_H^{C\alpha}$ rotates slower than it by the same amount, since the coupling constant $^1J(C,H)$ is positive. The small arrows in each of the diagrams c to e indicate the direction of rotation of the vectors relative to the rotating coordinate system. Diagrams c to f show only the x', y'-plane in each case.

After a time $\tau = [4J(C,H)]^{-1}$ the phase difference Θ, as given by Equation (8-2), is exactly $90°$ (c). At this instant a $180°$ pulse is applied to both the ^1H and ^{13}C nuclei. The phase of the $180°$ pulse in the ^1H channel is chosen so that the direction of the B_1 field is along the x'-axis. In the ^{13}C channel, however, it does not matter whether one uses a $180°_{x'}$ pulse or a $180°_{y'}$ pulse, since either of these will turn M_C from the $(+z)$ direction into the $(-z)$ direction.

The effects on $M_H^{C\alpha}$ and $M_H^{C\beta}$ can be explained in two stages. The $180°_{x'}$ pulse in the ^1H channel reflects both vectors through the x'-axis, without changing their directions of rotation (d). It would then be expected that the two components would be refocussed after an interval $\tau = [4J(C,H)]^{-1}$, as in the spin-echo experiment described in Section 7.3.2. However, the situation is altered radically by the $180°$ pulse in the ^{13}C channel. This pulse inverts the populations of the levels, both between N_1 and N_2 and between N_3 and N_4. At the same time the chloroform molecules with their ^{13}C nuclei in the α-state become those with the β-state, and vice-versa. This means that $M_H^{C\alpha}$ becomes $M_H^{C\beta}$ and $M_H^{C\beta}$ becomes $M_H^{C\alpha}$. Consequently, in the vector diagram the slower vector and the faster vector change places. The rotation arrows now point in the other direction (e). After a further delay time of $\tau = [4J(C,H)]^{-1}$ the phase difference is exactly $180°$ (f).

In the next stage of the pulse sequence the $90°_{y'}$ pulse rotates $M_H^{C\alpha}$ into the $(+z)$ direction and $M_H^{C\beta}$ into the $(-z)$ direction (g). Comparing this with the starting situation (a), the magnetization vector for the chloroform molecules with their ^{13}C nuclei in the β-state $(^{13}C_\beta HCl_3)$ has been rotated through $180°$, i. e. the

◄

Figure 8-15.
The INEPT experiment.
A: Pulse sequences in the ^1H and ^{13}C channels.
B: Vector diagrams for a two-spin AX system with A = ^1H and X = ^{13}C (example: $^{13}CHCl_3$). Diagrams a to g show the ^1H magnetization vectors $M_H^{C\alpha}$ and $M_H^{C\beta}$ in the rotating coordinate system x', y', z at the instants marked a to g in A; in c to f only the x', y'-plane is shown. Diagrams g' and h (in box) show the ^{13}C magnetization vectors M_C.

level populations are inverted, whereas for the other half of the molecules ($^{13}C_\alpha HCl_3$) nothing has changed. Figure 8-16 makes this clearer. The population ratios are exactly the same as in Figure 8-14 C, but in that case the situation was reached by means of a selective 180° pulse exciting only the A_1 transition.

It can be seen from Figure 8-16 that the change in populations also affects the directions of the ^{13}C magnetization vectors. $\boldsymbol{M}_C^{H\alpha}$ retains its original direction along the $(+z)$-axis, but $\boldsymbol{M}_C^{H\beta}$ is now in the $(-z)$ direction. This situation is shown in vector diagram g′ of Figure 8-15 B. (The frame around diagrams g′ and h is to indicate that these show ^{13}C magnetization vectors $\boldsymbol{M}_C$, whereas diagrams a to g show ^1H vectors $\boldsymbol{M}_H$.) The $90°_{x'}$ pulse in the ^{13}C channel serves simply to create transverse magnetization components which are observable (h). After data acquisition and Fourier transformation one obtains two enhanced ^{13}C signals as a result of the transfer of polarization from the "sensitive" protons to the "insensitive" ^{13}C nuclei, one with a positive and the other with a negative amplitude, as in the SPI method (Section 8.4.1, Fig. 8-14 C). The amplification factors in the INEPT experiment are given in terms of $\gamma(^1H)/\gamma(^{13}C)$ (≈ 4) by Equation (8-4), as in the SPI experiment. However, the INEPT pulse sequence has two important advantages:

- The signal enhancement is obtained using non-selective pulses, which avoids the need for exact frequency measurements on the ^{13}C satellites.
- An amplification is obtained for the signals of all the ^{13}C nuclei in a molecule which have identical or similar couplings $^1J(C,H)$ to a neighboring proton. This follows directly from the pulse sequence, in which the only variable is the delay time $\tau = [4J(C,H)]^{-1}$.

So far we have only considered CH groups, i.e. two-spin systems. How does the ^{13}C NMR spectrum change when the INEPT pulse sequence is applied to CH_2 or CH_3 groups i.e. three- or four-spin systems? Whereas in the one-dimensional ^{13}C NMR spectrum a CH_2 group gives a triplet with the intensity distribution 1:2:1, in the INEPT experiment the intensity of the middle line is, in the ideal case, zero (Fig. 8-17). The outer two lines are enhanced by the factor $2\gamma(^1H)/\gamma(^{13}C)$, one with a positive and the other with a negative amplitude. For a CH_3 group the INEPT experiment gives four lines of approximately equal intensities, two with positive and two with negative amplitudes, each enhanced by the factor $3\gamma(^1H)/\gamma(^{13}C)$ [5]. Since the C,H coupling constants in the CH, CH_2 and CH_3 groups of saturated compounds are very similar (125–150 Hz, see Section 3.3.1), a single INEPT experiment using an average value for τ is usually sufficient to give good results.

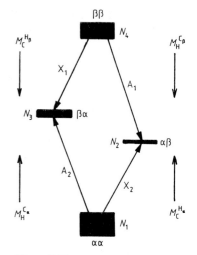

Figure 8-16.
Energy level scheme for a two-spin AX system ($^{13}CHCl_3$) in the INEPT experiment. The experiment inverts the populations of levels 2 and 4, which reverses the directions of the macroscopic magnetization vectors $\boldsymbol{M}_H^{C\beta}$ and $\boldsymbol{M}_C^{H\beta}$ compared with the equilibrium situation. In the ^{13}C NMR spectrum, as in the SPI method, two enhanced signals are observed, one positive and one negative (cf. Fig. 8-14 C).

Figure 8-18 C shows the INEPT spectrum of the neuraminic acid derivative **1**. Also shown for comparison are the relevant portions of the ^{1}H BB decoupled (Fig. 8-18 A, $\delta = 10$ to 110) and non-decoupled (Fig. 8-18 B) ^{13}C NMR spectra. A CH_3 quartet at $\delta \approx 23$ is easily recognized in the INEPT spectrum, with two positive and two negative peaks. So also are the two signals of a CH_2 group at $\delta \approx 40$, one with positive and the other with negative amplitude; the middle signal of the triplet has zero amplitude! The spectral region from $\delta = 50$ to 75 can be analyzed in a similar way. As this is a polarization transfer experiment, no enhanced signals are obtained for quaternary carbon nuclei, i.e. those with no directly attached hydrogen atoms. The "missing" signals are those of the quaternary carbons C_1 and C_2 and that of the carbamide group, at $\delta = 171.50$, 100.32 and 175.93 respectively (cf. Fig. 8-12 A).

The application of ^{1}H BB decoupling during data acquisition so as to simplify the spectrum is not possible, because of the opposite signs of the signal amplitudes within the multiplets; the signals would cancel each other. A way around this difficulty is to use the "refocussed" INEPT experiment. This differs from the "normal" INEPT experiment in that the data acquisition does not start until after a delay time of $2\Delta = 2[4J(C,H)]^{-1}$, with the insertion of additional 180° pulses in the ^{1}H and ^{13}C channels after half this delay time ($\Delta = [4J(C,H)]^{-1}$); the complete pulse sequence is shown in Figure 8-19 A. The purpose of these additional 180° pulses is to eliminate the effects of different chemical shifts during the time 2Δ (see Sections 7.3.2 and 8.3 and Fig. 8-10).

This extension of the pulse sequence can be understood from the vector diagrams in Figure 8-19 B. The evolution of the magnetization vectors M_C and M_H up to the instant g' (i.e. directly after the $90°_{y'}$ pulse in the ^{1}H channel) is as shown in Figure 8-15 a to g (or g'). Here we are concerned only with the vectors $M_C^{H\alpha}$ and $M_C^{H\beta}$, and the starting point is therefore diagram g' of Figure 8-15. In diagram g' (now in Fig. 8-19 B) $M_C^{H\alpha}$ lies along the $(+z)$ direction and $M_C^{H\beta}$ along the $(-z)$ direction, and the two vectors are therefore in antiphase. The $90°_{y'}$ pulse in the ^{13}C channel rotates $M_C^{H\alpha}$ into the $(-x')$ direction and $M_C^{H\beta}$ into the $(+x')$ direction. During this the phase relationship remains the same (h). As the rotation frequencies of the two vectors differ by $J(C,H)$ (Equation (8-1)), they move towards each other, and after a time $\Delta = [4J(C,H)]^{-1}$ their phase difference Θ is reduced to 90° (i). The effects of the $180°_{x'}$ pulses which are then applied simultaneously in the ^{1}H and ^{13}C channels will be considered in two stages for simplicity. The $180°_{x'}$ pulse in the ^{1}H channel interchanges the vectors $M_C^{H\alpha}$ and $M_C^{H\beta}$, so that they now begin to move apart (k). The $180°_{x'}$ pulse in the ^{13}C channel reflects both vectors through the x'-axis, without affecting the direction of rotation (l). During the following delay time Δ the two vectors

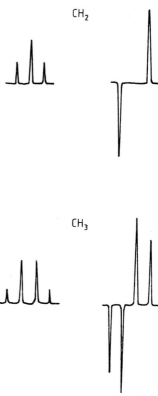

CH$_2$

CH$_3$

Figure 8-17.
Multiplets observed in the ^{13}C NMR spectrum for CH$_2$ and CH$_3$ groups (schematic).
Left: ^{13}C spectra without decoupling.
Right: INEPT spectra.

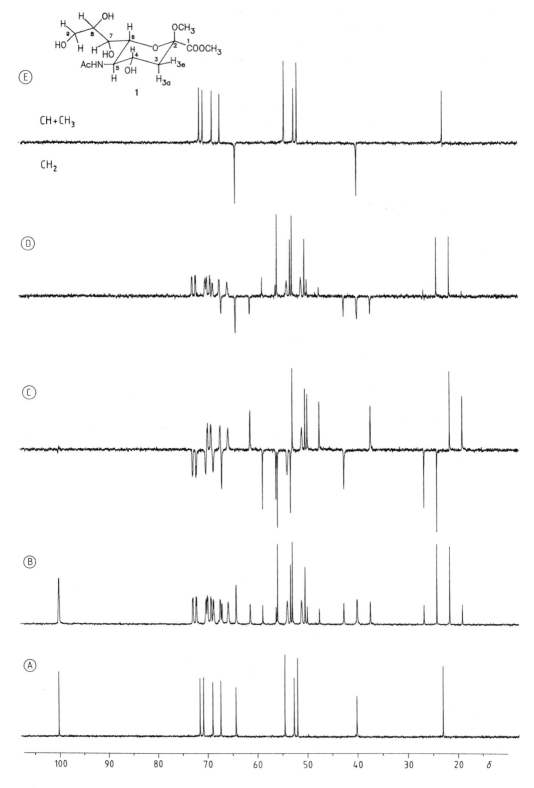

again move together (m). If we now record the FID (without BB decoupling!) we obtain, after Fourier transformation, two absorption signals with positive amplitudes. However, since the two peaks of the doublet have the same phase due to the refocussing, the BB decoupler can be switched on during the data acquisition. This simplifies the ^{13}C NMR spectrum so that it now consists of only a single absorption signal with positive amplitude, which is enhanced by the polarization transfer.

The delay time $\Delta = [4J(C,H)]^{-1}$ gives refocussing and maximum signal amplification only for CH groups, i.e. for two-spin AX systems. For a CH$_2$ group the delay time needed to give refocussing is $\Delta = [8J(C,H)]^{-1}$. In the case of a CH$_3$ group the four M_C vectors can no longer be exactly refocussed; the maximum possible refocussing and amplification is obtained when the value of Δ is about the same as that for a CH$_2$ group. Thus, for a compound which contains CH, CH$_2$ and CH$_3$ groups, the choice of Δ is essentially a compromise, since not only do these groups have different refocussing times in terms of $J(C,H)$, but also the coupling constants $J(C,H)$ are themselves different. A good compromise is to use, for example, a value $\Delta = 0.15[J(C,H)]^{-1}$.

In practice, however, a value of approximately $3[8J(C,H)]^{-1}$ is usually chosen for Δ. In the spectrum which results, not all the signals have positive amplitudes; those of CH and CH$_3$ groups are positive, while those of CH$_2$ groups are negative. Figure 8-18 D shows the spectrum of the neuraminic acid derivative **1** without BB decoupling. The triplets with negative amplitudes from the two CH$_2$ groups are clearly seen. The CH and CH$_3$ groups give doublets and quartets respectively, with positive amplitudes. In the INEPT spectrum with BB decoupling (Fig. 8-18 E) the two negative signals can immediately be assigned to the CH$_2$ groups, whereas the eight positive signals can arise from either CH or CH$_3$ groups. Thus, like the

◀

Figure 8-18.
Examples of INEPT experiments.
A: 50.3 MHz ^{13}C NMR spectrum of the neuraminic acid derivative **1** with ^{1}H BB decoupling ($\delta = 10$ to 110 region only).
B: Spectrum with C,H couplings, recorded by the gated decoupling technique (Section 5.3.2).
C: INEPT spectrum, recorded using the pulse sequence shown in Figure 8-15.
D: Refocussed INEPT spectrum, recorded using the pulse sequence shown in Figure 8-19 (without BB decoupling).
E: Refocussed INEPT spectrum with BB decoupling during data acquisition. Signals with positive amplitudes are assigned to CH or CH$_3$ groups, those with negative amplitudes to CH$_2$ groups. Quaternary carbon nuclei give no signals.
(Experimental conditions:
20 mg of the compound in 0.5 ml D$_2$O; 5 mm sample tube; 32 K data points; A: 1680 FIDs; total time approx. 1 h. B: 5008 FIDs; total time approx. 3 h. C: 976 FIDs; $\tau = 1.79$ ms; total approx. 66 min.
D: 1080 FIDs; $\tau = 1.79$ ms; $\Delta = 2.68$ ms; total time approx. 72 min.
E: 408 FIDs; $\tau = 1.79$ ms; $\Delta = 2.68$ ms; total time approx. 27 min.)

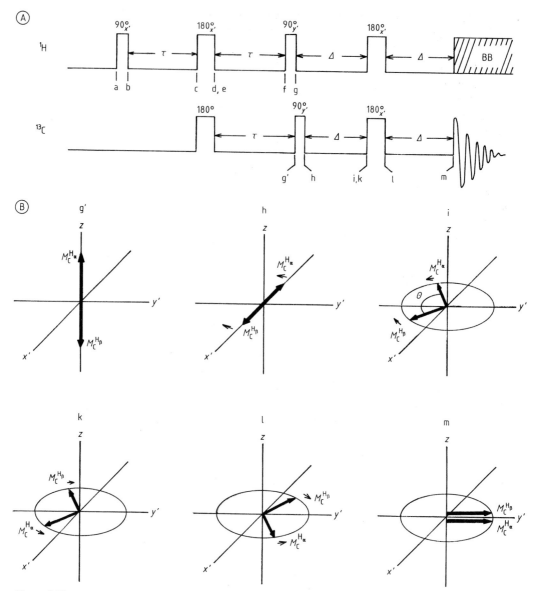

Figure 8-19.
The refocussed INEPT experiment.
A: Pulse sequences in the ^1H and ^{13}C channels.
B: Vector diagrams for a two-spin AX system with A = ^1H and X = ^{13}C (example: $^{13}CHCl_3$). The evolution of the ^1H and ^{13}C magnetization vectors up to the instant g' is as in Figure 8-15 B, and diagram g' here is identical to the previous g'. Diagrams h to m show the evolution of the vectors $M_C^{H\alpha}$ and $M_C^{H\beta}$ during the remainder of the pulse sequence A up to the instant m immediately before data acquisition.

J-modulated spin-echo experiment, these INEPT experiments, with and without BB decoupling, can be used as aids to recognizing the signals of CH, CH_2 and CH_3 groups in the spectrum. Quaternary carbon nuclei give no signals.

212

As mentioned at the beginning of this section, the INEPT method is not confined to the $^1H/^{13}C$ system. Particularly good results are obtained when the method is used to record ^{15}N spectra with ^1H BB decoupling, as the factor $\gamma(^1H)/\gamma(^{15}N)$ is approximately 10.

In order for an INEPT experiment to be successful, it is important that the relaxation times for the sensitive nuclei – protons in most cases – are not too short; in other words, the magnetization vectors M_H must not decay away too rapidly, since if they do there will be little or no transfer of polarization. However, proton relaxation times are usually long enough to ensure that an INEPT experiment can be carried out without difficulty.

8.5.3 The Reverse INEPT Experiment with Proton Detection [6–8]

The INEPT experiment makes it possible to amplify the signals of insensitive nuclides such as ^{13}C or ^{15}N so that they are easier to observe. Furthermore, by suitably designing the experiment one can make it easier to assign the signals. Equation (8-4) shows that the amplification factor depends on the ratio γ_A/γ_X, and is greatest when the transfer of polarization is from a sensitive nuclide (A) to an insensitive nuclide (X), i. e. from a nuclide with large γ (^1H) to one with a small γ (^{13}C or ^{15}N). Thus the experiments described in the previous section are regarded as the "normal" case. However, the INEPT experiment can equally well be carried out in reverse, by applying the initial pulse sequence in the insensitive nuclide channel and observing the signals of the sensitive nuclide. The ratio γ_A/γ_X in Equation (8-4) is then replaced by γ_X/γ_A, which is much smaller – only 0.25 instead of 4 with normal INEPT in the case of ^{13}C and ^1H! Nevertheless, despite this much smaller ratio the reverse INEPT experiment, and also several other analogous reverse procedures (Section 9.4.3), have an advantage due to the fact that one is observing the resonance of the sensitive nuclide (^1H). The signal intensity in the spectrum that one finally observes is not determined just by the ratio γ_X/γ_A, but depends crucially on the magnetogyric ratio γ_A and the resonance frequency of the observed nuclide.

Although the reverse INEPT experiment is not of very great practical importance, it too can be clearly described using vector diagrams, and it is an instructive exercise to analyze the differences between the normal and reverse INEPT procedures. For this purpose we again consider a two-spin AX system with A = ^1H and X = ^{13}C taking chloroform as our example, but in this case it is especially important to remember that only 1.1 % of $^{13}CHCl_3$ is present in chloroform, the main component being $^{12}CHCl_3$.

The pulse sequences (Fig. 8-20 I) are identical to those in Figure 8-15 A except that the roles of the ^1H and ^{13}C channels are reversed. Whereas in the normal INEPT experiment with $\tau = [4J(C,H)]^{-1}$ the pulse sequence results in antiparallel ^1H magnetization vectors $M_H^{C\alpha}$ and $M_H^{C\beta}$, in the reverse experiment it is the ^{13}C vectors $M_C^{H\alpha}$ and $M_C^{H\beta}$ that are antiparallel. It is unnecessary to repeat all the vector diagrams; instead we can simply carry over diagram g from Figure 8-15 and rename the vectors accordingly.

To understand how the antiparallel orientation of $M_C^{H\alpha}$ and $M_C^{H\beta}$ affects the vectors $M_H^{C\alpha}$ and $M_H^{C\beta}$, let us consider the energy level scheme for the two-spin AX system at the instant g. As before (in Figures 8-14 and 8-16) the thicknesses of the slabs represent the relative populations.

In the equilibrium state all four vectors $M_H^{C\alpha}$, $M_H^{C\beta}$, $M_C^{H\alpha}$, and $M_C^{H\beta}$ lie along the $(+z)$ direction. As already explained, the

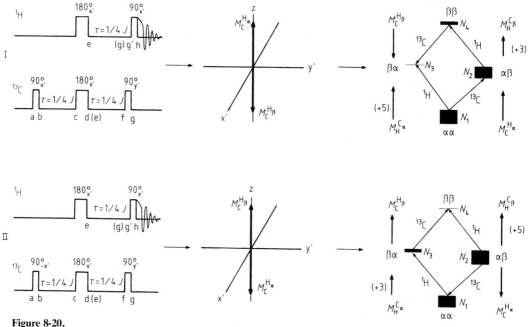

Figure 8-20.
The reverse INEPT experiment.
Pulse sequences, with vector diagrams and energy level schemes corresponding to the instant g in the sequences. Sequences I and II differ in the first of the 90° pulses in the ^{13}C channel; the effective r.f. field B_1 is in the $(+x')$ direction in I and in the $(-x')$ direction in II. The inversion of the level populations N_3 and N_4 (in I) or of N_1 and N_2 (in II) does not make the two M_H vectors antiparallel (as in normal INEPT) but makes them unequal in magnitude. (If, as in Fig. 8-14, the equilibrium state is represented by the numerical values $N_1 = 6$, $N_2 = 5$, $N_3 = 2$, $N_4 = 1$, then the population differences at the instant g are as shown in parentheses; these in turn are proportional to the magnitudes (lengths) of the magnetization vectors M_H.) For further details see text.

pulse sequence up to the instant g disturbs this equilibrium to put the two M_C vectors into an antiparallel configuration, with $M_C^{H\alpha}$ in the $(+z)$ direction and $M_C^{H\beta}$ in the $(-z)$ direction. This means that the populations of levels 1 and 2 have not altered from those in the equilibrium state, whereas those of levels 3 and 4 have become inverted. Therefore, at the instant g' $(= g)$ immediately before the $90°_{x'}$ signal generating pulse in the ^{1}H channel, the energy level populations are as in the scheme shown in Figure 8-20 I.

From this level scheme it can easily be seen that the polarization transfer corresponding to the ratio $\gamma_X/\gamma_A = \pm 0.25$ (Equation 8-4) does not lead to two antiparallel M_H vectors, but instead changes the lengths of the vectors to 1.25 and 0.75 times their initial values, respectively. If we again insert the numbers 6, 5, 2, and 1 to represent the populations N_1 to N_4, as in Figure 8-14, then we find that the lengths of the ^{1}H vectors in the equilibrium state are proportional to the population differences $+4$ and $+4$, whereas after the INEPT pulse sequence they are proportional to $+3$ and $+5$ (the numbers in parentheses in Fig. 8-20). In the following discussion we must therefore take into account the different lengths of the two M_H vectors.

Up to now we have been concerned only with the small fraction of chloroform molecules (1.1 %) that contain ^{13}C, while the majority (^{12}CHCl$_3$) have been tacitly ignored, as they are unaffected by the pulse sequence in the ^{13}C channel. However, all the pulses in the ^{1}H channel equally affect the protons in the ^{12}CHCl$_3$ molecules, and we must therefore take into account not only the vectors $M_H^{C\alpha}$ and $M_H^{C\beta}$ that are responsible for the ^{13}C satellites but also the macroscopic magnetization M_H of the ^{12}CHCl$_3$ molecules that make up 98.9 % of the sample. At the instant g' this vector lies along the $(-z)$ direction, since it has been turned into this orientation by the ("isotope-selective") 180° pulse in the ^{1}H channel. The fact that the magnitude of M_H is about 200 times greater than those of $M_H^{C\alpha}$ and $M_H^{C\beta}$ has an important bearing on the experiment. According to the energy level scheme (Fig. 8-20 I) $M_H^{C\alpha}$ is the longer of the two ^{1}H vectors and $M_H^{C\beta}$ is the shorter. The two vectors are parallel and both lie along the $(+z)$ direction, which means that before the application of the $90°_{x'}$ signal generating pulse in the ^{1}H channel they are antiparallel to M_H (Fig. 8-21 I). This $90°_{x'}$ pulse rotates them into the $(+y')$ direction, whereas M_H is rotated into the $(-y')$ direction, then the FID is recorded. Fourier transformation then yields the strong main signal (MS) with negative amplitude and the two ^{13}C satellites with positive amplitudes (Fig. 8-21 I). The great disadvantage of this experiment is that we are only interested in the two weak ^{13}C satellites, which differ only in their intensities, and not in the strong main signal. Nevertheless, by appropriately switching the phases of the pulses as described below, one can suppress the main signal, leaving only the satellites.

In this *phase cycle* one arranges that the first of the two 90° pulses in the ^{13}C channel acts alternately along the $(+x')$ direction and the $(-x')$ direction in successive half-cycles; in other words, the phase of this pulse is rotated by 180° from each half-cycle to the next. The overall pulse sequence is then as follows, where the phase information given in parentheses relates to the second half on the phase cycle:

^{1}H: $\qquad\qquad 180^{\circ}_{x'} - \tau - 90^{\circ}_{x'} - \text{FID}_{+(-)}$

^{13}C: $90^{\circ}_{x'(-x')} - \tau - 180^{\circ}_{x'} - \tau - 90^{\circ}_{y'}$

The first half-cycle with a $90^{\circ}_{x'}$ pulse (denoted by I in Figures 8-20 and 8-21) produces the changes in level populations described in detail above, giving ^{13}C satellites with different intensities. In the second half-cycle (II) with a $90^{\circ}_{-x'}$ pulse, $M_C^{H\beta}$ is in the $(+z)$ direction and $M_C^{H\alpha}$ in the $(-z)$ direction at the instant g. Consequently the populations N_1 and N_2 are inverted relative to the equilibrium state, whereas N_3 and N_4 are unchanged. As a result of this, $M_H^{C\beta}$ is now the longer of the two M_H vectors (see the energy level scheme, Fig. 8-20 II).

If now we take the two ^{1}H NMR spectra sketched in Figure 8-21 for the half-cycles I and II and subtract the second from the first, the main signal MS is eliminated, whereas the pairs of unequal satellites are replaced by two equally intense signals of opposite phases (with amplitudes $+2$ and -2). In

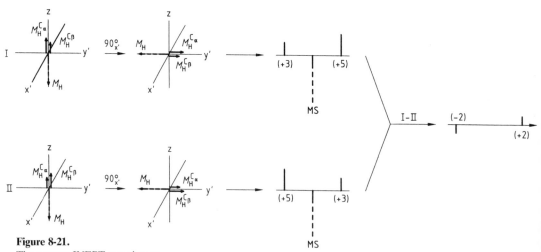

Figure 8-21.
The reverse INEPT experiment.
The state of the M_H vectors at the instants g' (immediately before the $90^{\circ}_{x'}$ signal generating pulse in the ^{1}H channel), and h (immediately after this pulse), together with the resulting ^{1}H NMR spectra for the sequences I and II of Figure 8-20. Example: ordinary chloroform consisting of 98.9 % ^{12}CHCl$_3$ and 1.1 % ^{13}CHCl$_3$. In the difference spectrum (I-II) the main signal (MS) from the ^{12}CHCl$_3$ molecules is eliminated, leaving the two ^{13}C satellites from the ^{13}CHCl$_3$ molecules, with opposite phases.

practice the differences between the successive FIDs of many such phase cycles ($FID_{+(-)}$ in above pulse sequence) are accumulated in the computer and the Fourier transformation is carried out only on completion.

As already explained in Section 8.5.2, the delay time $\tau = [4J(C,H)]^{-1}$ is strictly correct only for CH groups (AX spin systems), not for CH_2 and CH_3 groups. However, as a good compromise a value $\tau = [6J(C,H)]^{-1}$ is generally used in practice.

As in the normal INEPT experiment, one can obtain spectra with the satellite signals *in* phase instead of antiphase by slightly extending the pulse sequence. This is of interest when one wishes to eliminate the C,H coupling during the acquisition stage by broad-band decoupling in the ^{13}C channel. The pulse sequence for this is as follows:

^1H: $\qquad\qquad 180^\circ_{x'} - \tau - 90^\circ_{y'} - \varDelta - 180^\circ_{x'} - \varDelta - FID_{+(-)}$
^{13}C: $90^\circ_{x'(-x')} - \tau - 180^\circ_{x'} - \tau - 90^\circ_{y'} - \varDelta - 180^\circ_{x'} - \varDelta - BB$ decoupling

As compromise values one chooses $\tau = [6J(C,H)]^{-1}$ and $\varDelta = [4J(C,H)]^{-1}$. However, owing to the broad frequency range of the ^{13}C resonances, broad-band decoupling in the ^{13}C channel is not a straightforward matter. The same applies to experiments involving the $^{15}N/^1H$ system.

The example of the reverse INEPT experiment described here shows one way of suppressing an unwanted ^1H main signal, using a phase cycle with appropriate switching. However, there are a number of other ways of achieving the same result, and the principles of three of these are briefly outlined here.

- In one method the ^1H resonances are saturated before the pulse sequence is applied to the spin system (*presaturation*). This causes the main signal to disappear, whereas the ^{13}C satellites develop normally according to the pulse sequence. This method has the advantage that the ^{13}C signals are enhanced by the nuclear Overhauser effect (NOE).
- A second method is to use relaxation processes, by applying a preliminary pulse train known as the BIRD sequence [9] before the main pulse sequence of the experiment:

$90^\circ_{x'}(^1H) - 1/2J(C,H) - 180^\circ_{x'} (^1H,^{13}C) - 1/2J(C,H) - 90^\circ_{-x'}(^1H)$

This has the effect of rotating M_H into the $(-z)$ direction, whereas the vectors $M_H^{C\alpha}$ and $M_H^{C\beta}$ are both in the $(+z)$ direction. The main experiment is started after a delay time chosen such that the relaxing M_H vector passes through the zero point at that instant.

- Pulsed field gradients are now used in many types of two-dimensional experiments based on inverse detection. This enables one to choose the desired coherence pathway, so that the unwanted main peaks are suppressed and time-consuming phase cycles are avoided.

Finally it should be mentioned that a reverse INEPT experiment on the $^1H/^{13}C$ system yields an increase in intensity by a factor of about 8 compared with direct detection of the ^{13}C signal, thus shortening the required accumulation time by a factor of 64 for the same spectrum quality (S:N). On the other hand, compared with the normal $^{13}C/^1H$ INEPT experiment the gain in sensitivity is only a factor of 2. Nevertheless, this still corresponds to a time shortening by a factor of 4. In the case of the $^{15}N/^1H$ system the advantage in signal intensity and time saving is even greater than for $^{13}C/^1H$.

8.6 The DEPT Experiment [10, 11]

In interpreting ^{13}C NMR spectra it is very useful to know which signals belong to quaternary, CH, CH_2 and CH_3 carbon nuclei. In many cases this information can be obtained from a *J*-modulated spin-echo (APT) or refocussed INEPT experiment, but all these techniques have their weaknesses. For example, in the *J*-modulated spin-echo method one cannot normally distinguish between the signals of quaternary and CH_2 carbon nuclei, or between those of CH and CH_3 carbon nuclei. In the refocussed INEPT technique without BB decoupling the analysis of the spectra is often difficult in cases where signals are superimposed, whereas in the refocussed INEPT experiment with BB decoupling one cannot distinguish between the signals of CH and CH_3 groups.

Difficulties such as these do not arise with the DEPT technique (*distortionless enhancement by polarization transfer*). For this reason it is now one of the most important techniques available to the NMR spectroscopist.

Figure 8-22 A–D shows the same region ($\delta = 10$ to 110), recorded under four different sets of conditions, of the ^{13}C NMR spectrum of the methylketoside of *N*-acetyl-D-neuraminic acid methyl ester (**1**).

The ^{13}C spectrum with ^1H BB decoupling is shown at A; spectra B to D were recorded using the DEPT pulse sequence described below. It can be seen that these latter three are subspectra of A, which add together to give spectrum A, with the exception of one peak. Thus, sub-spectrum B contains only the signals of CH groups, sub-spectrum C only those of CH_2 groups, and D only those of CH_3 groups. With this technique the signals missing from the sub-spectra are those of quaternary carbon nuclei; in our example the missing signal is that of C-2 at $\delta = 100.32$.

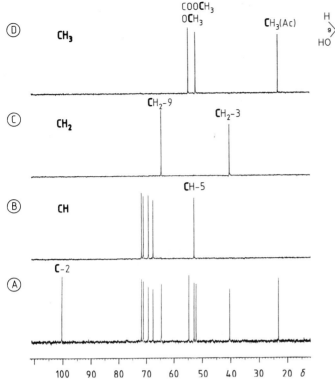

The DEPT experiment; for pulse sequence see text.
A: 100.6 MHz ^{13}C NMR spectrum of the neuraminic acid derivative **1** with ^{1}H BB decoupling (δ = 10 to 110 region only).
B: CH sub-spectrum: DEPT(90).
C: CH$_2$ sub-spectrum: DEPT(45) − DEPT (135).
D: CH$_3$ sub-spectrum: DEPT(45) + DEPT(135) −0.707 DEPT(90).
(*Experimental conditions:* 167 mg of the compound in 2.3 ml D$_2$O; 10 mm sample tube; 16 K data points; 32 FIDs for $\Theta_1 = 45°$ and $\Theta_3 = 135°$, 64 FIDs for $\Theta_2 = 90°$; $\tau = 3.57$ ms; total time approx. 12 min.)

The pulse sequence for the DEPT experiment is as follows:

^{1}H channel:
$$90^{°}_{x'} - \tau - 180^{°}_{x'} - \tau - \Theta_{y'} - \tau - \text{BB decoupling}$$
^{13}C channel:
$$90^{°}_{x'} - \tau - 180^{°} - \tau - \text{FID } (t_2)$$

The first part of this, in the ^{1}H channel, is identical to that for the INEPT experiment, except that the delay times τ for a two-spin AX system have the value $[2J(\text{C,H})]^{-1}$. The pulse $\Theta_{y'}$ with the variable pulse angle Θ is new, however. In the DEPT experiment Θ is chosen to be one of the angles $\Theta_1 = 45°$, $\Theta_2 = 90°$, or $\Theta_3 = 135°$, and three separate experiments are carried out with these values. During data acquisition the BB decoupler is switched on. In this case vector diagrams can only be used to show the effects of the pulse sequence on the spin system for some of the steps. In particular, such diagrams are no longer adequate for explaining the effect of the $\Theta_{y'}$ pulse. For this reason no diagrammatic explanation will be attempted here.

The manner whereby the three DEPT experiments are used to give the sub-spectra of Figure 8-22 is made clear from the curves in Figure 8-23, which show how the intensities of the

219

CH, CH$_2$ and CH$_3$ signals depend on the angle Θ. These curves were calculated from the following equations [11]:

CH: $\quad I = [\gamma(^1\text{H})/\gamma(^{13}\text{C})] \sin\Theta$

CH$_2$: $\quad I = [\gamma(^1\text{H})/\gamma(^{13}\text{C})] \sin2\Theta$ $\qquad\qquad$ (8-6)

CH$_3$: $\quad I = [3\gamma(^1\text{H})/4\gamma(^{13}\text{C})] (\sin\Theta + \sin3\Theta)$

It can be seen that the CH$_2$ and CH$_3$ curves pass through zero at the angle $\Theta_2 = 90°$, whereas the CH curve has a maximum at this point. Therefore the experiment with the value $\Theta_2 = 90°$ directly gives the *CH sub-spectrum:*
DEPT (90) (Fig. 8-22 B).

The *CH$_2$ sub-spectrum* (Fig. 8-22 C) is obtained by subtracting the spectrum recorded at the value $\Theta_3 = 135°$ from that recorded at $\Theta_1 = 45°$:
DEPT (45) – DEPT (135).

Finally the *CH$_3$ sub-spectrum* (Fig. 8-22 D) is obtained by combining the results from all three experiments as follows:
DEPT (45) + DEPT (135) -0.707 DEPT (90).

To obtain comparable absolute intensities the accumulation time in the experiment at $\Theta_2 = 90°$ must be twice that used for the other two.

The signals of the quaternary carbon nuclei can be identified by comparing the DEPT spectra with the BB-decoupled spectrum (Fig. 8-22 A). In this example, as Figure 8-22 shows only the spectral region from $\delta = 10$ to 110, the signals of two of the quaternary carbons are missing, namely C-1 at $\delta = 171.50$ and the acetamido carbon at $\delta = 175.93$ (cf. Fig. 8-12).

To carry out the experiment it is, in principle, necessary to know the value(s) of $^1J(\text{C,H})$, so as to determine the delay time τ for the pulse sequence. In practice this information is not usually available. However, it has been found from a large number of experiments that the results are not greatly influenced by the exact value of τ, and the DEPT pulse sequence can therefore be used even for systems containing widely differing coupling constants. For example, to obtain the spectra of **1** which are reproduced here, a value $\tau = 3.57$ ms was used. This corresponds to a coupling constant $^1J(\text{C,H}) = 140$ Hz, which is a typical value for carbon nuclei with sp^3 hybridization (Section 3.3.1).

Furthermore, it can be deduced from Figure 8-23 that for most routine practical applications it is sufficient to carry out only two experiments, at $\Theta_2 = 90°$ and $\Theta_3 = 135°$. As already explained, the DEPT(90) spectrum (recorded at $\Theta_2 = 90°$) contains all the signals of CH groups, while the DEPT(135) spectrum gives negative signals for carbon nuclei in CH$_2$ groups and positive signals for those in CH and CH$_3$ groups. Since the CH group signals have already been assigned from the DEPT(90) spectrum, which is invariably recorded in practice (Fig. 8-22 B), the remaining positive signals can be assigned to CH$_3$ groups.

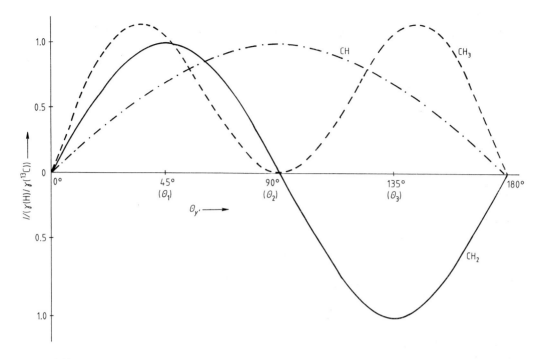

Figure 8-23.
DEPT experiment. Curves calculated from Equation (8-6) for the intensities of CH, CH₂ and CH₃ signals as functions of the pulse angle $\Theta_{y'}$; CH $\cdot-\cdot-\cdot-$, CH₂: ———, CH₃: ------.

Figure 8-24 shows the DEPT(135) spectrum (i.e. that recorded at $\Theta_3 = 135°$) for compound **1**. The five signals of the CH groups are marked by arrows. The information deduced from the DEPT experiment is summarized in Table 8-2.

For this compound we now know the chemical shifts of three CH₃ signals, two CH₂ signals, five CH signals, and lastly those of the three quaternary carbon nuclei. Using the rules given in Section 6.3 we can assign some of these signals without diffculty; the results of this are shown in Figure 8-22 and in the last column of Table 8-2. Several more of the signals have been assigned on the basis of experience and comparison with related compounds; however, we will not follow this through, as the next chapter will describe experiments which enable one to complete the assignments without recourse to these other sources of information.

The DEPT experiment can, of course, also be used to study other spin systems, e. g. those with coupled ¹H and ¹⁵N nuclei.

221

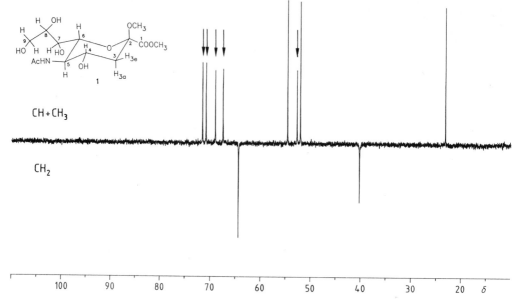

Figure 8-24.
DEPT(135) spectrum of the neuraminic acid derivative **1**, recorded using the pulse sequence given in the text, with $\Theta_{y'} = 135°$. The signals of the five CH groups, identified with the help of the DEPT(90) spectrum ($\Theta_{y'} = 90°$), are marked by arrows. The other three positive signals arise from CH_3 groups, and the two negative signals from CH_2 groups.
(Experimental conditions:
20 mg of the compound in 0.5 ml D_2O; 5 mm sample tube; 32 K data points; 300 FIDs; $\tau = 3.57$ ms; total time approx. 20 min.)

Table 8-2.
Partial assignment of the ^{13}C NMR signals of **1** from the results of the DEPT experiment.

δ [ppm]	CH_3	CH_2	CH	C	Assignment
23.2	×				CH_3 (Ac)
40.31		×			C-3
52.12	×				
52.83			×		C-5
54.65	×				
64.50		×			C-9
67.51			×		
69.18			×		
70.98			×		
71.67			×		
100.32				×	C-2
171.50[a]				×	
175.93[a]				×	

[a] Values from the complete spectrum (Fig. 8-12 A).

8.7 The Selective TOCSY Experiment [12–14]

The selective TOCSY (**TO**tal **C**orrelation **S**pectroscop**Y**) experiment, alternatively known as the HOHAHA (**HO**monuclear **HA**rtmann–**HA**hn) experiment, can best be described by considering an example. Figure 8-25 A shows the 500 MHz ^{1}H NMR spectrum of a dimethylated cellobiose (**3**). The assignments indicated here were determined from the results of an H,H-COSY measurement, an experiment that we will learn about in Chapter 9. A first-order analysis of the multiplets indicates that H-1 (the resonance of which is hidden under the HDO signal of the D$_2$O solvent) is coupled to H-2, while H-2 is coupled to both H-1 and H-3, H-3 is coupled to both H-2 and H-4, and so on. The couplings through four bonds are

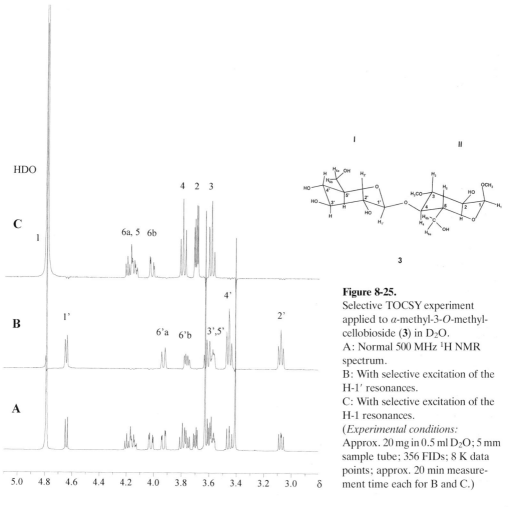

3

Figure 8-25.
Selective TOCSY experiment applied to α-methyl-3-*O*-methyl-cellobioside (**3**) in D$_2$O.
A: Normal 500 MHz ^{1}H NMR spectrum.
B: With selective excitation of the H-1′ resonances.
C: With selective excitation of the H-1 resonances.
(*Experimental conditions:*
Approx. 20 mg in 0.5 ml D$_2$O; 5 mm sample tube; 356 FIDs; 8 K data points; approx. 20 min measurement time each for B and C.)

223

in general too small to be resolved. Figures 8-25 B and C show spectra recorded by the method of the selective TOCSY experiment. By comparing spectra B and C with spectrum A we can see that B is the spectrum of glucose unit I (protons H-1' to H-6'), while C is that of glucose unit II (protons H-1 to H-6). Thus, the method selects from the total spectrum the resonances of a group of protons that belong to a common coupled system (see Section 4.2).

What is the principle of the selective TOCSY experiment, and what is its range of uses? Figure 8-26 shows the pulse sequence. It begins with a "soft pulse", i.e. one of low power (weak r.f. field B_1) and long duration (τ_p), which is adjusted to give a pulse angle θ [see Eq. (1-14)] of exactly 90°. The pulse is tuned to the resonance frequency of a particular proton. In the above example it was tuned first to the H-1' resonance ($\delta = 4.64$, Figure 8-25 B), then to the H-1 resonance ($\delta = 4.79$, Figure 8-25 C). Thus the pulse generates a transverse magnetization in the x', y' plane due solely to the selected protons. During the mixing phase following the pulse a "spin-lock" condition is applied, and polarization is transferred from the selected proton (here H-1' or H-1) to the rest of the protons in the coupled system, so that in the acquisition phase (the recording of the FID) we obtain signals from all the coupled protons. The nature and mechanism of the spin-lock is rather complicated, and to give a full explanation would go beyond the level of treatment in this book. A much simplified description is that during the application of the spin-lock the nuclei effectively experience only the weak radiofrequency field B_1, so that the chemical shift differences become extremely small and the scalar couplings are dominant. This results in mixing of the spin states, allowing a transfer of polarization. Thus, when a particular proton in the spin system (H-1 or H-1' in our example) has been excited selectively, the polarization becomes transferred to all the other protons in the system, not merely those with a direct scalar coupling to the excited proton.

The selective TOCSY procedure is widely used for analyzing the spectra of oligosaccharides and peptides, compounds whose proton spectra typically contain many overlapping multiplets. However, it must be emphasized that the method does not directly yield assignments; it only identifies those multiplets that belong to a common coupled spin system. The "range" of the experiment, i.e. the extent of the spin system that it can reveal, may be controlled to some degree by altering the spin-lock period. Thus, in the above example, if the spin-lock period is halved from 60 ms to 30 ms, the protons H-6 and H-6' are no longer detected.

The two-dimensional version of the TOCSY experiment will be described in Section 9.4.5.

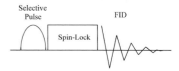

Figure 8-26.
Pulse sequence for the selective TOCSY experiment.

8.8 The One-Dimensional INADEQUATE Experiment [15]

Experiments which provide evidence for and measurements of C,C couplings are interesting in two ways; they give information both on:

- the structure of the carbon skeleton of the molecule, and
- which of the carbon nuclei are magnetically coupled to each other, and thus on connectivity relationships.

For structure determination exact J(C,C) values are needed, whereas for obtaining information on connectivities it is only necessary to have evidence of a coupling. Unfortunately C,C couplings can be measured only with difficulty. The reason for this is the low natural abundance of the ^{13}C isotope, which is only 1.1 %. This problem can be overcome by synthesizing ^{13}C-enriched compounds, but the newer spectroscopic techniques are aimed at achieving the same results without using expensive and time-consuming chemical methods.

Let us again remind ourselves what we normally observe in a ^{13}C NMR spectrum with ^{1}H BB decoupling: just single peaks! These arise from the 1.1 % of carbon atoms in the molecule which contain a ^{13}C nucleus. The peaks are singlets because there is a 98.9 % probability of the adjacent carbon nuclei being the ^{12}C isotope. We only observe a C,C coupling when two ^{13}C nuclei are connected through one, two or three bonds; the corresponding coupling constants are ^1J for $^{13}C - {}^{13}C$, ^2J for $^{13}C - C - {}^{13}C$ and ^3J for $^{13}C - C - C - {}^{13}C$. Whereas the probability of finding a ^{13}C nucleus at a given position in the molecule is already low, being only 1:100, that of finding two adjacent ^{13}C nuclei in the molecule is only 1:10 000! Owing to this low probability the signals of coupled carbon nuclei appear only as satellites of the main signals (Sections 1.6.2.9 and 3.7). In the simplest case of a molecule with two carbon atoms these satellites constitute a doublet whose intensity is 1.1 % of that of the main, singlet, peak.

We do not expect to find higher-order spectra, since the probability of finding three coupled ^{13}C nuclei as immediate neighbors is smaller by a further factor of 100.

The ^{13}C NMR spectrum of a carbon nucleus which is bonded to two other carbons, e. g. that of C-2 in the fragment $C^1 - C^2 - C^3$, consists of a singlet for $^{12}C^1 - {}^{13}C^2 - {}^{12}C^3$ with a relative intensity of about 98 %, together with two doublets for $^{13}C^1 - {}^{13}C^2 - {}^{12}C^3$ and $^{12}C^1 - {}^{13}C^2 - {}^{13}C^3$, each with an intensity of 1.1 % of the total intensity. Thus, for a quaternary carbon nucleus bonded to four other carbons, one should be able to find up to four superimposed doublets as ^{13}C satellites;

however, these will only be separate if the coupling constants $J(C,C)$ are all different in magnitude.

$^1J(C,C)$ values vary from 30 to 70 Hz, which means that the ^{13}C satellites due to couplings through one bond are well separated from the main signal. However, couplings through two and three bonds are an order of magnitude smaller, with the result that the satellites merge with the wings of the main peak and are therefore not detected. Additional problems can arise from side-bands caused by the rotation of the sample tube in the magnetic field. For these reasons it is in most cases impossible to analyze the ^{13}C satellites in normal ^{13}C NMR spectra.

The situation would be quite different if one could suppress the main signal – this is exactly what one achieves in the INADEQUATE experiment (*incredible natural abundance double quantum transfer*). The pulse sequence used in this technique is as follows:

$$90°_{x'} - \tau - 180°_{y'} - \tau - 90°_{x'} - \varDelta - 90°_{\varphi'} - FID(t_2)$$

As is already apparent from the name of the experiment, it involves a double quantum transfer. As the physical processes on which the method is based cannot be clearly described by a graphical presentation, we will not attempt to give vector diagrams for this pulse sequence, since the crucial steps would be missing. Let us remind ourselves of the selection rule stated in Section 1.4.1: according to this, transitions between two energy levels are forbidden if $\Delta m \neq 1$; this means that zero quantum, double quantum and multiple quantum transitions are forbidden and cannot be observed. However, it has been shown both theoretically and experimentally that although multiple quantum transitions cannot be observed, a train of selective pulses or, for example, the pulse sequence $90° - \tau - 90°$, can induce a *double quantum coherence* in a two-spin system. This coherence cannot be clearly visualized in any pictorial representation; it can only be described mathematically in terms of the off-diagonal elements of the density matrix. (In contrast, *single quantum coherence* is simply the bunching together of precessing individual spins which have the same phase (Fig. 1-12).)

In the one-dimensional INADEQUATE experiment the double quatum coherence is induced by the pulse sequence $90°_{x'} - 2\tau - 90°_{x'}$. The intermediate $180°_{y'}$ pulse which is also included serves only to refocus spins which have fanned out during the time τ as a result of chemical shift differences and field inhomogeneities. The $90°_{\varphi'}$ pulse which follows the above sequence after a short switching time of a few microseconds converts the magnetization vectors of the individual ^{13}C nuclei and of the AX spin systems into observable signals, i. e. into an FID. Fourier transformation gives the frequency spectrum, consisting of main and satellite signals. As a consequence of the double quantum coherence the main signals and the satellite

226

signals have different phases. By appropriately switching the phases of the pulses and of the receiver system – an operation whose details will not be described here – the main signal can be completely suppressed, leaving only the AX or AB satellite spectra. Of the two signals in the A and X parts of the spectrum, one has a positive and the other a negative amplitude.

Figure 8-27 B shows the one-dimensional INADEQUATE spectrum of the neuraminic acid derivative **1** which is already familiar from the DEPT experiment, together with the ^{1}H BB decoupled ^{13}C NMR spectrum (Fig. 8-27 A). The range covered by these partial spectra is the same as in Figure 8-22. The details only become apparent when the satellite signals are plotted on an expanded scale (Fig. 8-27 C). In this small spectral region ($\delta = 64$ to 72) we can recognize the two satellites for each signal, but the main signals are completely suppressed.

We know from the DEPT experiment that the signal at $\delta = 64.5$ arises from a ^{13}C nucleus in a CH$_2$ group (C-9), and the other four from ^{13}C nuclei in CH groups (C-4, C-6, C-7

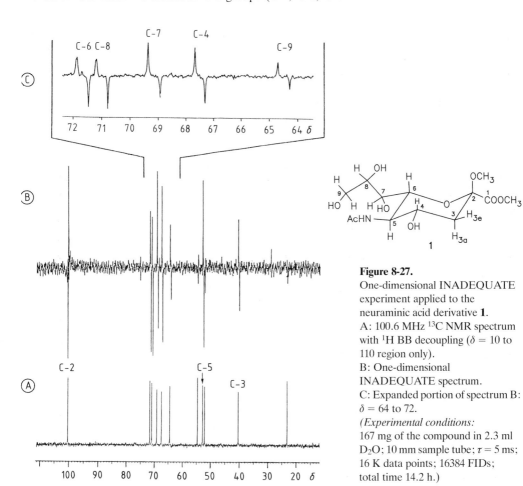

Figure 8-27.
One-dimensional INADEQUATE experiment applied to the neuraminic acid derivative **1**.
A: 100.6 MHz ^{13}C NMR spectrum with ^{1}H BB decoupling ($\delta = 10$ to 110 region only).
B: One-dimensional INADEQUATE spectrum.
C: Expanded portion of spectrum B: $\delta = 64$ to 72.
(*Experimental conditions:*
167 mg of the compound in 2.3 ml D$_2$O; 10 mm sample tube; $\tau = 5$ ms; 16 K data points; 16384 FIDs; total time 14.2 h.)

227

and C-8). Since each of the four CH carbons has two other carbon atoms as neighbors, we expect to find two doublets as satellites in each case. Evidently the C,C coupling constants to the two neighbors are equal or nearly so, with the result that the two doublets are superimposed. The same is also true for the other doublets in Figure 8-27 B, which have not been shown on an expanded scale. The measured coupling constants are listed in Table 8-3. Here the assignments of the signals have been assumed, even though we do not know all of these from the experiments described up to now. It can be seen that all the coupling constants fall within a comparatively narrow range (40.1 ± 3.5 Hz).

In order for the experiment to be successful, correct adjustment of the delay time τ is essential. According to theory the inducement of the double quantum coherence occurs with maximum efficiency for

$$\tau = \frac{2n + 1}{4J(C,C)} \qquad n = 0, 1, 2, 3 \qquad (8\text{-}7)$$

For measuring widely differing C,C coupling constants, e. g. where ^1J, ^2J and ^3J values are to be determined, it is best to observe the satellites by carrying out several experiments with different τ-values, unless it is possible for Equation (8-7) to be satisfied for several C,C coupling constants by inserting different values of n.

In addition to its use for determining coupling constants, the INADEQUATE technique can also help in assigning the signals of nuclei that are coupled to each other. However, this goal, which is of a more qualitative nature, can be achieved in a more elegant way by using the two-dimensional version of the INADEQUATE experiment (Section 9.5). In our example assignments are possible in only a few cases, as the coupling constants are not sufficiently different (Table 8-3).

Unfortunately the INADEQUATE experiment does not involve a transfer of polarization as do the methods described in Sections 8.5 and 8.6. Consequently, to record such a spectrum requires at least an overnight accumulation, even when using a high-field spectrometer and a highly concentrated sample (100 to 200 mg of sample in 2 ml of solvent). The accumulation time in our example was about 14 h. Because of the long accumulation times needed, such measurements are unlikely to become a routine procedure.

Table 8-3.
C,C coupling constants for **1** determined from the one-dimensional INADEQUATE spectrum.

J (C,C)	J [Hz]
J (2,3)	36.9 ± 0.2
J (3,4)	37.1 ± 0.2
J (4,5)	38.1 ± 0.8
J (5,6)	39.7 ± 0.7
J (6,7)	43.7 ± 0.7
J (7,8)	44.0 ± 0.5
J (8,9)	41.1 ± 0.2

8.9 Bibliography for Chapter 8

[1] R. Benn and H. Günther, *Angew. Chem. Int. Ed. Engl. 22* (1983) 350.

[2] S. L. Patt and J. N. Shoolery, *J. Magn. Reson. 46* (1982) 535.

[3] K. G. R. Pachler and P. L. Wessels, *J. Magn. Reson. 12* (1973) 337.

[4] G. A. Morris and R. Freeman, *J. Amer. Chem. Soc. 101* (1979) 760.

[5] R. K. Harris: *Nuclear Magnetic Resonance Spectroscopy. A Physicochemical View.* London: Pitman, 1983, p. 175.

[6] M. R. Bendall, D. T. Pegg and D. M. Doddrell, *J. Magn. Reson. 45* (1981) 8.

[7] R. Freeman, T. H. Mareci and G. A. Morris, *J. Magn. Reson. 42* (1981) 341.

[8] M. R. Bendall, D. T. Pegg, D. M. Doddrell and J. Field, *J. Magn. Reson. 51* (1983) 520.

[9] A. Bax and S. Subramanian, *J. Magn. Reson. 67* (1986) 565.

[10] D. M. Doddrell, D. T. Pegg and M. R. Bendall, *J. Magn. Reson. 48* (1982) 323.

[11] M. R. Bendall, D. M. Doddrell, D. T. Pegg and W. E. Hull: *DEPT booklet.* Bruker Analytische Meßtechnik, Karlsruhe-Rheinstetten, 1982.

[12] D. G. Davis and A. Bax, *J. Amer. Chem. Soc. 107* (1985) 7197.

[13] A. Bax and D. G. Davis, *J. Magn. Reson. 65* (1985) 355.

[14] H. Kessler, H. Oschkinat, C. Griesinger and W. Bermel, *J. Magn. Reson. 70* (1986) 106.

[15] A. Bax, R. Freeman and S. P. Kempsell, *J. Amer. Chem. Soc. 102* (1980) 4849.

Additional and More Advanced Reading

S. Berger and S. Braun: *200 and More NMR Experiments. A Practical Course.* Weinheim: Wiley-VCH, 2004.

A. E. Derome: *Modern NMR Techniques for Chemistry Research.* Oxford: Pergamon Press, 1987.

J. K. M. Sanders and B. K. Hunter: *Modern NMR-Spectroscopy. A Guide for Chemists.* Oxford: Oxford University Press, 1987.

F. J. M. van de Ven: *Multidimensional NMR in Liquids. Basic Principles and Experimental Methods.* New York: VCH Publishers, 1995.

9 Two-Dimensional NMR Spectroscopy

9.1 Introduction

All the NMR spectra described up to now have two dimensions: the abscissa which corresponds to the frequency axis, from which one reads off the chemical shifts, and the ordinate which gives the intensities of the signals. However, when we speak of a two-dimensional (2D) NMR spectrum we understand this to mean a spectrum in which both the abscissa *and* the ordinate are frequency axes, with the intensities constituting a third dimension.

If chemical shifts are plotted along one of the frequency axes and coupling constants along the other, we call this a *two-dimensional J-resolved spectrum*. On the other hand, if both axes give chemical shifts we have a *two-dimensional (shift)-correlated NMR spectrum*. From the practical NMR spectroscopist's point of view the most important 2D spectra of the latter type are those which show either ^{1}H vs. ^{1}H or ^{1}H vs. ^{13}C chemical shift correlations.

But developments did not stop at two dimensions! In some laboratories engaged in specialized NMR work, three- and multi-dimensional experiments are now carried out almost on a routine basis. The 3D spectra that they produce have three frequency axes, each of which may represent the resonances of a different nuclide; for example they could be ^{1}H and ^{13}C with either ^{15}N or ^{31}P. Alternatively one might have ^{1}H resonances plotted on two axes and those of a heteronuclide on the third. For sensitivity reasons the detection channel is always that of ^{1}H. In this chapter, however, we will confine our attention to two-dimensional experiments.

The two-dimensional methods are based on the couplings between nuclear dipoles. These need not necessarily be scalar couplings such as were treated in Chapter 3. As in the Overhauser effect, the interactions may also be of a dipolar type, i.e. through space. This opens up the possibility of investigating the geometrical structure of molecules. Another kind of two-dimensional technique is that based on the transfer of magnetization by chemical exchange; this can be used to study dynamic processes.

As was mentioned in the introduction to Chapter 8, these techniques represent a new generation of NMR experiments. So that their advantages and their great practical significance can be appreciated, the following discussion will deal with the basic principles of the methods that are currently the most important.

As before, we will confine our attention to ^{1}H and ^{13}C applications, but the techniques are in essence also applicable to other nuclides and nuclide combinations.

First we must familiarize ourselves with the basic idea underlying two-dimensional NMR spectroscopy. The possibility was first proposed in 1971 by J. Jeener. However, two-dimensional NMR spectroscopy only became a practical reality when R. R. Ernst et al. and R. Freeman et al. carried out their pioneering experiments. These two groups developed many different variants of 2D experiments, including not only those of an advanced and complex nature, but also others of simplified types.

Here we shall not go into the exact mathematical details of the phenomena, as these can be found in the literature [1].

9.2 The Two-Dimensional NMR Experiment

9.2.1 Preparation, Evolution and Mixing, Data Acquisition

In a normal pulsed NMR experiment (Section 1.5) the excitation pulse is followed immediately by the *data acquisition* (detection) *phase*, in which the interferogram or FID is recorded and the data are stored. In the experiments using complex pulse sequences which were described in Chapter 8 (INEPT, DEPT, INADEQUATE), the spin system undergoes a *preparation phase* before data acquisition. In two-dimensional NMR experiments these two phases are separated by an intermediate phase, that of *evolution and mixing*. We must first understand what effects this phase has on the spin system, and how it changes a one-dimensional NMR experiment into a two-dimensional.

The principle can be understood from a simple hypothetical experiment. For this purpose we will consider the ^{13}C NMR spectrum of a two-spin AX system with A = ^{1}H and X = ^{13}C, as in ^{13}CHCl$_3$. We will assume that, in contrast to the normal procedure, a variable delay time t_1 is inserted between the exci-

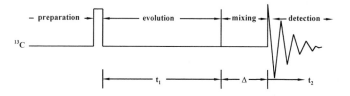

Figure 9-1.
The basic principle of a two-dimensional NMR experiment. The variable is the interval t_1, which can be of the order of milliseconds to seconds. During the period t_1 the system "evolves". Δ is the mixing time, a period during which the spin states are allowed to "mix", and is kept constant throughout the experiment. The choice of a value for Δ depends on the nature of the experiment.

tation pulse and the recording of the FID. Furthermore we assume that the ^{1}H BB decoupler is switched on only during the recording of the FID. The excitation pulse is assumed to be exactly adjusted to $90^{\circ}_{x'}$ (in contrast to the pulse angle of less than $30^{\circ}_{x'}$ which is typically used). The pulse sequence is shown in Figure 9-1.

Let us assume that we perform n experiments with different values of t_1, starting with $t_1 = 0$, then increasing the value by a constant amount (a few ms) from each experiment to the next. When the n interferograms undergo Fourier transformation with respect to t_2 we obtain n frequency domain spectra F_2; in our example of chloroform each of these is a single peak, as is familiar to us from the spectrum recorded in the normal way. If the decoupling were applied during the whole experiment, all n of the F_2 spectra would be identical, except for a decrease in intensity due to relaxation during the time t_1. However, as the BB decoupler is only switched on during data acquisition, the ^{13}C nuclei and the protons are coupled during the evolution phase. The vector diagrams (Fig. 9-2 A) show how this C,H coupling affects the single peak in the final spectrum.

After the $90^{\circ}_{x'}$ pulse ($t_1 = 0$, Fig. 9-2 a) the macroscopic magnetization M_C of the ^{13}C nuclei lies along the y'-direction. M_C can be resolved into two components $M_C^{H\alpha}$ and $M_C^{H\beta}$ corresponding to ^{13}C nuclei in chloroform molecules whose protons are in the α or β states (^{13}CH$_\alpha$Cl$_3$ and ^{13}CH$_\beta$Cl$_3$). This procedure was explained in detail in Section 8.3.

The two vectors $M_C^{H\alpha}$ and $M_C^{H\beta}$ rotate with different Larmor frequencies:

$$M_C^{H\alpha} \text{ with frequency } \nu_C - \frac{1}{2} J(C,H), \text{ and}$$
$$M_C^{H\beta} \text{ with frequency } \nu_C + \frac{1}{2} J(C,H). \tag{9-1}$$

Assuming that the coordinate system x', y' rotates with a frequency ν_C which is the average of the two Larmor frequencies, $M_C^{H\beta}$ rotates faster than the coordinate system by an amount $J(C,H)/2$ and $M_C^{H\alpha}$ slower than it by the same amount, since $^1J(C,H)$ is always positive. The small arrows in the figure indicate the directions of rotation of the vectors relative to the coor-

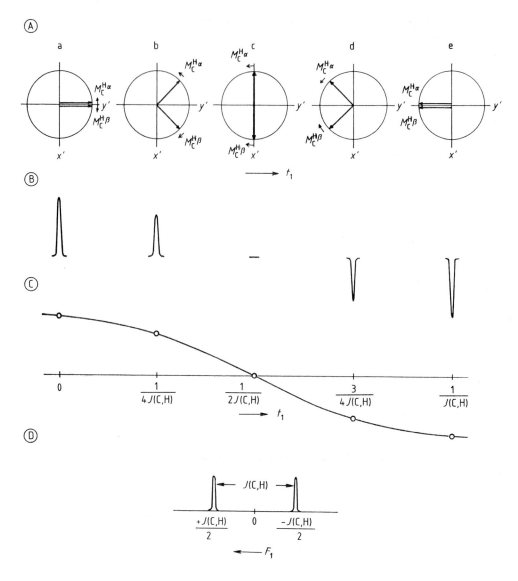

Figure 9-2.
A: Diagrams showing the evolution of the ^{13}C magnetization vectors for a two-spin AX system (with A = ^1H and X = ^{13}C) at different times t_1. The coordinate system x', y', z rotates with the frequency ν_C.
a: $t_1 = 0$;
b: $t_1 = [4J(C,H)]^{-1}$;
c: $t_1 = [2J(C,H)]^{-1}$;
d: $t_1 = 3[4J(C,H)]^{-1}$;
e: $t_1 = [J(C,H)]^{-1}$.
B: Fourier transformation with respect to t_2 (with ^1H BB decoupling) gives a ^{13}C NMR spectrum made up of single peaks whose amplitudes depend on t_1.
C: The curve shows that the amplitude is modulated at a frequency related to $J(C,H)$.
D: Fourier transformation with respect to t_1 yields two peaks with a separation $J(C,H)$.

234

dinate system. During the time t_1 the magnetization vectors $M_C^{H\alpha}$ and $M_C^{H\beta}$ rotate through the angles φ_α and φ_β as given by Equation (9-2).

$$\varphi_\alpha = 2\pi\left(v_C - \frac{1}{2}J(C,H)\right)t_1$$
$$\varphi_\beta = 2\pi\left(v_C + \frac{1}{2}J(C,H)\right)t_1$$

(9-2)

The phase difference Θ between the two is then:

$$\Theta = \varphi_\beta - \varphi_\alpha = 2\pi J(C,H)t_1 \qquad (9\text{-}3)$$

After a time $t_1 = [4J(C,H)]^{-1}$, which is of the order of milliseconds, the phase difference between the two vectors is exactly 90° (Fig. 9-2 b), after $t_1 = [2J(C,H)]^{-1}$ it is 180° (Fig. 9-2 c), and after $t_1 = [J(C,H)]^{-1}$ the two components are again in phase, but now their vectors lie along the $(-y')$ direction (Fig. 9-2 e).

How does this affect the appearance of the corresponding frequency spectra F_2? As the BB decoupler is switched on during the data acquisition, the C,H coupling is eliminated in all cases, and the F_2 spectra consist of singlets. However, this decoupling has no effect on the interaction between the A and X nuclei during the time t_1.

To understand this, let us consider the situation after a time $t_1 = [4J(C,H)]^{-1}$, when according to Equation (9-3) the phase difference is 90°. By switching on the BB decoupler we eliminate the C,H coupling, and as a result the cause ot the difference between the rotation frequencies of $M_C^{H\alpha}$ and $M_C^{H\beta}$ also disappears. Both vectors now rotate at equal rates with the average Larmor frequency v_C. However, although they are now rotating at the same rate, the 90° phase difference is retained. The amplitude of the signal induced in the receiver coil is proportional to the vector sum of $M_C^{H\alpha}$ and $M_C^{H\beta}$. At $t_1 = 0$ the two vectors are parallel, and the vector sum is therefore at its greatest, so that the signal is a maximum. At $t_1 = [2J(C,H)]^{-1}$ the component in the y'-direction is exactly zero, so no signal is observed. When $t_1 = [J(C,H)]^{-1}$ the signal is as large as at the beginning, except for the decay through relaxation, but its amplitude is now negative. In Figure 9-2 B the signals are shown schematically under their corresponding vector diagrams. We can see qualitatively that the F_2 spectra are amplitude-modulated by the coupling constant $J(C,H)$ (Fig. 9-2 C). A second Fourier transformation of the n F_2 spectra with respect to t_1 yields two frequencies, whose difference corresponds exactly to the coupling constant $J(C,H)$.

Thus the F_2 spectrum contains the chemical shifts δ (^{13}C), and the F_1 spectrum contains the coupling constants $J(C,H)$. In the example of chloroform every F_2 spectrum consists of a single peak, and the F_1 spectrum is a doublet whose splitting is equal to the coupling constant $J(C,H) = 209$ Hz, the frequencies of the two peaks being $+104.5$ and -104.5 Hz (Fig. 9-2 D).

$$S(t_1, t_2) \xrightarrow{\text{FT}} S(t_1, F_2) \xrightarrow{\text{FT}} S(F_1, F_2)$$

The question of how many experiments with different t_1-values are needed, which is of considerable practical importance, cannot be answered with complete generality. However, it should normally be possible to complete an experiment in one overnight run.

In principle we already have in Figure 9-2 an example of a simple two-dimensional J-resolved NMR experiment.

The mixing phase immediately follows the evolution phase, or can even interrupt it – often this occurs after a time $t_1/2$. It consists of additional pulses and a constant mixing time, although this mixing time can itself consist of several different defined time intervals. During the mixing phase the spin system can continue to evolve as in the evolution phase itself, but its contribution to the total evolution which occurs is constant for all the experiments carried out with different t_1-values.

9.2.2 Graphical Representation

We have seen that a double Fourier transformation gives a spectrum with two frequency axes, which we denote by the function $S(F_1, F_2)$. How can such a spectrum be displayed?

Two forms of graphical display are commonly used:
- the *stacked plot* and
- the *contour plot*.

In a *stacked plot* a series of F_2 spectra for different F_1-values are plotted one above another. To show the spectral features clearly, each trace is shifted over by a constant amount relative to the preceding one.

Figure 9-3 A shows the two-dimensional ^{13}C NMR spectrum of chloroform as a stacked plot. (The method of recording this spectrum is described in Section 9-3.) In this example, by projecting the signals onto the F_2-axis (abscissa) we obtain directly the δ-value of 77.7 for chloroform (^{13}CHCl$_3$). If instead we project the signals onto the F_1-axis, we obtain the corresponding multiplets, in this case simply a doublet with the splitting $J(C,H) = 209$ Hz, which is the coupling constant for chloroform. The middle trace in the diagram corresponds to the value $F_1 = 0$.

In the second mode of representation, the *contour plot*, the peaks in the stacked plot are viewed from above, then a section is taken at a certain height above the plane of the F_1- and F_2-axes, and the contours where this intersects the peaks are plotted. This has been carried out for the chloroform

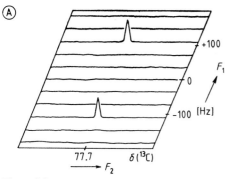

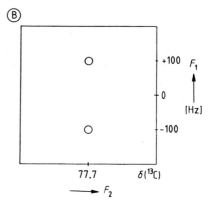

Figure 9-3.
Schematic representation of a two-dimensional NMR spectrum. A: *Stacked plot*; the F_2-axis corresponds to the usual frequency axis with the δ-scale of the one-dimensional NMR spectrum. The F_2 spectra for each of the different F_1-values are shown plotted one above another. For clarity the individual traces are shifted relative to each other by a constant increment.
B: *Contour plot*; the diagram shows a section in a plane at a fixed height through the peaks of the stacked plot. As the traces are not offset in this case, the F_1- and F_2-axes are at right angles.

spectrum in Figure 9-3 B. As we shall see, a contour plot is in many cases easier to interpret than a stacked plot.

It should be noted that the individual traces in Figure 9-3 A do not correspond to the different experiments. Rather must we regard the stacked plot (A) as a complete separate entity; it is in fact the result of a double Fourier transformation, as also is the contour plot (B).

9.3 Two-Dimensional *J*-Resolved NMR Spectroscopy

9.3.1 Heteronuclear Two-Dimensional *J*-Resolved NMR Spectroscopy [2]

In the experiment described in Section 9.2 a double Fourier transformation gives the chemical shifts δ along the F_2-axis and the C,H couplings along the F_1-axis. Since the coupling is between nuclei of different species we have here the simplest case, namely heteronuclear two-dimensional *J*-resolved NMR spectroscopy (also called heteronuclear J,δ-spectroscopy).

Of the many different variations of this experiment, we shall discuss here the gated decoupling method [3], which uses the pulse sequence shown in Figure 9-4 A. To understand how this pulse sequence affects a spin system, we will again take as our example the two-spin AX system of chloroform (^{13}CHCl$_3$).

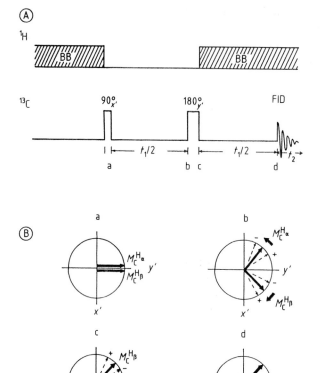

Figure 9-4.
Two-dimensional heteronuclear
J-resolved ^{13}C NMR spectroscopy.
A: pulse sequence.
B: Evolution of the transverse
^{13}C magnetization vectors M_C for a
two-spin AX system with A = ^{1}H
and X = ^{13}C in the rotating coordinate system (x', y'-plane shown).
$M_C^{H\alpha}$ and $M_C^{H\beta}$ are the ^{13}C magnetization vectors for molecules with
the proton in the α and β state
respectively. Diagrams a to d
correspond to the instants marked
in A.
The system "evolves" according to
the coupling constant J (C,H), with
the vectors simultaneously fanning
out due to field inhomogeneities.
The arrows outside the circle in b
indicate the direction of rotation of
each fan relative to the rotating
coordinate system; the '+' and '−'
signs indicate which spins in each
fan are rotating faster or slower
than the average. (In practice the
BB decoupler is switched on during
the *first* half of the evolution phase
(first $t_1/2$ delay period) and during
data acquisition, which has exactly
the same effect on the spectrum.
The version shown here makes the
explanation simpler.)

The experiment is derived from the spin-echo pulse sequence
($90^\circ_{x'} - \tau - 180^\circ_{x'} - \tau$ (echo); see Section 7.3.2). The sequence
differs from that in Figure 9-1 in having a $180^\circ_{y'}$ pulse at the
half-way point $t_1/2$ of the evolution period and ^{1}H BB decoupling during the second half of this period. The individual phases
can be understood by considering the vector diagrams shown in
Figure 9-4 B. Here we again use a coordinate system x', y', z
which rotates at the average Larmor frequency ν_C.

As shown in Figure 9-2, the $90^\circ_{x'}$ pulse in the ^{13}C channel turns
both magnetization vectors $M_C^{H\alpha}$ and $M_C^{H\beta}$ into the y'-direction
(Fig. 9-4 a). During the time $t_1/2$ the two vectors move apart,
$M_C^{H\beta}$ rotating faster than the coordinate system and $M_C^{H\alpha}$
slower. The arrows outside the circle indicate the directions of
relative motion of the vectors. At the same time the spins fan
out owing to the inevitable field inhomogeneities. (Here we
consider only the components of the individual spins in the x',
y'-plane.) Each of these groups of nuclear magnetic vectors contains some faster spins and some slower spins. The + and – signs
in Figure 9-4 b indicate which sides of the fans correspond to the
faster and which to the slower spins.

All the vectors are then reflected through the y'-axis by applying a $180_{y'}^{\circ}$ pulse. (This should formally reverse the directions of the rotation arrows and the plus and minus signs in the vector diagram, Fig. 9-4 c.) All the spins would then again come to a focus along the y'-direction after a further interval $t_1/2$ (giving a spin echo) if the ^{1}H BB decoupler were not switched on. However, the latter removes the C,H coupling which is the cause of the difference in frequencies, and the vectors that were previously $M_C^{H\alpha}$ and $M_C^{H\beta}$ now both rotate with the frequency ν_C, i.e. at the same rate as the coordinate system. However, their phase difference remains.

The frequency differences between the individual spins within the fans caused by field inhomogeneities are unaffected by the spin decoupling, and consequently during the second period $t_1/2$ the fans close up again (Fig. 9-4 d).

At the end of the total interval t_1 this process is complete, and data acquisition can begin. The signal induced in the receiver coil is proportional to the sum of the two vectors. The magnitude of this vector sum depends on the phase difference, which in turn depends on the magnitude of the C,H coupling constant $J(C,H)$. Thus the ^{13}C NMR signal is modulated by $J(C,H)$.

In the case of the chloroform molecule chosen as our example, Fourier transformation of the FID with respect to t_2 yields a singlet modulated by $J(C,H)$. A second Fourier transformation with respect to t_1 then gives a doublet along the direction of the F_1-axis, with a splitting $J(C,H)/2$; the splitting is only half the coupling constant because the system has only been allowed to evolve for half the time ($t_1/2$).

For molecules other than our example the F_2 spectrum contains as many singlets as there are non-equivalent carbon atoms in the molecule. All the signals $S\ (t_1, F_2)$ are modulated in the time dimension t_1 by the corresponding C,H coupling constants, and along the direction of the F_1-axis we therefore obtain the multiplets produced by the couplings; singlets for quaternary carbons, doublets for CH groups, triplets for CH_2 groups, and quartets for CH_3 groups. Each multiplet is centered on zero frequency.

Figure 9-5 shows, in the form of a stacked plot, a two-dimensional J-resolved 100.6 MHz ^{13}C NMR spectrum, recorded by this technique, of the compound already used as an example in Chapter 8, the methylketoside of N-acetyl-D-neuraminic acid methyl ester (1). The same spectrum is shown in Figure 9-6 as a contour plot.

The spectrum shown at the top edge of both these two-dimensional NMR spectra, in the direction parallel to the F_2-axis, is the projection of the multiplets, and corresponds to the one-dimensional ^{13}C NMR spectrum with ^{1}H BB decoupling.

By looking at the multiplet structure of each signal in the direction of the F_1-axis, we can immediately determine whether

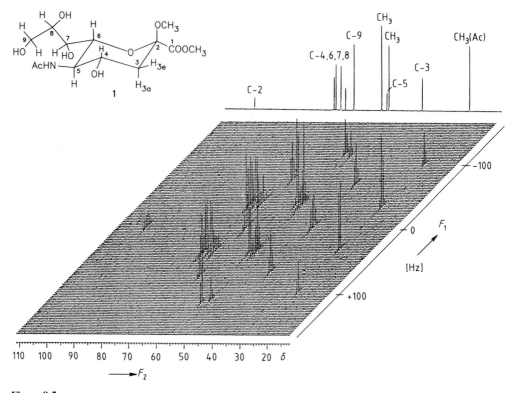

Figure 9-5.
Stacked plot showing the two dimensional heteronuclear *J*-resolved 100.6 MHz ^{13}C NMR spectrum of **1**. The projection of the multiplets onto the F_2-axis is shown along the top edge, and corresponds to the ^{13}C spectrum with BB decoupling. The multiplet splittings along the F_1-direction allow one to determine how many hydrogen atoms are directly bonded to each carbon nucleus. The peak separations in the multiplets correspond to $J(C,H)/2$, as the system has only been allowed to evolve for a time $t_1/2$. The signals of the two quaternary carbons of the carboxyl and acetamido groups are not shown; they are at $\delta = 171.5$ and 175.93.

(*Experimental conditions:*
167 mg of the compound in 2.3 ml D_2O; 10 mm sample tube; 128 measurements with t_1 altered in 1.56 ms increments; each measurement with 48 FIDs and 4 K data points; total time 4.5 h.)

it arises from a CH_3, CH_2 or CH group or from a quaternary carbon nucleus. In the F_2 spectrum the signals have been marked with their assignments (see top of diagram) so far as these are known with certainty at present (see also Table 8-2). By measuring the separation of two adjacent lines in each multiplet we obtain a value for $^1J(C,H)/2$. A rough comparison of the features in the contour plot confirms the statement already made in Section 8.5.2 that the coupling constants are very similar.

When the two methods of presentation are compared it is evident that the contour plot can be measured and interpreted more easily and clearly than the stacked plot.

240

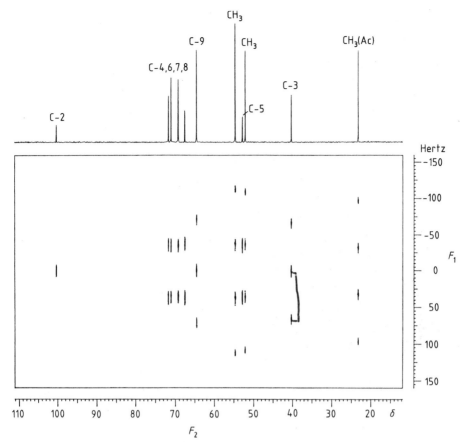

Figure 9-6.
Contour plot of the same results as in Figure 9-5.

9.3.2 Homonuclear Two-Dimensional *J*-Resolved NMR Spectroscopy

^{1}H NMR spectra of large molecules such as steroids, peptides or oligosaccharides, or of mixtures of compounds with similar structures, are often incapable of being analyzed owing to severe overlapping of the resonances, even with the help of the techniques described in Chapter 6. Nevertheless, the required information is contained in the spectral parameters such as the chemical shifts and coupling constants. It was shown in Section 9.3.1 that for complex compounds such as **1** such parameters can be obtained from a two-dimensional heteronuclear NMR experiment. The following questions now arise: do similar methods exist for homonuclear spin systems? – and how does a spin system consisting only of coupled protons behave if we perform a spin-echo experiment (Fig. 9-4) analogous to the above heteronuclear experiment? Since no procedure equivalent to the BB decoupling of the previous experiment is possible in the homonuclear system, the H,H coupling is in this case effective throughout the whole experiment.

We will discuss the experiment by taking the two-spin AX system as an example. The spin-echo pulse sequence used is sketched in Figure 9-7 A, and is as follows:

$$90^\circ_{x'} - t_1/2 - 180^\circ_{x'} - t_1/2 - \text{FID } (t_2)$$

The variable quantity is t_1. Figure 9-7 B shows by means of vector diagrams how this pulse sequence affects the spin system and its magnetization vectors.

The $90^\circ_{x'}$ pulse turns the macroscopic magnetization vectors $\mathbf{M_A}$ and $\mathbf{M_X}$ of the protons A and X into the y'-direction. Of the two magnetization vectors we will consider only $\mathbf{M_A}$. As on several previous occasions, we will again separate $\mathbf{M_A}$ into two components $M_A^{X\alpha}$ (where the X protons are in the α state) and

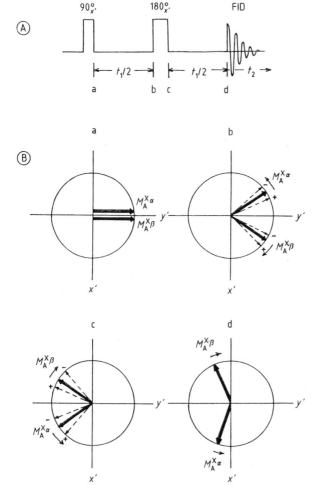

Figure 9-7.
Two-dimensional homonuclear *J*-resolved NMR spectroscopy.
A: Pulse sequence.
B: Evolution of the magnetization vectors in the rotating coordinate system for a two-spin AX system, in which A and X are both protons. Only the magnetization vectors $M_A^{X\alpha}$ and $M_A^{X\beta}$ for the A nuclei are shown. These vectors fan out as a result of field inhomogeneities; the faster spins are indicated by the '+' sign and the slower ones by the '−' sign. The arrow outside the circle indicates the direction of rotation of the fan as a whole relative to the coordinate system. The vector diagrams correspond to the instants a to d shown in the pulse sequence A.

$M_A^{X\beta}$ (where the X protons are in the β state), as shown in Figure 9-7 a.

Like the individual spins, these two vectors rotate about the z-axis with different frequencies:

$$v_A^{X\alpha} = v_A - \frac{1}{2}J(A,X) \quad \text{and} \quad v_A^{X\beta} = v_A + \frac{1}{2}J(A,X)$$

Here v_A is the Larmor frequency of the A protons in the absence of a coupling to X. We also assume that the coordinate system is rotating at this frequency. Which of the two vectors rotates fastest is determined simply by the sign of the coupling constant $J(A,X)$. In our example we will assume that the sign is positive, which means that $M_A^{X\beta}$ rotates faster than the coordinate system and $M_A^{X\alpha}$ slower. After a time $t_1/2$ a phase difference has developed between the two magnetization vectors (Fig. 9-7 b). In addition each vector has fanned out as a result of field inhomogeneities. The plus and minus signs indicate which sides of the fan correspond to the faster and the slower spins respectively.

The $180_{x'}^{\circ}$ pulse which is now applied reflects the magnetization vectors through the x'-axis. If this were a heteronuclear system in which X was some nucleus other than a proton, the directions of rotation relative to the coordinate system would remain unchanged (Section 7.3.2). In the homonuclear case, however, the $180_{x'}^{\circ}$ pulse acts also on the X-protons.

To understand this step we depart from our usual procedure by considering in this case not the macroscopic magnetization vector M_X but a single nucleus X. Each individual nucleus always has a magnetic moment component μ_z along the z-direction. The $180_{x'}^{\circ}$ pulse reverses this z-component, so that an X-nucleus in the α state becomes one in the β state and vice versa. For the macroscopic sample this means that the $180_{x'}^{\circ}$ pulse changes $M_A^{X\alpha}$ into $M_A^{X\beta}$ and $M_A^{X\beta}$ into $M_A^{X\alpha}$. From vector diagram c of Figure 9-7 B we see that after the $180_{x'}^{\circ}$ pulse the two vectors continue to move apart. After a further period $t_1/2$ the phase difference has doubled.

However, the $180_{x'}^{\circ}$ pulse also has another effect. During the second $t_1/2$ interval the fanning out caused by field inhomogeneities is reversed, since although the $180_{x'}^{\circ}$ pulse reverses the directions of rotation of the magnetization vectors relative to the coordinate system, within each of the fans the faster spins are still the fastest, as their higher frequency arises from the external field and not from the coupling. Figure 9-7 d shows the final situation in the spin system before data acquisition.

The phase difference between the two magnetization vectors depends on the value of t_1 chosen for the experiment, and, more importantly, on the magnitude of the coupling constant $J(A,X)$.

In practice one records interferograms (FIDs) for a range of different t_1-values at constant increments of a few ms. Fourier transformation of an individual FID gives a frequency spectrum F_2 which contains information on both chemical shifts and coupling constants. A second Fourier transformation with respect to t_1 yields another frequency spectrum F_1; as in the heteronuclear case this contains only the multiplets produced by the couplings, and thus gives the coupling constants. However, the multiplets appear on lines which are tilted relative to the F_2-axis at an angle of 45°, if both dimensions have the same scale. By manipulating the data one can make all the signals of a multiplet, belonging to one proton with a chemical shift δ, appear on a line perpendicular to the F_2-axis. If this 2D spectrum is then projected onto the F_2-axis we obtain signals only at the positions corresponding to the chemical shifts of the different protons in the molecule. The resulting spectrum thus consists only of singlets, and corresponds to a fully decoupled ^{1}H NMR spectrum which is entirely analogous to a ^{13}C NMR spectrum with ^{1}H BB decoupling. Such a spectrum can provide an useful aid to making assignments or determining δ-values for large molecules.

As an example Figure 9-8 shows the two-dimensional *J*-resolved ^{1}H NMR spectrum of the neuraminic acid derivative **1** with which we are already familiar.

At the top of the contour diagram are shown the projection of the multiplets onto the F_2-axis (Fig. 9-8 A) and, for comparison, the normal (one-dimensional) 400 MHz ^{1}H NMR spectrum (Fig. 9-8 B). On the latter spectrum the assignments of the signals are indicated, so far as these can be deduced from their positions and splitting patterns.

In the contour diagram one can recognize, along the direction of the F_1-axis, the same multiplet patterns as are seen in the normal one-dimensional spectrum. These can be seen even more clearly if the individual multiplets are plotted out separately, as has been done for five of them in Figure 9-9. From these extracts it is evident that the resolution is not significantly different from that in the one-dimensional 400 MHz spectrum.

The region around $\delta = 3.9$ has been omitted from this treatment; this part of the spectrum is confused due to the overlapping of the strong methyl peak of the ester group with the multiplets of four of the protons.

This experiment is widely used for analyzing spectra which contain many overlapping multiplets (giving best results when these are first-order). However, this method too has its limitations, as we were forced to recognize in our example where the strong methyl signal overlapped with other resonances in the region between about $\delta \approx 3.8$ and 4.0. In the sections which follow we will meet other techniques which allow assignments to be made even in cases such as this.

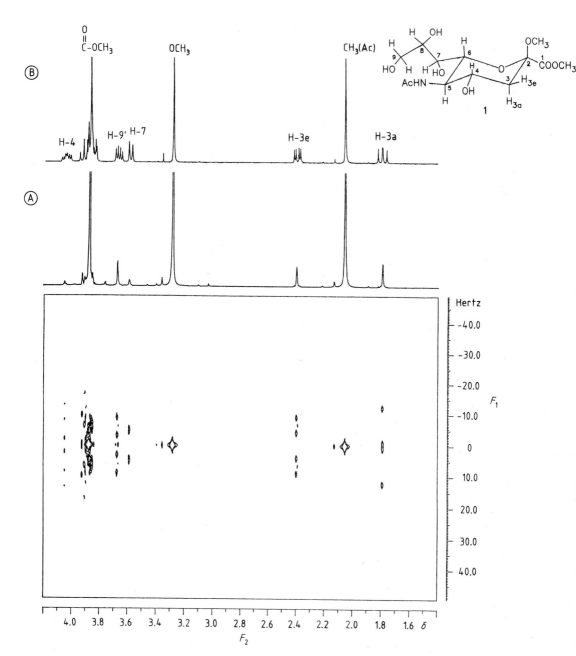

Figure 9-8.

Contour plot of the two-dimensional homonuclear *J*-resolved 400 MHz ^{1}H NMR spectrum of **1**.

A: Projection of the 2D spectrum onto the F_2-axis. This is, in effect, a "decoupled" ^{1}H NMR spectrum.

B: Normal 400 MHz ^{1}H NMR spectrum of **1**.

(*Experimental conditions:*

20 mg of the compound in 0.5 ml D_2O; 5 mm sample tube; 128 measurements with t_1 altered in 5.06 ms increments; each measurement with 48 FIDs and 4 K data points; total time 4.2 h.)

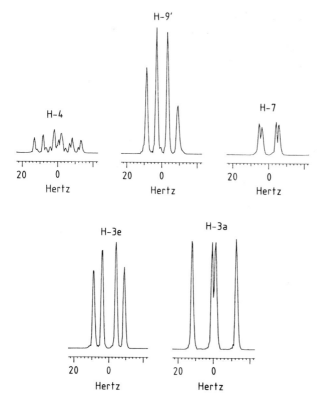

Figure 9-9.
Sections parallel to the F_1-axis through the 2D spectrum of Figure 9-8, for the H-3a, H-3e, H-4, H-7 and H-9' multiplets.

9.4 Two-Dimensional Correlated NMR Spectroscopy

Many assignment problems can be solved in an elegant way with the help of two-dimensional (*shift*) *correlated* NMR spectroscopy. The practical methods to be treated below will begin in Section 9.4.1 with the heteronuclear case, the HETCOR (**HET**eronuclear **COR**relation) or C,H-COSY (**CO**rrelated **S**pectroscop**Y**) experiment, as this can be described using simple classical vector diagrams. First we will learn about the basic principles by considering a simple hypothetical experiment, and will then see how the experimental arrangement can be made more sophisticated for practical use. For simplicity the effects of relaxation and field inhomogeneities will be neglected. After that, in Section 9.4.2, we will discuss the homonuclear version of the experiment, H,H-COSY.

The disadvantages of observing ^{13}C resonances compared with those of ^1H have already been explained in Chapter 1 (Section 1.6.3.2). The same arguments apply, of course, to those two-dimensional methods in which ^{13}C serves as the directly observed nuclide. The development of the inverse (or reverse) procedures has been an important step forward in this respect. Thus, alongside the original C,H-COSY experiment there is the inverse method, H,C-COSY, in which the observed nuclide is ^{1}H. This method will be described in Section 9.4.3, and the fundamental reasons for the advantages of inverse methods will be explained.

In names such as C,H-COSY and H,C-COSY, the order in which the two nuclides (here H and C) are given follows the now accepted convention that the first named is the observed nuclide. Thus, C,H-COSY is a method in which one observes ^{13}C resonances, whereas in H,C-COSY one observes ^{1}H resonances.

In the last three sections (9.4.4 to 9.4.6) we will learn about some further techniques that until recently would have been regarded as quite specialized, but which have now become part of the standard armory of NMR spectroscopy.

9.4.1 Two-Dimensional Heteronuclear (C,H)-Correlated NMR Spectroscopy (HETCOR or C,H-COSY)

We again start with the two-spin AX system of the chloroform molecule (^{13}CHCl$_3$), and apply the pulse sequence shown in Figure 9-10 A [5–7].

The vector diagrams of Figure 9-10 B show how the applied pulse sequence $90^{\circ}_{x'} - t_1 - 90^{\circ}_{x'}$ affects the ^{1}H macroscopic magnetization vectors $\boldsymbol{M}_{\mathrm{H}}^{\mathrm{C\alpha}}$ and $\boldsymbol{M}_{\mathrm{H}}^{\mathrm{C\beta}}$. Here $\boldsymbol{M}_{\mathrm{H}}^{\mathrm{C\alpha}}$ and $\boldsymbol{M}_{\mathrm{H}}^{\mathrm{C\beta}}$ represent the magnetization vectors for chloroform molecules whose ^{13}C nuclei are in the α and β states respectively. (The procedure of resolving the magnetizsation into these two vectors has already been explained in Section 8.5.2.)

The first $90^{\circ}_{x'}$ pulse turns both magnetization vectors from the z-direction into the y'-direction (Fig. 9-10 b). During the subsequent evolution phase t_1 both vectors rotate with their Larmor frequencies:

$$v_{\mathrm{H}} - \frac{1}{2} J(\mathrm{C,H}) \quad \text{and} \quad v_{\mathrm{C}} + \frac{1}{2} J(\mathrm{C,H}) \qquad (9\text{-}4)$$

where v_{H} is the Larmor frequency in the absence of coupling, i. e. the proton resonance frequency of ^{12}CHCl$_3$.

The angles φ_α and φ_β through which the vectors $\boldsymbol{M}_{\mathrm{H}}^{\mathrm{C\alpha}}$ and $\boldsymbol{M}_{\mathrm{H}}^{\mathrm{C\beta}}$ travel in the time t_1 are:

$$\varphi_\alpha = 2\pi \left(v_{\mathrm{H}} - \frac{1}{2} J(\mathrm{C,H}) \right) t_1$$
$$\varphi_\beta = 2\pi \left(v_{\mathrm{H}} + \frac{1}{2} J(\mathrm{C,H}) \right) t_1 \qquad (9\text{-}5)$$

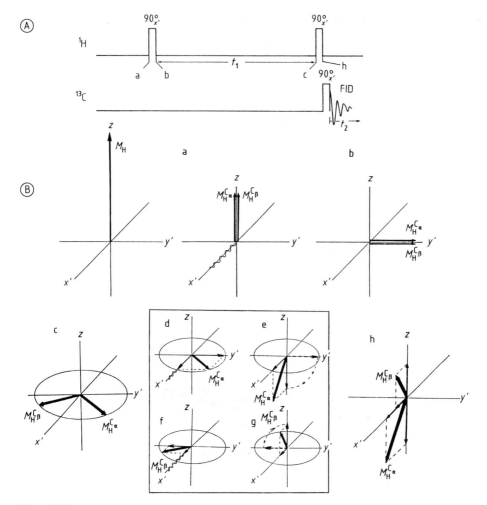

Figure 9-10.

Two-dimensional C,H-correlated NMR spectroscopy.

A: Pulse sequence.

B: Diagrams showing the evolution of the ^{1}H magnetization vectors $M_H^{C\alpha}$ and $M_H^{C\beta}$ for two-spin AX system (A = ^{1}H, X = ^{13}C) in the rotating coordinate system x', y', z. The vector diagrams a to c correspond to the instants marked in A. Diagrams d and e (for $M_H^{C\alpha}$) and f and g (for $M_H^{C\beta}$) show how these vectors are affected by the second $90_{x'}^\circ$ ^{1}H pulse. Diagram h shows the situation immediately before the $90_{x'}^\circ$ signal generating pulse in the ^{13}C channel.

The phase difference Θ depends only on the time t_1 and on J(C,H):

$$\Theta = \varphi_\beta - \varphi_\alpha = 2\pi J(\text{C,H})\, t_1 \qquad (9\text{-}6)$$

For $t_1 = [4J(\text{C,H})]^{-1}$, $\Theta = 90°$, and
for $t_1 = [2J(\text{C,H})]^{-1}$, $\Theta = 180°$.

Equations (9-4) to (9-6) correspond exactly to Equations (9-1) to (9-3), the only difference being that here we are concerned with ^{1}H instead of ^{13}C resonances.

248

Figure 9-10 c shows the situation after an (arbitrary) evolution time t_1, which we assume to be in the region of several ms. As the precession frequencies of $M_H^{C\alpha}$ and $M_H^{C\beta}$ do not coincide with the frequency of the rotating coordinate system, these two vectors are now inclined to the y'-axis at angles which depend on the frequency differences. Let us first concentrate our attention on the vector $M_H^{C\alpha}$. This has components in the $(+y')$ and $(+x')$ directions (Fig. 9-10 d). The second $90_{x'}^{\circ}$ pulse in the ^{1}H channel now turns the y'-component into the $(-z)$ direction, whereas the x'-component is unaffected. The new direction of the magnetization vector $M_H^{C\alpha}$ is determined by the components in the x'- and $(-z)$-directions; in the example drawn here (Fig. 9-10 e) it is in the front lower quadrant of the x', z-plane. The same thing happens to the y'-component of $M_H^{C\beta}$, except that this is rotated into the $(+z)$ direction, so that after the $90_{x'}^{\circ}$ pulse $M_H^{C\beta}$ is in the front upper quadrant of the x', z-plane, as given by vector addition of the x'- and z-components (Fig. 9-10 f and g). However, for the rest of the discussion we are not concerned with the vectors $M_H^{C\alpha}$ and $M_H^{C\beta}$ themselves, but only with their z-components (Fig. 9-10 h). These longitudinal magnetization components are proportional to the population differences between the energy levels 1 and 3 (for $M_H^{C\alpha}$) and between 2 and 4 (for $M_H^{C\beta}$). From this we can draw two conclusions:

- The pulse sequence $90_{x'}^{\circ} - t_1 - 90_{x'}^{\circ}$ has altered the level populations from those at the start; at the instant which we chose in Figure 9-10, the population of level 3 is even greater than that of the lowest-lying level 1. In an extreme case the situation at the time t_1 is exactly as shown in Figure 8-14 B.
- The state reached by the spin system depends on the time t_1, i. e. on the angles φ_α and φ_β covered by the vectors (Equation 9-5). These are in turn determined also by the Larmor frequence ν_H and the coupling constant $J(C,H)$.

Up to now we have considered only the magnetization vector M_H of the protons. What effects does the pulse sequence applied to the ^{1}H channel have on the ^{13}C NMR spectrum? Let us look again at the situation shown in Figure 9-10 h, which can be described using an energy level scheme like that in Figure 8-14 B. The intensity of the ^{13}C NMR signal is determined by the level populations after the second $90_{x'}^{\circ}$ pulse, which means that we are concerned here with a transfer of polarization or magnetization, such as we met earlier in the SPI and INEPT experiments (Section 8.5). Here, however, the ^{13}C NMR signal is not amplified by a constant factor as in the experiments described there, but is modulated as a function of t_1 by the Larmor frequencies of the protons. Unfortunately this situation can no longer be illustrated diagrammatically in a clear way. The mathematical analysis gives the result that the magnetization vectors $M_H^{C\alpha}$ and $M_H^{C\beta}$, and therefore also $M_C^{H\alpha}$ and $M_C^{H\beta}$, which are connected with them as a consequence of the energy level scheme (see Fig. 8-14), are always changed by the same amount, but with opposite signs.

The $90_{x'}^{\circ}$ signal generating pulse in the ^{13}C channel turns these two longitudinal vectors into the $(+y')$ and $(-y')$ directions respectively. During the data acquisition phase t_2 they precess at the frequencies corresponding to the X_2 and X_1 transitions, giving the FID signal in the receiver coil. Fourier transformation with respect to t_2 gives two ^{13}C signals along the F_2-axis, whose modulation depends on t_1 and on the resonance frequencies of the protons. If we now record n spectra with different values of t_1, and carry out a second Fourier transformation with respect to t_1, we obtain a two-dimensional sepctrum with four signals, two of which have negative amplitudes (Fig. 9-11). In this 2D spectrum the ^1H resonances are displayed along the F_1-axis and the ^{13}C resonances along the F_2-axis. The coordinates of the four signals are: (A_1, X_1); (A_1, X_2); (A_2, X_1); (A_2, X_2). Here the symbols A and X stand for the frequencies of the corresponding ^1H and ^{13}C transitions, but the results are usually given as δ-values.

The spectrum parallel to the F_2-axis corresponds to the one-dimensional coupled ^{13}C spectrum, and that along the F_1-axis to the one-dimensional coupled ^1H spectrum. In a molecule as simple as chloroform containing only two coupled nuclei, this gives spectra that are quite clear. For larger molecules, however, the interpretation becomes much more difficult, or even impossible. The experiment must therefore be modified for practical use.

The first question might be: could we switch on the ^1H BB decoupler during data acquisition, so as to make the doublets in the ^{13}C NMR spectrum become singlets? Unfortunately this is not possible! Decoupling would completely eliminate the signals of interest, for we must recall that in each experiment, after the $90_{x'}^{\circ}$ ^{13}C signal generating pulse, the vectors $M_C^{H\alpha}$ and $M_C^{H\beta}$ are equal in magnitude but opposite in phase, and therefore the signal amplitudes have opposite signs. It is only when there is no decoupling that two signals are observed, as the frequencies are then different. Switching on the BB decoupler would eliminate the coupling which is the cause of the difference in frequencies. The two magnetization vectors would then completely cancel each other and no signal would be induced in the receiver coil.

A different situation arises, however, if we insert a delay time $\Delta_2 = [2J(C,H)]^{-1}$ between the $90_{x'}^{\circ}$ excitation pulse in the ^{13}C channel and the acquisition of the FID (Fig. 9-12). During the time Δ_2 the angle swept through by the faster magnetization vector $M_C^{H\beta}$ is exactly 180° greater than that for $M_C^{H\alpha}$, and the two vectors are again in phase, although they continue to

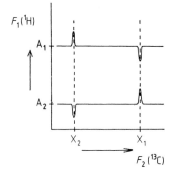

Figure 9-11.
Schematic two-dimensional C,H-correlated NMR spectrum of a two-spin AX system (for pulse sequence see Fig. 9-10). The two signals along the F_2-direction correspond to the one-dimensional ^{13}C NMR spectrum without decoupling, except that the signals have opposite signs. Along the F_1-direction is seen the doublet of the ^1H NMR spectrum with the C,H coupling (the ^{13}C satellites, also with opposite signal amplitudes).

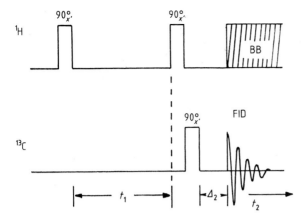

Figure 9-12.
Extended pulse sequence for
simplifying the two-dimensional
C,H-correlated NMR spectrum of
Figure 9-11. After the $90^\circ_{x'}$ pulse in
the ^{13}C channel there is a delay
$\Delta_2 = [2J\,(C,H)]^{-1}$ before the start
of data acquisition and simulta-
neous ^{1}H BB decoupling.

precess at different rates. If at this instant the BB decoupler is
switched on, both vectors precess at the same rate from this
point on. After the Fourier transformation of the FID with
respect to t_2 we find that the ^{13}C NMR spectrum (F_2) consists
of only one signal at ν_C. In the modified experiment one records
spectra for n different t_1-values by this method. The values are
increased by a constant amount each time, this increment
being of the order of a few ms. As a result of the polarization
transfer caused by the pulse sequence in the ^{1}H channel, the n
^{13}C NMR signals thus obtained are modulated by the proton
resonances. A second Fourier transformation with respect to t_1
gives the two-dimensional spectrum, which now consists of
only two signals with the coordinates (A_1, X) and (A_2, X)
(Fig. 9-13).

Finally, the two-dimensional spectrum, which in our example
now consists of two lines, can be reduced to just one signal by
using the pulse sequence shown in Figure 9-14 A. The new
features here are the 180° pulse in the ^{13}C channel after a time
of exactly $t_1/2$, and the delay time Δ_1 before the second $90^\circ_{x'}$
pulse in the ^{1}H channel.

Let us again try, so far as possible, to illustrate and understand the experiment by means of vector diagrams in a rotating
coordinate system (Fig. 9-14 B). The first $90^\circ_{x'}$ pulse turns both
the ^{1}H magnetization vectors $M_H^{C\alpha}$ and $M_H^{C\beta}$ into the y'-direc-
tion (Fig. 9-14 a). Owing to their different Larmor frequencies
$\nu_H - J(C,H)/2$ and $\nu_H + J(C,H)/2$, the two vectors move
apart. After the time $t_1/2$ the phase difference is $\Theta = \pi J(C,H)t_1$
(Equation 9-6, Fig. 9-14 b). The 180° pulse in the ^{13}C channel
reverses the identities of the ^{13}C nuclei in the α and β states, so
that $M_H^{C\alpha}$ becomes $M_H^{C\beta}$ and $M_H^{C\beta}$ becomes $M_H^{C\alpha}$. In the diagram
the vector which is rotating fastest (indicated by the thicker of
the two arrows outside the circle) is now following the slower
one (Fig. 9-14 c). After a further interval $t_1/2$ $M_H^{C\beta}$ has caught
up with $M_H^{C\alpha}$ and the two are again in phase (Fig. 9-14 d). The

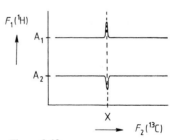

Figure 9-13.
Schematic two-dimensional C,H-
correlated NMR spectrum of a two-
spin AX system (pulse sequence as
in Fig. 9-12). The 2D spectrum is
reduced to two signals with opposite
signs; their separation along the F_1
frequency axis is equal to $J\,(C,H)$.

251

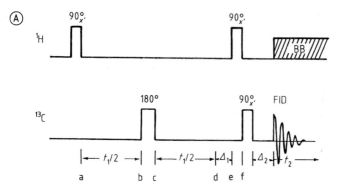

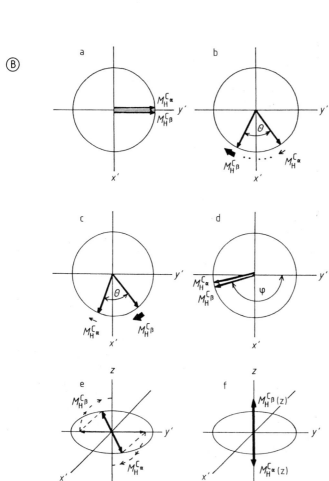

Figure 9-14.
A: Pulse sequence for a two-dimensional C,H-correlated NMR experiment which reduces the 2D spectrum of a two-spin AX system to only one peak.
B: The vector diagrams a to f show the positions of the ^{1}H magnetization vectors $M_H^{C\alpha}$ and $M_H^{C\beta}$ or their z-components (f) at the instants indicated in A; in diagrams a to d only the x', y'-plane is shown.

total angle through which the vectors have traveled and the angle φ depend only on the Larmor frequency for the protons with no coupling to the ^{13}C nuclei.

The same situation could also have been arrived at by continuous C,H decoupling, e. g. by ^{13}C BB decoupling in the ^{13}C

252

channel – not as previously in the ^{1}H channel – but from an experimental point of view this procedure would have a number of disadvantages, which we cannot go into here.

A $90^\circ_{x'}$ pulse applied in the ^{1}H channel immediately following the period t_1 would result in no polarization transfer, and consequently no modulation of the ^{13}C NMR signal. However, if an additional delay Δ_1 is inserted before the $90^\circ_{x'}$ ^{1}H pulse, $M_H^{C\alpha}$ and $M_H^{C\beta}$ move apart again. After a time $\Delta_1 = [2J(C,H)]^{-1}$ the phase difference is 180°. The $90^\circ_{x'}$ ^{1}H pulse which now follows turns the y'-components of these two vectors into the $(+z)$ and $(-z)$ directions respectively (Fig. 9-14 e and f). This is the step which produces the polarization. Its magnitude depends only on the angle φ. If the vectors are exactly along the y'-direction before the pulse, the polarization is at a maximum, whereas if they are along the x'-direction the polarization is zero. The angle covered by the vectors in the time t_1 is a function of the Larmor frequency ν_H of the "decoupled" protons. The system continues its evolution during the time Δ_1, but this contribution is constant for all the spectra that are recorded with different t_1-values. The polarization situation on which the ^{13}C signal intensities depend is therefore determined solely by the proton Larmor frequency ν_H.

The remaining stages are analogous to those in the previous experiment (Fig. 9-12). The $90^\circ_{x'}$ pulse in the ^{13}C channel tips the ^{13}C magnetization vectors into the $(+y')$ and $(-y')$ directions, and after a further delay time $\Delta_2 = [2J(C,H)]^{-1}$ the vectors $M_C^{H\alpha}$ and $M_C^{H\beta}$ are in phase. The ^{1}H BB decoupling which is switched on at this instant eliminates the C,H coupling during the data acquisition. The first Fourier transformation with respect to t_2 gives a signal at ν_C. If one records a series of spectra with $\Delta_1 = \Delta_2 = [2J(C,H)]^{-1}$ and different t_1-values, the signal intensities are modulated by the frequency ν_H. A second Fourier transformation with respect to t_1 therefore gives a two-dimensional spectrum (F_1, F_2) which contains only one signal with the coordinates (ν_H, ν_C) (Fig. 9-15). According to theory the signal intensity is a maximum for $\Delta_1 = [2J(C,H)]^{-1}$. Also it is amplified by the transfer of magnetization from the sensitive nucleus ^{1}H to the less sensitive nucleus ^{13}C.

The interval between the first period t_1 and the beginning of the data acquisition period t_2 is sometimes called the *mixing time*. The same experiment can also be applied to multiple spin systems.

Figure 9-16 shows the two-dimensional spectrum of glutamic acid (**2**). In the ^{13}C NMR spectrum at the top edge of the figure we see the three peaks belonging to the carbon nuclei that have directly-bonded protons. The two quaternary carbon nuclei of the carboxyl groups do not give correlation peaks, and therefore do not appear. The normal 500 MHz ^{1}H NMR spectrum is shown at the left-hand edge.

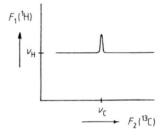

Figure 9-15.
Schematic two-dimensional C,H-correlated NMR spectrum of a two-spin AX system (pulse sequence as in Fig. 9-14). The 2D spectrum which is obtained consists of only one signal with the coordinates (ν_H, ν_C).

253

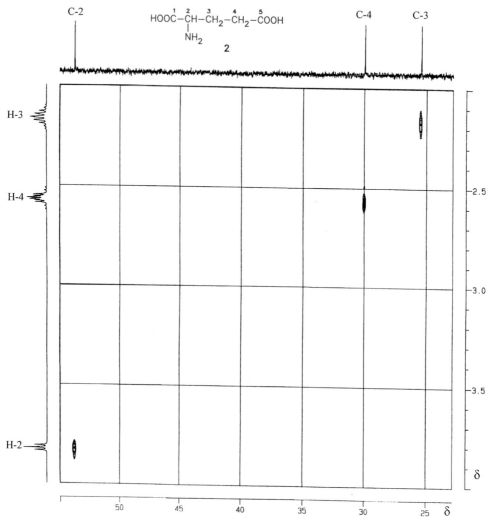

Figure 9-16.
Two-dimensional C,H-correlated spectrum of glutamic acid (**2**) in D_2O. The one-dimensional 500 MHz
^{1}H NMR spectrum is shown at the left-hand edge and the 125 MHz ^{13}C NMR spectrum at the top edge.
(*Experimental conditions:*
approx. 20 mg in 0.5 ml D_2O; 5 mm sample tube; 128 measurements with t_1 altered in 312 μs increments; each
measurement with 16 FIDs and 1 K data points; total time approx. 1 h.)

As the ^{1}H resonances can be unambiguously assigned from
their chemical shifts (see also Section 9.4.2), we can proceed
from this to assign the ^{13}C resonances of carbon nuclei 2–4 by
looking at the correlation peaks. A particular advantage of this
technique is that even in complicated molecules, such as are
often encountered in biochemistry and natural products
chemistry, there is little overlapping of the correlation peaks,
as it combines the large chemical shifts of ^{13}C NMR spectroscopy with those of ^{1}H NMR spectroscopy.

254

In Sections 9.4.3 and 9.4.4, again taking compounds **1** and **2** as our examples, we will see how correlation experiments can also be performed by a reverse procedure. These methods, together with the use of pulsed field gradients, lead to a great reduction in the time needed to complete a measurement, and consequently the "normal" C,H-COSY procedure described above has nowadays been largely superseded by the more time-efficient reverse methods.

By using the pulse sequence shown in Figure 9-14 A and suitably adjusting the delay times Δ_1 and Δ_2, it is possible to obtain correlation information for carbon and hydrogen nuclei separated by two or three bonds. For example, to detect correlations between nuclei with a coupling constant $J(C,H)$ of 10 Hz, which is fairly typical for $^2J(C,H)$ or $^3J(C,H)$, one must set Δ_1 and Δ_2 both equal to $1/[2J(C,H)] = 50$ ms. In practice, however, this procedure is seldom chosen; instead the HMBC method described in Section 9.4.4 is preferred, as it yields the same information in a much shorter time. The use of the HMBC method will be illustrated by an example when we come to the relevant section.

9.4.2 Two-Dimensional Homonuclear (H,H)-Correlated NMR Spectroscopy (H,H-COSY; Long-Range COSY)

The two-dimensional homonuclear (H,H)-correlated NMR experiment yields NMR spectra in which ^1H chemical shifts along both frequency axes are correlated with each other [8]. This technique has become known as COSY (*correlated spectroscopy*). It is based on the pulse sequence $90^\circ_{x'} - t_1 - \Theta_{x'}$ (Fig. 9-17).

First we will describe the COSY experiment with $\Theta_{x'} = 90^\circ_{x'}$, and we will consider its application to a two-spin AX system, where A and X are both protons with a coupling constant $J(A,X)$. Thus the pulse sequence in this case is : $90^\circ_{x'} - t_1 - 90^\circ_{x'}$.

Compared with the heteronuclear case there is an important difference here, because the first $90^\circ_{x'}$ pulse turns both magnetization vectors, M_A and M_X, into the direction of the y'-axis. Due to the coupling $J(A,X)$ the A nuclei have two macroscopic magnetization vectors, $M_A^{X\alpha}$ and $M_A^{X\beta}$, according to whether the X nucleus is in the α or the β state. Similarly we also have to take into consideration two M_X vectors, $M_X^{A\alpha}$ and $M_X^{A\beta}$. These four vectors rotate in the x', y'-plane around the z-axis with the frequencies $\nu_A \pm J(A,X)/2$ and $\nu_X \pm J(A,X)/2$.

During the time t_1, which is the variable in the COSY experiment, the four magnetization vectors fan out in the x', y'-plane as a result of their different frequencies.

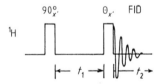

Figure 9-17.
Pulse sequence for the two-dimensional homonuclear H,H-correlated NMR experiment COSY. The variable is t_1. The pulse angle Θ is usually $90°$ or $45°$, or occasionally $60°$.

At the instant t_1 each of these vectors has components in the x'- and y'-directions. The second $90°_{x'}$ pulse which now follows turns each of the y'-components into the $(+z)$ or $(-z)$ direction. Associated with this step there is a transfer of polarization. The amount of magnetization that is transferred depends on the situation in the spin system at the instant t_1, and therefore on the Larmor frequencies v_A and v_X and on $J(A,X)$.

The x'-components of the magnetization vectors, which have a similar dependence on the situation to which the system has evolved, and which continue to rotate in the x', y'-plane, give a FID which, after Fourier transformation with respect to t_2, yields a four-line AX-type spectrum with the frequencies:

$$v_A + \frac{1}{2} J(A,X) \quad (A_1) \qquad v_A - \frac{1}{2} J(A,X) \quad (A_2)$$

$$v_X + \frac{1}{2} J(A,X) \quad (X_1) \qquad v_X - \frac{1}{2} J(A,X) \quad (X_2)$$

These frequencies correspond to the transitions labeled A_1, A_2, X_1 and X_2 in Figure 8-14.

The signals as functions of t_1 are modulated with these four frequencies. The second Fourier transformation with respect to t_1 therefore gives a two-dimensional spectrum with four groups, each containing four signals. Two of these groups are centered around the positions (v_A, v_A) and (v_X, v_X), and are called the *diagonal peaks*, while the other two are centered around (v_A, v_X) and (v_X, v_A), and are called the *correlation peaks* or *cross peaks*. Thus the diagonal peaks and the cross peaks form the corners of a square. The point of importance for us is that cross peaks always occur when two nuclei, in this case A and X, interact with each other through a *scalar coupling*.

Within a group the separation in each dimension (F_1 and F_2) between two adjacent signals is exactly equal to the coupling constant $J(A,X)$. The projection of such a COSY spectrum onto the F_1- or F_2-axis therefore corresponds to the one-dimensional ^{1}H NMR spectrum.

In Figure 9-18 a spectrum of this kind for the two-dimensional AX spin system is shown schematically as a contour plot. Here only the absolute values of the signals (the "magnitudes spectrum", see Section 1.5.4) are shown. This form of presentation is usually chosen in practice, since in the COSY experiment the phases of the diagonal and cross peaks always differ by 90°. Therefore, if one corrects the phase so that the cross peaks appear as absorption signals, the diagonal peaks will appear as dispersion signals.

If we look only at the cross peaks we find that, in both the horizontal and vertical directions, we have absorption signals with alternating positive and negative amplitudes, i.e. the signals are in antiphase. For multi-spin systems these phase relationships between the cross peaks provide an aid to assign-

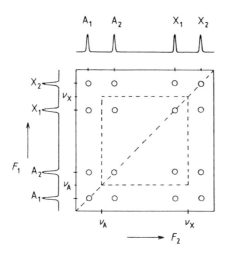

Figure 9-18.
Schematic representation of a
COSY experiment on a two-spin
AX system in which A and X are
protons. The signal amplitudes are
shown here as absolute values. In
an actual spectrum the peaks on the
diagonal are dispersion signals,
while the correlation peaks (cross
peaks) are absorption signals with
alternating signs. The diagonal
peaks of a pair of mutually coupled
nuclei and their cross peaks form
the corners of a square.

ing and determining coupling constants. However, it can also
happen that superimposed signals with opposite phases cancel
each other. Since in most cases one is only interested in correlations, it is usual to dispense with phase-sensitive detection when
recording the COSY spectrum, and instead to display the signals
in "phase-corrected" form. It is usually preferred, as has already
been explained above, to calculate only the absolute values of
the signals. This method was also used for Figures 9-19 to 9-21.

If a proton is coupled to more than one neighboring proton,
the diagonal peak forms a corner of several squares. This
makes it possible to identify the positions of the resonances of
the coupled nuclei, even in complicated spectra. Thus the
COSY experiment is an important aid in assigning ^{1}H resonances. It is far superior to decoupling experiments, as a *single*
experiment gives connectivity information on *all* the coupled
nuclei.

As a simple example Figure 9-19 shows the 500 MHz COSY-
90 spectrum of glutamic acid (**2**). The spectrum at the upper and
left edges of the diagram is the normal one-dimensional
500 MHz ^{1}H NMR spectrum. Of the three multiplets which can
be seen, that at $\delta \approx 3.8$ can be assigned on the basis of its
position and intensity to the proton on C-2.

In the COSY spectrum we find three signals on the diagonal,
which correspond to the three multiplets in the normal one-
dimensional spectrum. From these diagonal peaks and the
cross peaks we can draw two squares, which enable one to see
immediately which multiplets belong to the mutually coupled
protons. Since the protons on C-3 are coupled both to the proton on C-2 and to the two protons on C-4, the multiplet which
is to be assigned to the C-3 protons forms a corner of two
squares.

By chosing appropriate experimental conditions it is often
possible to also distinguish small, long-range couplings.

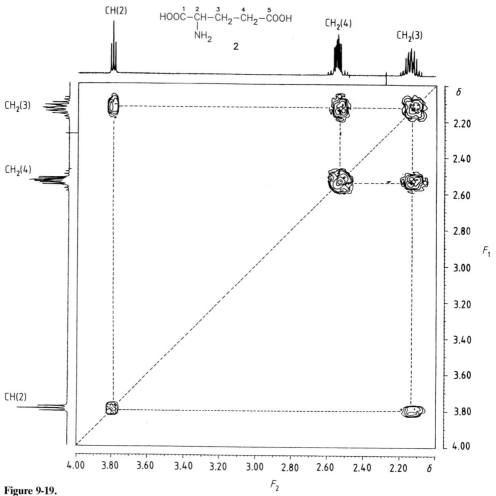

Figure 9-19.
500 MHz COSY-90 spectrum of glutamic acid (**2**) shown as a contour plot. At the left-hand edge and at the top is the one-dimensional ^{1}H NMR spectrum with assignments. The diagonal and cross peaks joined by dashed construction lines indicate which protons have a mutual scalar coupling. The diagonal peak of the two protons on C-3 forms a corner of two squares, as these protons are coupled both to the proton on C-2 and to those on C-4. (*Experimental conditions:*
10 mg of the compound in 0.5 ml D_2O; 5 mm sample tube; 256 measurements with different values of t_1, each measurement with 16 FIDs; digital resolution 2.639 Hz/data point.)

A disadvantage of the method is that for larger molecules the COSY contour diagram becomes confused. In particular, when the chemical shift differences $\Delta\delta$ between the coupled nuclei are small the cross peaks, which are close to the diagonal, are difficult to recognize, as this is the region where the strong diagonal peaks with their broad wings interfere.

If instead of the second $90^\circ_{x'}$ pulse one applies a pulse with a smaller angle $\Theta_{x'}$, the spectrum becomes simpler. The transfer of magnetization takes place preferentially, so that some of the signals within the cross and diagonal peaks are reduced in inten-

258

sity more than others. However, the smaller pulse angle reduces the sensitivity. A good compromise is to use an angle $\Theta_{x'} = 45^\circ_{x'}$.

Let us take as an example the 400 MHz COSY-45 spectrum of our test compound, the neuraminic acid derivative **1**. Figure 9-20 shows the region from $\delta = 1.4$ to 4.2, while Figure 9-21 shows that from $\delta = 3.4$ to 4.2. The projection of the signals onto the F_2-axis is shown along the top edge in each case; on the left-hand side of each diagram (F_1-axis) is the one-dimensional 400 MHz ^{1}H NMR spectrum. A comparison shows that the resolution in the spectrum obtained by projection is only slightly inferior to that in the normal spectrum.

To analyze the COSY spectrum we must form squares from the diagonal and cross peaks. Starting from the unambiguously assigned signals of H-3a and H-3e, the cross peaks (Fig. 9-20) immediately indicate the position of the H-4 multiplet in the region $\delta = 4.0$ to 4.05. This multiplet forms the corner of a further square (Fig. 9-21), from which we can determine the position of the signal due to H-5. Drawing the next square presents some difficulty; however, from the expanded region of the spectrum it appears that the chemical shift of H-6 may be only slightly different from that of H-5.

To continue the analysis of the spectrum it is best to start now from another signal which is definitely assigned. In our case a suitable signal is the doublet at $\delta = 3.6.$, assigned to H-7. From the cross peaks we find the position of another coupled neighbor; this could in principle be either H-6 or H-8. Without additional information we cannot decide between these two. However, we know from other experiments that the coupling constant $J(6,7)$ is very small. It therefore seems likely that the cross peak will lead us to the signal of H-8. Starting from H-8 we can now find H-9', then H-9.

As this example shows, one needs several assigned signals as starting points for the analysis of a spectrum. In our example these were the signals of H-3 and H-7. From these it was possible to locate the signals of all the remaining protons, although there is still some uncertainty in the region from $\delta = 3.85$ to 3.95, due to the overlapping of the multiplets from H-5, 6, 8 and 9 and the methyl signal of the ester group. The results are shown in Table 9-1.

Another result which is evident from this spectrum is that for non-coupled protons – in this case those of the methyl group – only signals on the diagonal are found.

Table 9-1.
Summary of the procedure followed in analyzing the two-dimensional H,H-correlated NMR spectrum of **1**.

starting point	newly assigned
H-3a/H-3e	H-4
	H-5
	H-6*
H-7	H-8*
	H-9*
	H-9'

* The experiment gives only approximate positions for these signals.

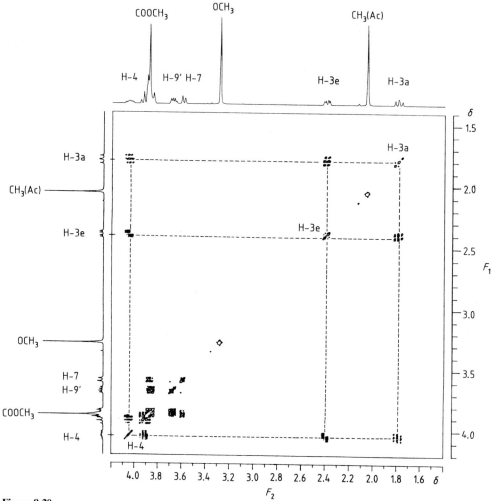

Figure 9-20.

400 MHz COSY-45 spectrum of the neuraminic acid derivative **1**. At the top edge is shown the projection of the 2D spectrum onto the F_2-axis, while at the left-hand edge is the one-dimensional 400 MHz ^{1}H NMR spectrum. For reasons of clarity only three squares have been drawn in, linking the signals of H-3a, H-3e and H-4. (*Experimental conditions:*

20 mg of the compound in 0.5 ml D_2O; 5 mm sample tube; 512 measurements with t_1 altered in 632 μs increments; each measurement with 32 FIDs and 2 K data points; total time 15.4 h.)

There are many variants of the COSY experiment. One such is *long-range COSY*. In this a fixed delay Δ, usually about 0.1 to 0.4 s, is introduced before and after the second $90^\circ_{x'}$ pulse, so that the sequence becomes:

$$90^\circ_{x'} - t_1 - \Delta - 90^\circ_{x'} - \Delta - \text{FID}$$

The effect of this delay is to allow the development of correlation effects also for very weakly coupled protons, so that in the

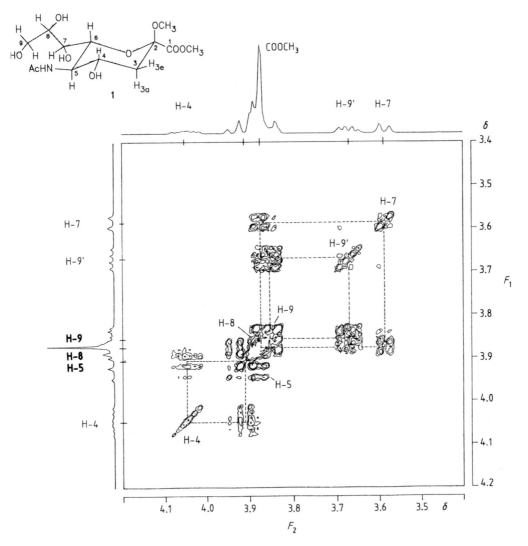

Figure 9-21.
Expanded plot of a portion of Figure 9-20. Starting from the signals of H-4, H-7 and H-9′, which are already assigned with certainty as shown in the top spectrum, the cross peaks lead us to the chemical shifts to **H-5**, **H-8** and **H-9**.

two-dimensional spectrum we find correlation peaks for protons that show no detectable scalar couplings in the one-dimensional ^{1}H NMR spectrum. Figure 9-22 shows a long-range COSY spectrum of glutamic acid. On comparing this with Figure 9-19 we notice two additional, slightly weaker, peaks at $\delta \approx (3.8, 2.6)$ and $\delta \approx (2.6, 3.8)$. These indicate a coupling through four bonds between proton H-2 and the two protons H-4, which would not be detectable under the usual conditions. The long-range COSY method can provide help in assigning complex spectra.

261

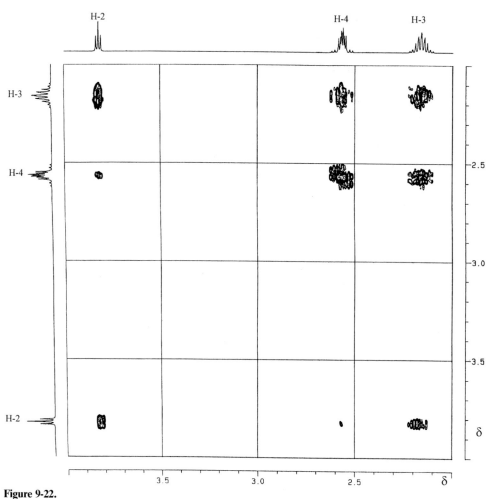

Figure 9-22.

500 MHz long-range H,H-COSY spectrum of glutamic acid (**2**). The one-dimensional ^{1}H spectrum is shown at the top and on the left. Compared with the COSY-90 spectrum (Fig. 9–19) two new signals at approximately (2.6,3.8) and (3.8,2.6) have now appeared. These provide evidence of a correlation between the protons on C-2 and C-4; in the one-dimensional 500 MHz spectrum it is not possible to detect a coupling between these protons. (*Experimental conditions:*

approx. 20 mg in 0.5 ml D$_2$O; 5 mm sample tube; 128 measurements with t_1 altered in 600 μs increments; each measurement with 8 FIDs and 1 K data points; total time approx. 40 min.)

9.4.3 Reverse Two-Dimensional Heteronuclear (H,C)-Correlated NMR Spectroscopy (HSQC; HMQC)

Methods in which one observes the *insensitive* nuclide are usually very time-consuming. Two-dimensional heteronuclear (C,H)-correlated NMR spectroscopy (Section 9.4.1) suffers from this disadvantage, even though it is such a useful method. In Section 8.5.3 of the previous chapter we saw that by reversing the INEPT method so that one observes the *sensitive* nuclide, ^{1}H, the required amount of sample or the accumulation time can be greatly reduced. The common feature of all these *reverse* procedures is that one first generates phase coherence in the channel corresponding to the insensitive nuclide (e. g., ^{13}C or ^{15}N) and this is then transferred to the sensitive nuclide (usually ^{1}H), whose resonance is then observed. Such reverse methods with proton detection are also available for recording (H,C)-correlated spectra. It must be said at the outset that the resulting spectra yield no new information compared with the normal (C,H)-COSY method, but the detection sensitivity is considerably increased, allowing one to use smaller samples and/or shorter accumulation times. Here we will focus attention on a form of reverse two-dimensional (H,C)-correlated experiment that was proposed by Bodenhausen and Ruben [10], as we can use vector diagrams that we have already met earlier to understand the action of the pulse sequence on a two-spin AX system. The experiment is known as the HSQC (heteronuclear single quantum coherence) method, both in the literature and in the computer programs supplied by instrument manufacturers. The overall pulse sequence is as follows:

^{1}H:
$$90^\circ_{x'}-\tau-180^\circ_{x'}-\tau-90^\circ_{y'}-t_1/2-180^\circ_{y'}-t_1/2-90^\circ_{x'}-\tau-180^\circ_{x'}-\tau-\text{FID}$$
$$(t_2)$$

^{13}C:
$$\tau-180^\circ_{x'}-\tau-90^\circ_{x'}\quad-\quad t_1\quad-\quad 90^\circ_{x'}-\tau-180^\circ_{x'}-\tau-\text{BB}$$

The strategy of this experiment involves a first step in which an ^{1}H polarization corresponding to the magnetization M_H is transferred, by means of a normal INEPT pulse sequence with its associated amplifying effect, to the ^{13}C nuclei (magnetization M_C). In the second step the magnetization vectors M_C are allowed to develop for a time t_1 (which is changed incrementally), then in the final step the resulting ^{13}C polarization (coherence) is transferred back to the protons by a reverse INEPT sequence, and lastly the ^{1}H resonance is recorded. Rather than taking the ^{15}N/^{1}H system as an example, as in the original publication [10], we will again consider here the

^{13}C/^{1}H system (example: chloroform). Lastly we will, as before, illustrate it by showing its application to the neuraminic acid derivative **1**.

An important question in any reverse experiment is: how can the unwanted main signal be suppressed so that we are left only with the signals of the ^{13}C-coupled protons – the ^{13}C satellites? To begin by answering this question, let us first consider the magnetization vector M_H of the protons in the ^{12}CHCl$_3$ molecules that make up 98.9 % of the chloroform sample, i. e. the protons that are not coupled to ^{13}C and are responsible for the main signal in the spectrum. On the basis of the relationships explained in Section 8.2 it can easily be shown that at the end of the pulse sequence described above there is no observable component of the proton magnetization M_H. For this we need only consider the pulses in the ^{1}H channel, since those in the ^{13}C channel have no effect on M_H. The first $90^{\circ}_{x'}$ pulse turns M_H into the direction of the y'-axis in the coordinate system that is rotating with the frequency ν_H. The $180^{\circ}_{x'}$ pulse then reflects M_H in the x,z plane so that it is along the $(-y')$ direction. The next two pulses, the $90^{\circ}_{y'}$ and the $180^{\circ}_{y'}$, leave M_H unchanged, then the second $90^{\circ}_{x'}$ pulse turns M_H into the $(+z)$ direction. After the interval τ the next $180^{\circ}_{x'}$ pulse then turns it into the $(-z)$ direction. In the ideal case, therefore, there is no transverse magnetization component that could be detected. Thus, our discussion of the effects of the pulse sequence can now concentrate solely on the vectors $M_C^{H\alpha}$ and $M_C^{H\beta}$.

The first part of the pulse sequence, up to the start of the first $90^{\circ}_{x'}$ pulse in the ^{13}C channel, is identical to that in the normal INEPT experiment. Therefore, immediately before this pulse the two vectors $M_C^{H\alpha}$ and $M_C^{H\beta}$ are in antiphase along the $(+z)$ and $(-z)$ directions, and after the $90^{\circ}_{x'}$ pulse they are still in antiphase, but now along the $(+y')$ and $(-y')$ directions (see Section 8.5.2, Fig. 8-11, vector diagrams a–h). During the variable time interval t_1 the two M_C vectors develop in the same way as already described in Section 9.4.1 for the M_C vectors, except that here the two vectors are antiparallel. The $180^{\circ}_{y'}$ pulse in the ^{1}H channel ensures that the vectors $M_C^{H\alpha}$ and $M_C^{H\beta}$, which are rotating with different frequencies, are again exactly antiparallel at the end of the time t_1. The condition reached by the spin system, in particular the positions of the two antiparallel vectors $M_C^{H\alpha}$ and $M_C^{H\beta}$ in the x',y' plane of the rotating frame, depends on the time t_1 and the Larmor frequencies of the ^{13}C nuclei. The second $90^{\circ}_{x'}$ pulse in the ^{13}C channel at the end of the time t_1 turns the y' components of the two ^{13}C vectors (which depend on t_1) into the z direction. This alters the level populations, and also therefore the vectors $M_H^{C\alpha}$ and $M_H^{C\beta}$. Thus, polarization has been transferred back from ^{13}C to ^{1}H, and this part of the pulse sequence therefore corresponds to a *reverse* INEPT operation. The two M_H vectors $M_H^{C\alpha}$ and $M_H^{C\beta}$

are, like the M_C vectors, in antiphase; also, even more importantly, they are modulated by the ^{13}C resonances. The $90^\circ_{x'}$ signal generating pulse in the ^1H channel tips $M_H^{C\alpha}$ and $M_H^{C\beta}$ into the $(+y')$ and $(-y')$ directions, respectively, and they will become refocussed after the interval 2τ, provided that τ is exactly equal to $[4J]^{-1}$. The simultaneous 180° pulses in the ^1H and ^{13}C channels after the time τ serve to cancel frequency-dependent phase shifts and any loss of signal intensity due to field inhomogeneities. Since $M_H^{C\alpha}$ and $M_H^{C\beta}$ are in phase at the time of signal acquisition, one can switch on the broad-band decoupler during the acquisition and thus eliminate the C,H coupling.

In practice, as in all two-dimensional techniques, one performs n individual measurements with different values of t_1, which is altered by constant increments of a few microseconds, starting from zero and continuing up to several milliseconds. Fourier transformation of each of the FIDs with respect to t_2 gives the frequency spectrum F_2 containing the information about ^1H chemical shifts and coupling constants. A second Fourier transformation with respect to t_1 then yields the two-dimensional spectrum with correlation peaks at the positions (ν_C, ν_H). Figure 9-23 shows the two-dimensional (H,C)-correlated spectrum of glutamic acid (**2**), recorded by this method. The region of the spectrum shown here is the same as in Figure 9-16, but with the axes interchanged: the 500 MHz ^1H NMR spectrum appears along the top edge and the 125 MHz ^{13}C NMR spectrum at the left-hand edge. (In heteronuclear experiments it is usual to show the spectrum of the directly observed nuclide at the top.) The only resonances not included in this region are those of the quaternary carbon nuclei belonging to the two carboxyl groups ($\delta = 173.7$ and 177.0; see the complete spectrum in Figure 9-25). Both spectra yield the same information, namely the correlations between directly bonded carbon and hydrogen atoms. However, the reverse experiment has the great advantage that it can be performed in a much shorter time.

The HSQC experiment suffers from the considerable disadvantage that it requires pulses of precisely 90° and 180°, and is very sensitive to small errors in these settings. This problem is largely avoided in the HMQC (heteronuclear multiple quantum coherence) method devised by Bax et al. [11,12]. As in most experiments that use complex pulse sequences, the mechanism cannot be explained simply by vector diagrams, and therefore the pulse sequence is not given here. The important feature of the method is that the unwanted main peaks arising from protons attached to ^{12}C atoms are suppressed by suitable phase cycles. Figure 9-24 shows a spectrum of the neuraminic acid derivative **1** recorded by this technique. The 500 MHz ^1H NMR spectrum appears along the upper edge, while at the left-hand edge is a portion of the 125 MHz

^{13}C NMR spectrum; the signals of the three quaternary carbon nuclei present in the complete spectrum are not included here.

Some of the resonances in the ^{1}H and ^{13}C spectra can immediately be assigned with confidence on the basis of chemical shifts and multiplicities. For the ^{1}H spectrum these are the three methyl signals, together with H-3a, H-3e, and – with slightly more difficulty – H-4 and H-7. Those in the ^{13}C spectrum are the methyl signal of the *N*-acetyl group, and C-3, C-5 and C-9. The

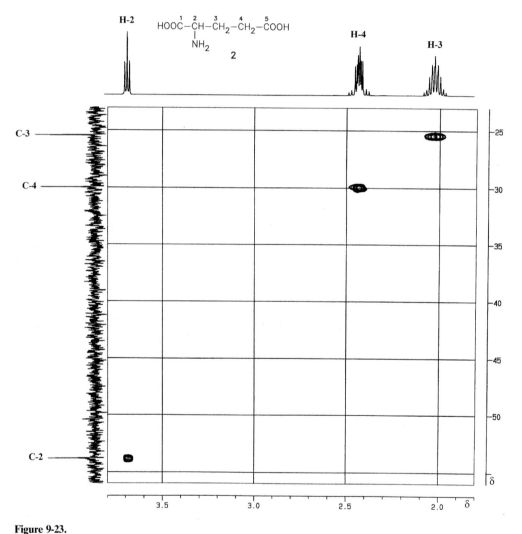

Figure 9-23.
Two-dimensional H,C-correlated spectrum of glutamic acid (**2**) in D$_2$O, recorded by the HSQC method with ^{1}H as the observed nuclide. The one-dimensional 500 MHz ^{1}H NMR spectrum is shown at the top edge and a portion of the 125 MHz ^{13}C NMR spectrum at the left-hand edge (only the resonances of the two quaternary ^{13}C nuclei C-1 and C-5 are missing).
(*Experimental conditions:*
approx. 20 mg in 0.5 ml D$_2$O; 5 mm sample tube; 128 measurements with t_1 altered in 44 μs increments; each measurement with 4 FIDs and 0.5 K data points; total time approx. 20 min.)

two OCH₃ groups can also be recognized from their chemical shifts, but they cannot be assigned unambiguously.

If we continue the analysis by starting from the ¹H resonances that have been definitely assigned, there is no diffculty in using the correlation peaks to identify the corresponding ¹³C resonances. In addition to the signals that were already assigned, this gives the signals of the two OCH₃ groups and of C-4 and C-7.

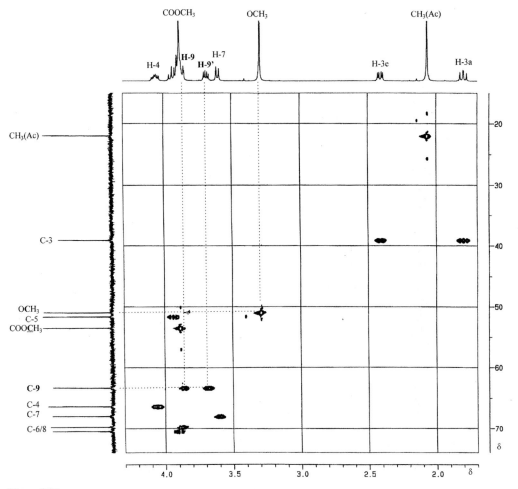

Figure 9-24.
Two-dimensional H,C-correlated spectrum of the neuraminic acid derivative **1**, recorded by the HMQC method with ¹H as the observed nuclide. The one-dimensional 500 MHz ¹H NMR spectrum is shown at the top edge and a portion of the 125 MHz ¹³C NMR spectrum at the left-hand edge (only the resonances of the quaternary carbon nuclei are missing). The signals that can be assigned with confidence are marked accordingly. The dashed construction lines show the analysis procedure for two examples. The new assignments thus obtained are indicated by bold type.
(*Experimental conditions:*
approx. 20 mg in 0.5 ml D₂O; 5 mm sample tube; 256 measurements with t_1 altered in 20 μs increments; each measurement with 8 FIDs and 1 K data points; total time approx. 80 min.)

Conversely, if we now begin with the ^{13}C signals that have been assigned, we can recognize, in addition to the ^1H resonances already assigned, those of H-5, H-9 and H-9'. C-3 and C-9 each give two correlation peaks, as these carbon nuclei are each bonded to two diastereotopic hydrogen atoms.

Examples:

○ In Figure 9-24 the process of analysis is shown for just two examples, since to show more would make the spectrum confusing. In the first example we start from the ^1H signal of the OCH$_3$ group and thereby find the corresponding ^{13}C signal for this group; in the second we start from the ^{13}C signal of C-9 and find the positions of the ^1H signals of the two methylene protons on C-9.

With these newly assigned signals the analysis of the ^1H and ^{13}C NMR spectra is almost complete. The only assignments still undecided are those for H-6, H-8, C-6 and C-8. Even the two-dimensional H,C-correlated NMR spectrum gives no information on this point, as the positions of the resonances for these nuclei are not sufficiently different in either the ^1H or the ^{13}C spectrum. In practice one would normally regard the problem as now being solved, since nobody would base a structural identification on differences as small as these. The results are summarized in Table 9-2.

An even more elegant and efficient method is the gradient-selected HMQC technique (gs-HMQC), in which field gradients are used to select the required coherence paths. This means that the 2D spectrum consists only of the signals of interest, with no peaks from protons attached to ^{12}C atoms. Thus the phase cycles used to suppress the main signals (from ^1H attached to ^{12}C) in the normal HMQC experiment are no longer necessary, so that less time is needed to complete the measurement than in an HSQC or normal HMQC experiment. The spectrum of glutamic acid (**2**) shown in Figure 9-25 A was obtained by the gs-HMQC method in a total time of only five minutes. The detailed action of the sequence of radiofrequency and gradient pulses in this experiment could not be adequately explained here without getting more deeply involved in the theory of spin vector interactions.

9.4.4 The Gradient-Selected (gs-)HMBC Experiment

In the C,H- and H,C-correlated spectra described in Sections 9.4.1 and 9.4.3 there are no correlation peaks between protons and quaternary carbon atoms (C$_q$), since these methods are designed to detect only the larger couplings $^1J(C,H)$. Thus, for

Table 9-2.

Summary of the procedure followed in analyzing the two-dimensional (H,C)-correlated NMR spectrum of **1**.

starting point	newly assigned
H-4	C-4
H-7	C-7
OCH$_3$ (ketoside)	OCH$_3$
OCH$_3$ (ester)	OCH$_3$
C-5	H-5
C-9	H-9
C-9	H-9'

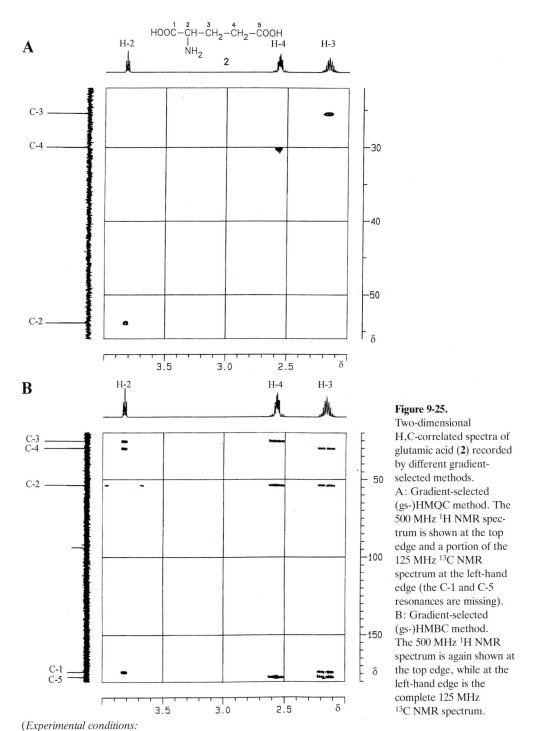

Figure 9-25.
Two-dimensional H,C-correlated spectra of glutamic acid (**2**) recorded by different gradient-selected methods.
A: Gradient-selected (gs-)HMQC method. The 500 MHz ^{1}H NMR spectrum is shown at the top edge and a portion of the 125 MHz ^{13}C NMR spectrum at the left-hand edge (the C-1 and C-5 resonances are missing).
B: Gradient-selected (gs-)HMBC method. The 500 MHz ^{1}H NMR spectrum is again shown at the top edge, while at the left-hand edge is the complete 125 MHz ^{13}C NMR spectrum.

x

(*Experimental conditions:*
approx. 20 mg in 0.5 ml D_2O; 5 mm sample tube; for A: 128 measurements with t_1 altered in 47 μs increments; each measurement with 2 FIDs and 1 K data points; total time approx. 10 min; for B: 256 measurements with t_1 altered in 23 μs increments; each measurement with 8 FIDs and 2 K data points; total time approx. 70 min.)

x

269

example, the spectrum of glutamic acid does not enable one to assign the ^{13}C NMR signals of the two carboxyl groups at $\delta = 173.7$ and 177.0.

Bax and Summers [13] suggested an experiment that would enable one to make assignments in cases where ^{13}C and ^1H nuclei are coupled through two or more bonds and thus have smaller *J*-values. This is called the HMBC (heteronuclear multiple bond correlation) method. It will be described here by again taking glutamic acid (2) as our example, then we will also apply it to our other model compound, the neuraminic acid derivative 1.

The spectrum shown in Figure 9-25 B was recorded by the gradient-selected (gs-)HMBC method without $^{13}C\{^1H\}$ decoupling, which is a further variant of the original gs-HMBC experiment. The one-dimensional 500 MHz ^1H NMR spectrum is shown at the upper edge, and the complete 125 MHz ^{13}C NMR spectrum ($\delta = 30$ to 180) is at the left-hand edge.

To analyze this spectrum we start from our existing knowledge of the unambiguously assigned ^{13}C and ^1H signals of C-2, C-3 and C-4 and their directly attached protons (see Figure 9-25 A). Evidently the ^{13}C signals at $\delta = 173.7$ and $\delta = 177.0$ (Figure 9-25 B, left-hand edge) belong to C-1 and C-5, but to decide which signal belongs to which of these two carbon nuclei we need to use the information given by the HMBC experiment, as explained below.

- We start with the H-2 signal, and look for correlation peaks on a vertical line at $\delta = 3.82$ (lower scale in Figure 9-25 B). As expected, we find peaks for C-3 and C-4, attributed to geminal and vicinal C,H couplings respectively, with an order of magnitude in the region of 10 Hz. There is a further correlation peak at $\delta = 173.7$, which must correspond to the quaternary carbon nucleus of one of the carboxyl groups. We expect H-2 to show an appreciable coupling [$^2J(C,H)$] to C-1, but not one to C-5, which is five bonds distant. Therefore the signal at $\delta = 173.7$ can be assigned unambiguously to C-1 (see sketch).

- If we now start from the H-3 signal at $\delta = 2.16$, we find correlation peaks for C-4, C-2 and the carbon nuclei of both carboxyl groups (see sketch).

- Lastly, in the case of H-4 at $\delta = 2.55$ we find correlation peaks for C-3, C-2 and the carbon nucleus with $\delta = 177.0$, and therefore we can assign the latter to the other carboxyl group ^{13}C nucleus C-5 (see sketch).

It can be seen that H-2 has two further correlation peaks which correspond to C-2. Peaks such as these, for directly bonded hydrogen and carbon nuclei, should in general be suppressed, but that does not always occur for all the protons in the molecule, and in this case the peaks for H-2 have not been suppressed. In such cases one obtains a doublet in the 2D spectrum, corresponding to the positions of the ^{13}C satellites in the ^1H NMR spectrum, with a separation $^1J(C,H) \approx 130$ to 150 Hz. Because of this doublet structure one can very easily distinguish between such one-bond correlations and those for two or three bonds. This is also the reason for performing the experiment without $^{13}C\{^1H\}$ decoupling.

Figure 9-26 shows the spectrum of our model compound, the neuraminic acid derivative **1**, recorded by the same method. As in Figure 9-25, the one-dimensional ^1H and ^{13}C NMR spectra are shown at the upper and left edges respectively. In both these spectra the assignments obtained from the DEPT, H,H-COSY and H,C-COSY spectra are indicated. In the case of the ^1H NMR spectrum, however, the information is incomplete because of overlapping between the multiplets for the protons H-6, H-8 and H-9 and the signal of the OCH$_3$ group, and therefore these cannot be used as starting-points for our analysis of the 2D spectrum. The only remaining uncertainties in the ^{13}C assignments are between those of C-6 and C-8, and between those of the two quaternary carbon nuclei, namely C-1 and the C=O carbon of the acetyl group.

Here, rather than working through the complete analysis of the spectrum, we will look at a few key steps as examples to show how one sets about such an analysis, including firming up the assignments of the ^{13}C resonances mentioned above. The sketches in the margin should help to explain the procedure. (Additional details can be found in Table 9-3).

• The proton H-3a ($\delta = 1.8$, see lower scale) gives correlation peaks for the carbon nuclei at the 5-, 4- and 2-positions and the one with $\delta(^{13}C) = 171.5$ (right-hand scale). Thus, the latter nucleus can only be C-1 (see sketch).
• For H-3e ($\delta = 2.5$) we find correlation peaks for C-5, C-4 and C-2, but not for C-1! H-3e and C-1 are a vicinal pair, and therefore the coupling between them depends critically on the steric configuration, as explained in Chapter 3. Evidently $^3J(H$-$3e,C$-$1)$ happens to be practically zero in this case.
• H-9′ has correlation peaks for C-7 and for the signal at $\delta(^{13}C) = 70.98$, which we have up to now only been able to assign to either C-8 or C-6 (see Figure 9-26). The correlation with H-9′ means that we can now assign it unambiguously to C-8. Therefore the signal at $\delta(^{13}C) = 71.7$ must belong to C-6 (see sketch).

None of the techniques that we looked at earlier yielded correlation peaks relating the protons and ^{13}C nuclei of an OCH$_3$ group to other nuclei linked via the oxygen atom, as the couplings were too small. This restriction does not apply to the HMBC method, as we shall see below for **1**.

Table 9-3.
Summary of the procedure followed in analyzing the two-dimensional (gs-)HMBC spectrum of **1**.

starting point	^{13}C NMR signals assigned	$\delta(^{13}C)$ [ppm]
H-3a	C-1	171.50
	C-2	100.32
	C-4	67.51
	C-5	52.83
H-4	C-3	40.31
	C-5	52.83
H-5	C-6	71.67
H-7	C-5	52.83
	C-8	70.98
	C-9	64.50
H-9′	C-7	69.18
	C-8	70.98
OCH$_3$ (ketoside)	C-2	52.12
OCH$_3$ (ester)	C-1	54.65
CH$_3$ (Ac)	CH$_3$ (d)	23.20
	C=O (ac)	175.93

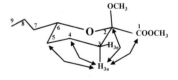

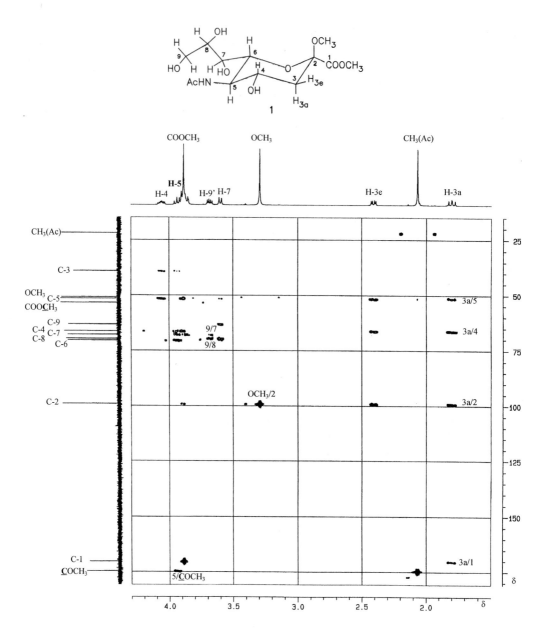

Figure 9-26.

Two-dimensional H,C-correlated spectrum of the neuraminic acid derivative **1**, recorded by the gradient-selected (gs-)HMBC method. The 500 MHz ^{1}H NMR spectrum is shown at the top edge and the 125 MHz ^{13}C NMR spectrum at the left-hand edge. The assignments already known are shown on each spectrum, and in addition the assignments newly arrived at by this method are shown in bold type. The analysis is carried out by considering vertical alignments starting from known ^{1}H resonances and horizontal alignments starting from known ^{13}C resonances (see text). The numbers beside the correlation peaks indicate the hydrogen and carbon nuclei to which they correspond, in that order, separated by a slash.

(*Experimental conditions:*

approx. 20 mg in 0.5 ml D$_2$O; 5 mm sample tube; 256 measurements with t_1 altered in 23 μs increments; each measurement with 16 FIDs and 1 K data points; total time approx. 2 h.)

- The protons of the ketoside OCH$_3$ group ($\delta = 3.28$) show a strong correlation peak for C-2, and those of the COOCH$_3$ (ester) group ($\delta = 3.87$) show correlation peaks for both C-1 and C-2 (see sketch).
- If we look along a horizontal line from the signal of the quaternary carbon nucleus of the NCOCH$_3$ group at $\delta = 175.93$, we find correlation peaks for H-5 ($\delta = 3.92$) and the protons of the methyl group at $\delta = 2.05$ (see sketch).
- Starting now from the C-6 signal, we find correlation peaks in the region $\delta = 3.90$ to 3.97; these belong to H-5 and H-8 (see sketch).

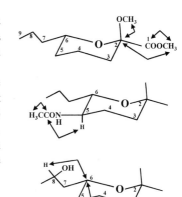

In the examples that we have looked at we sometimes chose a proton resonance as a starting point for the analysis and sometimes a ^{13}C resonance. In practice the choice depends very much on which signals have already been unambiguously assigned.

9.4.5 The TOCSY Experiment

In Section 8.7 we learned about the selective TOCSY or HOHAHA experiment. By selectively exciting the resonance of a single proton in the molecule we were able to identify the resonances of all other protons belonging to the same coupled spin system. The development of that one-dimensional method actually came later than the introduction by Braunschweiler and Ernst [15] of the two-dimensional TOCSY experiment, known simply as TOCSY.

The TOCSY pulse sequence is similar to that of the normal H,H-COSY experiment, except that here the second 90° pulse is replaced by a spin-lock stage. The sequence is:

$$90° - t_1 - \text{spin-lock} - \text{FID}$$

As in the COSY experiment, one performs a series of n measurements with increasing values of t_1. With regard to the exact nature of the spin-lock stage, the very much oversimplified explanation given in Section 8.7 will have to suffice for our present purpose.

Figure 9-27 shows the 500 MHz TOCSY spectrum of the cellobiose derivative **3**, which we met earlier in connection with the selective TOCSY experiment discussed in Section 8.7. The one-dimensional spectrum of the ring protons is shown at the top and at the left-hand edge, with assignments as indicated. As in the COSY spectrum, the diagonal peaks correspond to those of the normal 1D spectrum, whereas the off-diagonal peaks show the correlations between protons that belong to the same coupled spin system. For example, if we look along the horizontal line through the H-1' resonance at $\delta = 4.64$ (right-hand scale), we find six correlation peaks in addition to the diagonal peak. These show correlations between H-1' and all six protons of the glucose unit I, namely H-2' ($\delta = 3.07$), H-3' ($\delta = 3.60$), H-4' ($\delta = 3.45$), H-5' ($\delta = 3.58$), H-6'a ($\delta = 3.92$) and H-6'b ($\delta = 3.77$), these δ-values being read off on the lower scale.

If we now look along the horizontal line corresponding to H-1 ($\delta = 4.78$) in the glucose unit II, we find correlation peaks for H-2 to H-5, and a very weak one for H-6b. No correlation peak for H-6a is detectable under the conditions chosen for the measurement. The "range" within the molecule for detecting a correlation is critically dependent on the length of the mixing period during which the spin-lock is applied. By adjusting this time (typically in the range from 10 to several hundred milliseconds) one can choose between a short-range spectrum, which corresponds to the normal COSY spectrum, or a "total correlation" spectrum.

We can go on to deduce other unambiguous correlations from Figure 9-27, and from all these accumulated pieces of information we can identify all the resonances belonging to the two glucose units. For a given problem one must decide whether the two-dimensional TOCSY method is the best approach or whether the required information can be obtained more quickly and simply from the one-dimensional selective TOCSY method. The decision will depend on the nature of the problem and of the compound being investigated. In the above example the selective TOCSY spectra shown in Figure 8-25 are undoubtedly easier to interpret, quickly yielding the necessary information. However, it must be mentioned again here that TOCSY spectra do not yield assignments just by direct inspection.

As well as being very useful for determining the structures of oligosaccharides, TOCSY experiments have many applications in the field of peptides, since here the correlation peaks enable one to identify the protons belonging to the individual amino acids (see Section 13.3).

There are many variants of the two-dimensional TOCSY experiment described above. One that deserves a special mention is the *gradient-selected TOCSY* experiment; provided that the spectrometer incorporates the necessary facilities, this makes it possible to considerably reduce the time needed for a measurement.

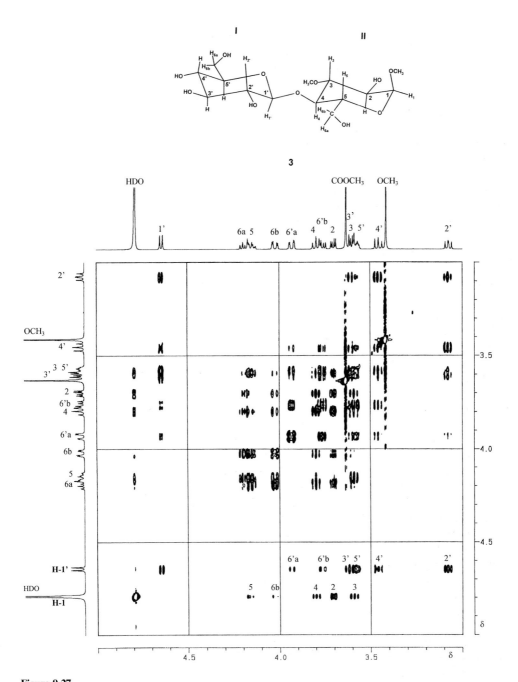

Figure 9-27.

Two-dimensional TOCSY spectrum of α-methyl-3-O-methylcellobioside (**3**) in D_2O. The 500 MHz ^1H NMR spectrum is shown at the top and left-hand edges with assignments. The H-1 resonance is hidden under the HDO signal of the solvent. Only the cross-peaks on the horizontal lines starting from H-1 and H-1' are labeled; on each line these indicate protons belonging to a common coupled spin system (rings I and II, see formula).

(Experimental conditions:

approx. 20 mg in 0.5 ml D_2O; 5 mm sample tube; 256 measurements with t_1 altered in 400 μs increments; each measurement with 8 FIDs and 1 K data points; total time approx. 1.5 h.)

275

9.4.6 Two-Dimensional Exchange NMR Spectroscopy: The Experiments NOESY [16], ROESY [17, 18] and EXSY [19, 20]

In all the two-dimensional NMR techniques discussed up to now the transfer of magnetization takes place between nuclei that have a scalar coupling. In addition there are two further mechanisms whereby magnetization can be transferred:

- Dipole-dipole interactions through space; this effect is familiar to us as the nuclear Overhauser effect (NOE, Chapter 10). The corresponding experiments are called NOESY [**N**uclear **O**verhauser **E**nhancement (or **E**ffect) **S**pectroscop**Y**] and ROESY [**R**otating frame **O**verhauser **E**nhancement (or **E**ffect) **S**pectroscop**Y**].
- Chemical exchange processes (Chapter 11), for example:

$$A \xrightarrow{\quad k \quad} X.$$

In this case, if the A-nuclei are polarized, there is a transfer of polarization from A to X with the rate constant k. The corresponding experiment is called EXSY (**EX**change **S**pectroscop**Y**).

Of these procedures the NOESY and ROESY experiments are of greatest importance, and we shall therefore look at it in some detail. However, we will consider here only the homonuclear case (i.e. where A and X are both protons).

The pulse sequence for NOESY is shown in Figure 9-28 A. In addition to the r.f. pulses there are two field gradient pulses G_1 and G_2, the importance of which will be explained later.

The behavior of the magnetization vectors is analyzed in Figure 9-28 B, in which the vector diagrams describe the situations at the six instants a to f. It is assumed that the two protons A and X belonging to the same molecule do not have a scalar coupling, but are close neighbors. Initially the macroscopic magnetization vectors M_A and M_X lie along the z-direction (diagram a), which is also the direction of the applied field B_0. The $90^\circ_{x'}$ pulse turns these vectors into the direction of the y'-axis (b). During the interval t_1 (which is changed by increments as in the two-dimensional experiments described earlier), M_A and M_X precess about the z-axis, and move apart due to their different Larmor frequencies (c). The second $90^\circ_{x'}$ pulse turns both the y'-components existing at the instant c until they lie along the z-axis (in either the positive or the negative direction). In the vector diagram shown here it is assumed that both components are turned into the positive z-direction, but have different magnitudes. In the subsequent interval Δ (the mixing time) the

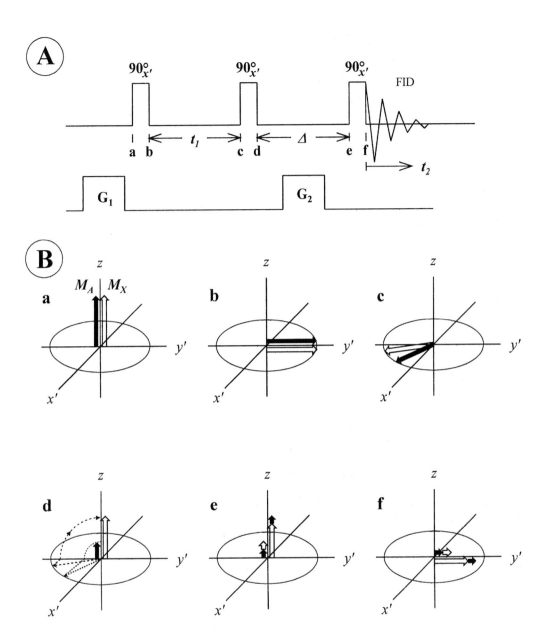

Figure 9-28.

The NOESY experiment.

A: pulse sequence.

B: vector diagrams showing the behavior of the ^{1}H magnetization vectors M_A and M_X. The protons A and X are assumed to be in the same molecule, spatially close together but with no scalar coupling between them. In the 2D experiment t_1 is altered by successive increments. The mixing time Δ is several seconds. The field gradient pulses G_1 and G_2 are identical in magnitude, direction and duration. In the vector diagrams e and f the small contrasting arrows superimposed at the ends of the vectors represent the portion of the magnetization that is transferred from the nuclei of the other sort by cross-polarization during the mixing time Δ.

system undergoes relaxation, and at the same time some polarization is transferred from M_A to M_X and vice-versa, a process known as cross-relaxation or cross-polarization. In diagram e (Figure 9-28 B) the two small superimposed contrasting arrows represent schematically the portion of the polarization that has been transferred. This produces the desired polarization for A and X, since M_A now contains a small proportion of the original M_X and vice-versa. The final $90°_{x'}$ pulse turns the resulting magnetization vectors into the y'-direction (diagram f), so that the FID can be recorded. The Fourier transformations with regard to t_2 and t_1 yield the two-dimensional spectrum. This shows signals at (ν_A, ν_A) and (ν_X, ν_X) (the diagonal peaks) and, more importantly for our purpose, at (ν_A, ν_X) and (ν_X, ν_A). These *correlation peaks* indicate the spatial proximity of the nuclei A and X because, as we shall see from the detailed treatment in Chapter 10, the nuclear Overhauser effect depends on dipole–dipole interactions, which fall off with distance in inverse proportion to r^6. The diagonal peaks at (ν_A, ν_A) and (ν_X, ν_X) are of no interest to us. On the contrary, they are a nuisance, as they are stronger than the correlation peaks. However, they can be considerably reduced by suitably chosen phase cycles. For the system discussed here we can use the phase cycle described earlier in Section 8.5.3, whereby the phase of the first 90° pulse is made to alternate, so that it acts first along the x'-direction then along the $(-x')$-direction.

It now remains to consider the effect of the two field gradient pulses G_1 and G_2. As already mentioned, the NOESY experiment consists of a series of individual measurements with increasing values of t_1. Sufficient time (approximately $5T_1$) must be allowed between these individual measurements for relaxation of the spin system, which means that one must insert a long delay time, thereby increasing the time needed for the entire experiment. If the next measurement is started too early, some transverse magnetization will remain from the previous one, causing artefacts in the two-dimensional spectrum. The effect of the field gradient pulse G_1 is to disperse this residual magnetization, as described in Section 8.2.3.

The purpose of the second field gradient pulse G_2 is similar to that of the first. In the NOESY experiment described here it can be seen (Figure 9-28 B, diagram e) that during the mixing period Δ we are only concerned with longitudinal magnetization components M_z. These are then transformed into detectable (transverse) magnetization by the third $90°_{x'}$ pulse. Any transverse magnetization still present during the mixing period can cause artefacts and undesirable additional contributions to the correlation peaks. The details of how this can occur need not be considered here; it is enough to add that the field gradient pulse G_2 disperses the unwanted transverse magnetization. To carry out the experiment successfully it is essential to choose an appropriate value for the mixing time Δ. For the NOESY experiment it should be of the same order as the spin–lattice relaxation time T_1.

Because through-space (dipole–dipole) interactions between protons are strongly dependent on the internuclear distance, the ability to detect them by the presence of correlation peaks is of great practical importance, as it gives valuable information about the stereochemistry of the molecule. In most cases this is of a qualitative nature – to determine internuclear distances accurately is very difficult. The method has many important applications, especially when used in combination with other two-dimensional or multidimensional techniques. One such area of application is in determining the structures of peptides, proteins and oligosaccharides. We will meet an example in Chapter 10 on the nuclear Overhauser effect (see Figure 10-6 and in Section 13.3). The disadvantage of the NOESY experiment is that the relative signs of the amplitudes of the cross peaks depend on the rotational correlation times τ_C. Thus, with large values of τ_C, which are typically found for large molecules or for highly viscous liquids, the diagonal peaks and cross peaks have the same (positive) sign, whereas with short correlation times τ_C, which apply for smaller molecules and low-viscosity solutions, their signs are opposite. Between these extremes there exists a critical correlation time τ_C^{crit}, for which the NOE effect is zero, causing the cross peaks to disappear. This is most likely to occur for molecules with a molar mass in the region 1000 to 3000 Da.

That problem is overcome in the ROESY experiment, for which one uses the pulse sequence:

$$90^0_{\phi 1} - t_1 - \text{spin lock (phase } \phi_2) - \text{FID}$$

(See also Figure 10-3). As an example, we again consider an NOE experiment on a system of two spins A and X, which are not scalar-coupled. If we assume that the 90^0 pulse is applied in the direction of the x'-axis ($90^0_{x'}$ pulse), it generates transverse magnetization vectors M_A and M_X along the direction of the y'-axis. During the time t_1 these transverse magnetization vectors develop in the x,y plane in accordance with their different Larmor frequencies. Following this period, a strong constant r.f. field is applied to the spin system for a defined time τ_m. The strength of this spin-lock field is sufficient to ensure that it contains both the Larmor frequencies ν_A and ν_X, and is therefore effectively in resonance with both M_A and M_X. The phase of the r.f. field is chosen so that it is shifted by exactly 90^0 from that of the $90^0_{x'}$ pulse, which in our example is along the direction of the y' axis in the rotating frame. The result of this is to fix the components of M_A and M_X that are along the y' direction, an effect known as "spin-locking". Under ideal conditions these components remain locked in direction throughout the period τ_m, although decaying with the transverse relaxation time constant T_2.

The magnetization components $M_A(y')$ and $M_X(y')$ have the same frequency in the rotating frame, and therefore also the same transition energy. Therefore they are capable of exchanging energy by cross-relaxation, like two identical tuning-forks with the same vibrational frequency. However, in contrast to the NOE effect, where cross-polarization occurs between longitudinal magnetization components, here it is between transverse magnetization components.

The experiment is performed by incrementing t_1, and the ROESY spectrum is obtained by Fourier transformation with respect to t_1 and t_2. The resulting spectrum differs from the NOESY spectrum in only one respect, namely that the diagonal peaks and the cross peaks always have opposite sign. The relative signs no longer depend on τ_C, and the form of the spectrum is independent of the spectrometer frequency.

It can be seen that the ROESY pulse sequence is identical in principle to that for the TOCSY experiment. The only difference is in the strength of the r. f. field used for the spin lock, which is here only about one-fifth of that used in the TOCSY experiment. Consequently, no TOCSY signals should appear in the ROESY spectrum. However, in cases where they do appear, they are of opposite sign to the ROESY cross peaks and are therefore easy to recognize.

In the EXSY experiment one uses the same pulse sequence as in NOESY. However, whereas in the NOESY experiment the exchange of polarization between A and X during the mixing period Δ occurs by cross-polarization, in the EXSY experiment it occurs through chemical exchange. The efficiency of the polarization transfer depends on the rate constant k for the chemical exchange process. To obtain useful results the mixing time Δ must be of the same order as k^{-1}. We will meet an example of an EXSY spectrum in Chapter 11 (Section 11.3.7, Figure 11-7).

With this method too one is usually content to obtain a qualitative result, in this case evidence for a chemical exchange. It is scarcely practicable to measure rate constants, as the relationships involved are too complicated. The difficulties include the following: firstly, during the evolution phase not only the A-spins but also the X-spins undergo development, secondly the spin system undergoes further relaxation during the mixing phase Δ, and thirdly only that fraction of the polarization which is transferred from A to X during the time Δ contributes to the modulation of the X-signals.

9.5 The Two-Dimensional INADEQUATE Experiment

[21–23]

The one-dimensional INADEQUATE technique was treated in detail in Section 8.8. At this point it would be useful to re-read that section, since everything which appears there relating to the basic approach, theoretical background, experimental procedure, and results obtained, applies here also. The difference between the one- and two-dimensional INADEQUATE pulse sequences is that in the two-dimensional version the short switching time Δ is replaced by a variable interval t_1:

$$1\,D: 90^\circ_{x'} - \tau - 180^\circ_{y'} - \tau - 90^\circ_{x'} - \Delta - 90^\circ_{\Phi'} - \text{FID}\ (t_2)$$
$$2\,D: 90^\circ_{x'} - \tau - 180^\circ_{y'} - \tau - 90^\circ_{x'} - t_1 - 90^\circ_{\Phi'} - \text{FID}\ (t_2)$$

The first part of the 2D pulse sequence (up to and including the second $90^\circ_{x'}$ pulse) again serves to establish the double quantum coherence. During the time t_1 this double quantum coherence evolves, and is then converted by the $90^\circ_{\Phi'}$ pulse into single quantum transitions which can be observed, and which correspond to the ^{13}C satellites. If the value of τ is correctly chosen, the Fourier transformation of the FID with respect to t_2 gives, as in the one-dimensional experiment, the satellite spectrum; in other words, for each pair of directly adjacent ^{13}C nuclei A and X one obtains two doublets. The central signals, which would correspond to the isolated ^{13}C nuclei, are suppressed. The mathematical analysis shows that the satellite signals, as functions of t_1, are modulated by the sum of the resonance frequencies of the two coupled ^{13}C nuclei $(\nu_A + \nu_X)$. (This situation cannot be shown diagrammatically.)

The two-dimensional experiment consists of n measurements with different t_1-values, the increment between successive measurements being a few µs. The second Fourier transformation of all the F_2 spectra with respect to t_1 gives a two-dimensional spectrum in which the F_1-coordinates of the signals correspond to the double quantum frequencies $(\nu_A + \nu_X)$, and the F_2-coordinates are the frequencies ν_A and ν_X. However, owing to the nature of the experiment the frequencies along the F_1-axis do not give $(\nu_A + \nu_X)$ directly; instead they give a frequency $\nu_{meas} = \nu_A + \nu_X - 2\nu_1$, where ν_1 is the frequency of the pulsed r.f. source. The ν_{meas}-values are in a frequency range extending up to a few kHz. Since these double quantum frequencies are of no interest to us, we will not consider them further. The important point is that at *one* particular F_1-value we find the satellite spectra of *both* the coupled ^{13}C nuclei, i. e. both doublets appear at the same F_1-value. Usually the actual value of the coupling constant $J(C,C)$ is of no importance here; the object of the experiment is to determine the connectivities in the molecule, i. e. to identify in the ^{13}C spectrum the signals of nuclei that are coupled and to assign them. The following example clarifies this better than a mere description.

Figure 9-29 shows the two-dimensional 100.6 MHz ^{13}C INADEQUATE spectrum of the neuraminic acid derivative **1**, in the range from $\delta = 10$ to 110. The spectrum at the top edge of the figure corresponds to the projection of the two-dimensional spectrum onto the F_2-axis. Each carbon nucleus gives a doublet consisting of the two satellite signals. The main signals would be at the centers of these doublets.

○ We begin the analysis by starting from the signal at $\delta = 100.32$, which is assigned unambiguously to the quaternary carbon C-2, and is marked ② in Figure 9-29. At the same F_1-value we find another signal ③ with $\delta = 40.31$ (on the F_2-axis), and this indicates a coupling to C-3. In the spectrum ② and ③ are joined by a horizontal dashed line. The signal at $\delta = 40.31$ is therefore assigned to C-3. At this same δ-value (on the F_2-axis), but at a different F_1-frequency, we find another doublet, which leads us to the C-4 signal ④. We can now continue this procedure and thus assign all the signals unambiguously. Even the difficult question of C-6 and C-8 can now be resolved: the doublet for C-6 is easily found from its coupling to C-5 ($\delta = 53$), and that for C-8 from its coupling to C-9.

The interpretation can be even further simplified by separately recording the AB or AX satellite spectra for each of the relevant F_1-values, resulting in Figure 9-30. It can be seen that the resolution along the F_1-axis is not quite good enough to completely separate the satellites of the three pairs 2/3, 6/7 and 7/8. However, the 2/3 pair can easily be assigned straight away from the chemical shifts, and the remaining two pairs also can be readily identified from the difference in intensities between their two AX spectra.

An additional aid to assignment is the fact that the centers of all the pairs of doublets joined by horizontal lines in Figure 9-29 lie on a straight line.

Up to this point in our discussion of the two-dimensional INADEQUATE technique, the experimental details have scarcely been mentioned, as it was intended only to emphasize the types of information obtainable by this method. To conclude, however, it is necessary to mention, alongside the many advantages offered by this new technique, its major drawback. Since, in contrast to other types of two-dimensional experiments, there is no enhancement of signal intensities by polarization transfer, 2D-INADEQUATE measurements are extremely time-consuming. Even with the most sensitive spectrometers and high sample concentrations (200–500 mg in 2 ml of solvent!), and with optimally adjusted parameters (90° and 180° pulse angles and the correct value of τ), at least one over-

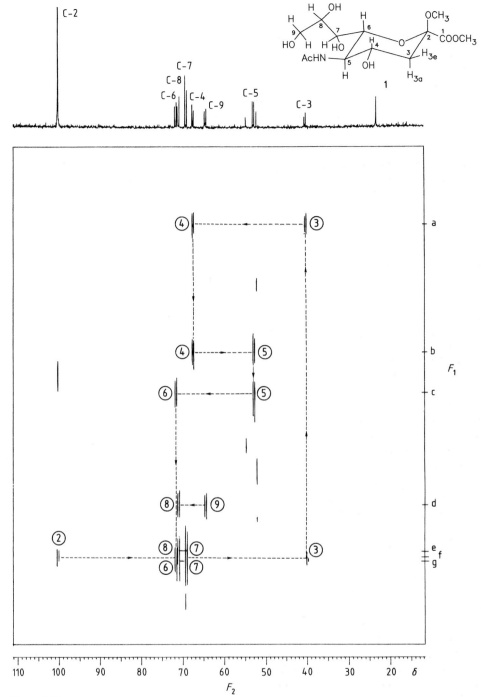

Figure 9-29.
A portion (δ = 10 to 110) of the two-dimensional 100.6 MHz ^{13}C INADEQUATE spectrum, with ^{1}H BB decoupling, of the neuraminic acid derivative **1**. The projection of the 2D spectrum onto the F_2-axis is shown at the top. The horizontal construction lines link the doublets of coupled pairs of ^{13}C nuclei. The identification numbers alongside the signals correspond to the numbering of the carbon atoms in **1**. Starting from C-2 and following the horizontal and vertical construction lines in the direction of the arrows leads to unambiguous assignments of all the ^{13}C resonances.

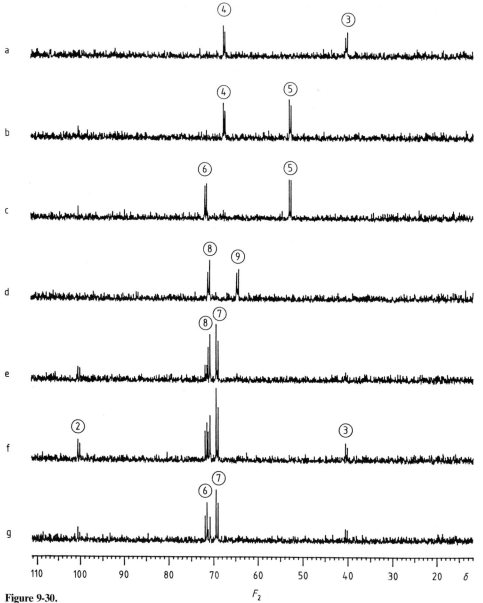

Figure 9-30.

Individual F_2-spectra for a series of different F_1-values from the 2D-INADEQUATE experiment on **1**. The traces a to g have been chosen so that in each case an AB- or AX-type satellite spectrum for a coupled pair of ^{13}C nuclei is visible. The F_1-values for spectra a to g are indicated on the axis in Figure 9-29. It is evident from traces e, f and g that the resolution along the F_1-axis is not quite adequate to completely separate the 2/3, 6/7 and 7/8 pairs.

(*Experimental conditions:*

167 mg of the compound in 2.3 ml D_2O; 10 mm sample tube; 128 measurements with t_1 altered in 50 μs increments; each measurement with 576 FIDs; total time 66 h.)

night run per measurement is needed. Nevertheless, in many cases the 2D-INADEQUATE spectrum makes it possible to assign all the signals from this one experiment, as our example illustrates.

Furthermore, developments to improve the method are still far from coming to an end.

9.6 Summary of Chapters 8 and 9

Table 9-4 summarizes the types of information that can be obtained using the various techniques described in Chapters 8 and 9. This makes it easier to compare the different methods.

Table 9-4.
Types of information obtained from different multiple pulse experiments.

Experiment	Nuclides observed	Types of information and applications
One-dimensional (1 D) techniques		
J-modulated spin-echo (attached proton test, APT)	^{13}C	CH and CH_3 carbon nuclei give positive signals, quaternary and CH_2 carbon nuclei give negative signals (an aid to assignment).
INEPT	$^1H/^{13}C$	The INEPT pulse sequence is used as a component of many two- and multidimensional experiments (example: HSQC).
DEPT	^{13}C	Tells how many hydrogen atoms are directly bonded to a carbon nucleus: CH, CH_2, CH_3. Disadvantage: no signals from quaternary carbon nuclei.
Selective TOCSY	^1H	Allows one to identify all the protons belonging to a common coupled spin system.
1D-INADEQUATE	^{13}C	C,C coupling constants.
Two-dimensional (2 D) techniques		
Heteronuclear J-resolved ^{13}C NMR spectroscopy	^{13}C	C,H coupling constants, number of directly bonded hydrogen atoms (as in DEPT).
Homonuclear J-resolved ^1H NMR spectroscopy	^1H	Determining δ-values in complicated spectra, identifying the peaks of a multiplet.
H,H-COSY	^1H	Assigning signals in complicated spectra.
Long-range COSY	^1H	Assigning signals of protons separated by four or more bonds where the couplings are small.
H,C-COSY	$^1H/^{13}C$	Assigning signals in the ^1H and ^{13}C spectra, starting from known signals.
Heteronuclear multiple bond correlations HMBC	$^1H/^{13}C$	Assigning ^1H and ^{13}C signals on the basis of $^2J(C,H)$- and $^{2+n}J(C,H)$-values.
2D-TOCSY	^1H	Allows one to identify all the protons belonging to a common coupled spin system.
NOESY, ROESY	^1H	Gives evidence for spatial proximity of nuclei.
EXSY	^1H	Qualitative evidence of exchange processes.
2D-INADEQUATE	^{13}C	Assigning signals by detecting couplings between adjacent ^{13}C nuclei.

9.7 Bibliography for Chapter 9

[1] R. R. Ernst, G. Bodenhausen and A. Wokaun: *Principles of Nuclear Magnetic Resonance in One and Two Dimensions*. Oxford: Claredon Press, 1986.

[2] L. Müller, A. Kumar and R. R. Ernst, *J. Chem. Phys. 63* (1975) 5490.

[3] G. Bodenhausen, R. Freeman and D. L. Turner, *J. Chem. Phys. 65* (1976) 839.

[4] W. P. Aue, J. Karhan and R. R. Ernst, *J. Chem. Phys. 64* (1976) 4226.

[5] A. A. Maudsley, L. Müller and R. R. Ernst, *J. Magn. Reson. 28* (1977) 463.

[6] G. Bodenhausen and R. Freeman, *J. Magn. Reson. 28* (1977) 471.

[7] R. Freeman and G. A. Morris, *J. Chem. Soc. Chem. Commun.* (1978) 684.

[8] W. P. Aue, E. Bartholdi and R. R. Ernst, *J. Chem. Phys. 64* (1975) 2229.

[9] A. Bax and R. Freeman, *J. Magn. Reson. 44* (1981) 542.

[10] G. Bodenhausen and D. J. Ruben, *Chem. Phys. Lett. 69* (1980) 185.

[11] L. Müller, *J. Amer. Chem. Soc. 101* (1979) 4481.

[12] A. Bax, R. H. Griffey and B. L. Hawkins, *J. Magn. Reson. 55* (1983) 301.

[13] A. Bax and M. F. Summers, *J. Amer. Chem. Soc. 108* (1986) 2093.

[14] W. Wilker, D. Leibfritz, R. Kerssebaum and W. Bermel, *Magn. Reson. Chem. 31* (1993) 287.

[15] L. Braunschweiler and R. R. Ernst, *J. Magn. Reson. 53* (1983) 521.

[16] J. Jeener, B. H. Meier, P. Bachmann and R. R. Ernst, *J. Chem. Phys. 71* (1979) 4546.

[17] A. A. Bothner-By, R. L. Stephens, J.-M. Lee, C. D. Warren, R. W. Jeanloz, *J. Amer. Chem. Soc. 106* (1984) 811.

[18] A. Bax, D. G. Davis, *J. Magn. Reson. 63* (1985) 207.

[19] R. Willem, *Prog. Nucl. Magn. Reson. Spectrosc. 20* (1987) 1.

[20] K. G. Orrell, V. Sik and D. Stephenson, *Prog. Nucl. Magn. Reson. Spectrosc. 22* (1990) 141.

[21] A. Bax, R. Freeman and S. P. Kempsell, *J. Amer. Chem. Soc. 102* (1980) 4849.

[22] A. Bax, R. Freeman, T. A. Frenkiel and M. H. Levitt, *J. Magn. Reson. 43* (1981) 478.

[23] A. Bax, R. Freeman and T. A. Frenkiel, *J. Amer. Chem. Soc. 103* (1981) 2102.

Additional and More Advanced Reading

S. Berger and S. Braun: *200 and More NMR Experiments, A Practical Course*. Weinheim: Wiley-VCH, 2004.

W. R. Croasmun and R. M. K. Carlson (Eds.): *Two-Dimensional NMR-Spectroscopy. Applications for Chemists and Biochemists*. New York: VCH Publishers, 1994, 2nd Edition.

A. Derome: *Modern NMR Techniques for Chemistry Research*. Oxford: Pergamon Press, 1987.

R. R. Ernst, G. Bodenhausen and W. Wokaun: *Principles of Nuclear Magnetic Resonance in One and Two Dimensions*. Oxford: Clarendon Press, 1986.

H. Kessler, M. Gehrke and C. Griesinger: Two-Dimensional NMR Spectroscopy: Background and Overview of the Experiments, *Angew. Chem. Int. Ed. Engl. 27* (1988) 490.

J. K. M. Sanders and B. K. Hunter: *Modern NMR Spectroscopy. A Guide for Chemists*. Oxford: Oxford University Press, 1987.

F. J. M. van de Ven: *Multidimensional NMR in Liquids. Basic Principles and Experimental Methods*. New York: VCH Publishers, 1995.

10 The Nuclear Overhauser Effect

10.1 Introduction

In connection with the use of ^1H decoupling techniques in ^{13}C NMR spectroscopy, reference has already been made several times to the nuclear Overhauser effect or the nuclear Overhauser enhancement (NOE). This can produce an increase of up to 200 % in the intensities of ^{13}C NMR signals when using ^1H broad-band (BB) decoupling to suppress C,H couplings. This useful side-effect of decoupling contributes appreciably to making possible the routine measurement of ^{13}C NMR spectra.

However, the observation of NOE is by no means confined to heteronuclear systems; indeed the results obtained from NOE measurements on homonuclear systems, especially in ^1H NMR spectroscopy, are very important in structure determination. Three examples will illustrate this.

○ Anet and Bourn in 1965 described the following decoupling experiment [1]: In the "semiclathrate" compound **1** the two protons H^A and H^B are very close together. If the resonance of H^A is saturated the intensity of the H^B signal increases by 45 %. (Here and in all subsequent examples the arrow indicates the effect of the saturated nucleus on the nucleus being observed.)

○ The second example concerns dimethylformamide (**2**), in which the two methyl groups are non-equivalent owing to hindered rotation about the C − N bond (Fig. 11-1, Chapter 11). Two methyl signals are therefore found at $\delta = 2.79$ and 2.94, together with a singlet at $\delta \approx 8.0$ for the formyl proton. However, it is not obvious which methyl signal belongs to which group. If one now saturates the methyl signal at $\delta = 2.94$, the intensity of the formyl proton signal increases by 18 %. When instead the other methyl signal is saturated, a decrease of 2 % is observed.

○ In an analogous experiment with 3-methylcrotonic acid (**3**), values very similar to those for dimethylformamide (**2**) are found.

What is the origin of this effect? What causes the changes in signal intensities under these experimental conditions? Can the effects be quantified? What applications are possible?

These questions will be answered in the following sections. We will also be returning to the two-dimensional experiment NOESY, which we met in Section 9.4.6.

10.2 Theoretical Background

The experimental results described can be explained by the theory of the Overhauser effect, originally developed to account for the mutual effects observed between the magnetic moments of nuclei and electrons. To understand the basic principle and to explain clearly the conditions under which the effect occurs, our derivation will initially be restricted to a one-dimensional NOE experiment for a system of two spins. We will then go on to consider a multi-spin system. This theoretical section will end with a brief look at the two-dimensional version of the experiment, highlighting the basic differences compared with the one-dimensional version.

10.2.1 The Two-Spin System

The NOE changes the intensities of the signals! This very briefly summarizes all the experimental results. Let us remind ourselves of what determines the intensity of an NMR signal. Section 1.4.1 showed that it is proportional to the difference in populations of the two energy levels between which the nuclear resonance transition occurs. To make this clear Figure 10-1 again shows the energy level scheme for a two-spin system AX; A and X may be protons – as in our examples – or they may be nuclei of different species. However, we assume that they do not have a scalar coupling, i.e. $J_{AX} = 0$. This simplifying assumption is not essential in principle, but it makes it easier to understand the effect.

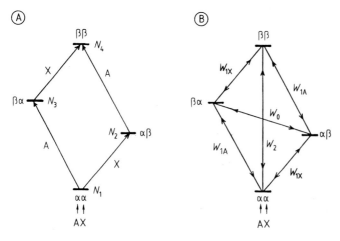

Figure 10-1.
Energy level schemes for a two-spin AX system with $J_{AX} = 0$.
A: The frequencies of the two A transitions are equal, as are those of the two X transitions. N_1 to N_4 are the populations of the energy levels.
B: The probabilities W for the six allowed relaxation transitions are shown. Those with a subscript 1 correspond to single quantum transitions, while W_0 corresponds to a zero quantum transition and W_2 to a double quantum transition.

The transitions between levels 1 and 3 and between levels 2 and 4 are transitions of the A-nucleus, while those between 1 and 2 and between 3 and 4 are those of the X-nucleus. These transitions are allowed and observable (Section 4.3). As we have assumed J_{AX} to be zero, the A- and X-nuclei each give a singlet in the NMR spectrum, the intensities of these being determined by the population differences between the relevant energy levels in the equilibrium situation.

The NOE experiment, which involves continuously saturating a transition of one nucleus (A, for example), evidently results in population ratios which no longer correspond to the equilibrium situation, since otherwise the signal intensities for the X-nucleus would not change. As with any kind of perturbation, the system tries to restore the equilibrium by spin-lattice relaxation. As will be explained in Section 7.2.1, the relaxation of the spin system occurs predominantly via a dipolar mechanism. According to the theory, the NOE and the dipole-dipole relaxation mechanism are intimately connected.

Let us now consider Figure 10-1 B. Instead of the NMR transitions between the energy levels as shown in A, this shows all the possible and theoretically allowed relaxation processes, with their transition probabilities W.

The four probabilities W_1 correspond to the single quantum transitions which constitute the spin-lattice relaxation processes, and which we have already met in connection with measurements of T_1. The transitions 4–1 and 3–2 are new, however; W_2 and W_0 denote the probabilities for relaxation of the spin system via double quantum or zero quantum transitions respectively, i.e. transitions for which Δm is 2 or zero. These cannot be excited by electromagnetic radiation; they are spectroscopically forbidden and therefore cannot be observed in the NMR spectrum (Section 1.4.1). However, both are allowed in relaxation. According to theory it is even possible for W_2 to be greater than W_1! Furthermore, W_2 and W_0 are determined almost entirely by dipole-dipole relaxation. In cases where non-dipolar mechanisms contribute appreciably to the relaxation, these mainly affect W_1.

Having thus set out the basic principles, we will now try, using Figure 10-2, to develop a qualitative understanding of the NOE for the example of the two-spin AX system (A = ^{1}H; X = ^{13}C).

In Figure 10-2 the thicknesses of the "slabs" indicate the relative populations occupying the different energy levels. Diagram A shows the initial condition, where the occupancy numbers or populations N_1 to N_4 correspond to equilibrium. If we now saturate the A-transitions, whose frequencies are equal (since $J_{AX} = 0$), levels 1 and 3 become equally populated, as also do levels 2 and 4, i.e. we have $N_{1'} = N_{3'}$ and $N_{2'} = N_{4'}$. In diagram B the contributions which have been transferred from N_1 and N_2 to N_3 and N_4 are indicated by black slabs.

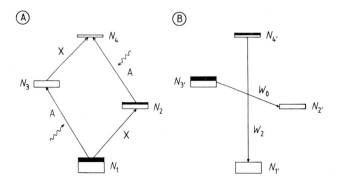

Figure 10-2.
Changes occurring in the population ratios in an NOE experiment. The energy level scheme A shows the initial situation, and B shows the situation when the two A transitions of equal energy are saturated. The relative thicknesses of the slabs represent the populations N_1 to N_4 and $N_{1'}$ to $N_{4'}$. Of the transition probabilities for relaxation only W_2 and W_0, corresponding to double quantum and zero quantum transitions, are shown.

The changes in the populations have no effect on the total intensity of the X signal, as this depends only on the total magnetization M_X, and M_X is not affected by the saturation of the A-transitions. Through the allowed relaxation processes with the probabilities W_2 and W_0 a new equilibrium with a new population distribution is established. By comparing diagrams A and B we can see in a qualitative way how the two relaxation processes labeled W_2 and W_0 affect the signal intensities. It will be noticed that $N_{1'} < N_1$ and $N_{4'} > N_4$. It follows from this that the population ratio $N_{1'}/N_{4'}$ is smaller than the equilibrium ratio N_1/N_4. The relaxation process labeled W_2 therefore tries to increase $N_{1'}$ at the expense of $N_{4'}$. This causes an increase in the population differences $N_{1'} - N_{2'}$ and $N_{3'} - N_{4'}$ which determine the intensity of the X-transitions. This is synonymous with an increase in signal intensity, i.e. an amplification. Conversely the relaxation process labeled W_0 tries to increase $N_{2'}$ at the expense of $N_{3'}$. This reduces the differences $N_{1'} - N_{2'}$ and $N_{3'} - N_{4'}$ thereby tending to reduce the signal intensity. The combination of these two (opposing) relaxation mechanisms determines the amplification observed as the nuclear Overhauser effect. Depending on the magnitudes of the contributions from W_2 and W_0, the intensity can therefore increase or decrease, but can also become zero. But when can we expect W_2 to predominate, and when W_0?

In small molecules with short correlation times τ_C, W_2 predominates, and one therefore observes a positive NOE. On the other hand, in molecules with long correlation times τ_C – for example in macromolecules – W_0 has the greatest effect. This dependence of the transitions probabilities W_2 and W_0 on the correlation times can be understood if one reflects that the fluctuating magnetic fields needed to induce the double quantum transition must contain frequencies close to the sum of the Larmor frequencies ν_A and ν_X, whereas for the zero quantum transition much lower frequencies are needed. This result has an important practical consequence: the NOE signal amplification depends on the magnetic field strength! Since the introduction of high-field spectrometers it has become much more common to find "negative amplifications".

From the close connection between dipole-dipole relaxation and NOE we can understand why the NOE depends on the distance between the nuclei, since the dipolar coupling decreases in inverse proportion to the sixth power of the distance.

10.2.2 Enhancement Factors

As explained above, the NOE amplification factor depends on the correlation time τ_C, in other words on how rapidly the molecules are rotating. Making the reasonable assumptions that the nuclei undergo rapid reorientation (short τ_C), and that their relaxation occurs entirely by a dipolar mechanism, the theory gives a fractional increase η as in Equation (10-1):

$$\eta = \frac{\gamma_A}{2\,\gamma_X} \quad \begin{matrix} \longleftarrow \text{A saturated} \\ \longleftarrow \text{X observed} \end{matrix} \qquad (10\text{-}1)$$

In this experiment the X-nuclei are observed while the resonance of the A-nuclei is saturated. γ is the magnetogyric ratio. The coefficient η gives the fractional increase in a signal due to the NOE. The resulting total intensity I of the enhanced X-signal is given by:

$$I = (1 + \eta)\, I_0 \qquad (10\text{-}2)$$

where I_0 is the original intensity.

Using the magnetogyric ratios given in Table 1-1 (Chapter 1) we can calculate from Equation (10-1) the fractional increase coefficients η for different nuclear combinations. The values thus obtained are maximum figures which are not fully realized in practice.

Examples:
○ What is the fractional increase η for ^{13}C NMR signals measured under ^{1}H BB decoupling conditions?
From Table 1-1 we find $\gamma\,(^1\text{H}) \approx 4\,\gamma\,(^{13}\text{C})$. Hence $\eta \approx 2$ (to be exact, 1.98). This means that, as already indicated earlier, the ^{13}C NMR signals are increased by a maximum of 200%. i.e. their intensity increases from 1 to 3 units.
○ What enhancement factor should be expected if one were to observe ^{1}H NMR signals while simultaneously saturating the resonances of the ^{13}C nuclei? Such an experiment is in practice unrealistic unless carried out on a ^{13}C-enriched sample. From Equation (10-1) (assuming 100% enrichment) we calculate $\eta = 0.125$; i.e. the ^{1}H signals would increase by 12.5%.

From the above two experiments, namely ^{13}C$\{^1$H$\}$ and ^{1}H$\{^{13}$C$\}$, we reach the following conclusion:
In a heteronuclear NOE experiment one should, so far as possible, saturate the more sensitive nucleus (that with the largest γ-value) and observe the less sensitive nucleus.

○ What is the maximum possible fractional increase η in a homonuclear NOE experiment? Such experiments are, almost without exception, only of practical importance in ^1H NMR spectroscopy. Putting $\gamma_A = \gamma_X$ in Equation (10-1) gives $\eta = 0.5$. Thus the maximum increase in intensity is 50%.

○ What is the effect of a negative value of γ? A practical situation where this arises is when observing ^{15}N resonances with simultaneous ^1H BB decoupling. As the γ-values of ^1H and ^{15}N differ by a factor of nearly 10, this gives, according to Equation (10-1), the unusually large value of about -5 for the maximum fractional increase η, and, moreover, it is negative! This negative sign can lead to complications. If as a result of molecular or experimental factors the fractional increase accidentally turns out to be -1, then according to Equation (10-2) the ^{15}N NMR signal disappears! This negative sign also explains why ^{15}N NMR spectra often contain negative signals despite having been recorded with correct phase settings.

10.2.3 Multi-Spin Systems

Up to now we have not considered how the NOE affects the signal intensities in multi-spin systems. In connection with these, two results of the theory deserve particular mention.

● In the homonuclear case – which is the only one of practical importance – the maximum possible fractional increase η is 0.5.

● In multi-spin systems negative contributions to the NOE can occur; these have nothing to do with the relaxation process labeled as W_0 in the earlier discussion.

How is this latter condition possible? Let us remind ourselves once more that a signal becomes enhanced – or reduced – when some process occurs in the spin system which leads to new level populations different from those at equilibrium. In our case this happened as a result of the decoupling. Let us assume that we have a three-spin system $A-B-C$. Each pair of adjacent nuclei will have a dipole-dipole coupling, but there will be no appreciable dipolar coupling between A and C. If we saturate the A transitions, the populations involved in the B transitions will be altered, and the intensities of the B signals will be increased. Now, however, the dipolar interaction between B and C is affected by the new spin populations which deviate from the equilibrium condition. Consequently, saturation of A causes an NOE through the indirect path via B, which also alters the intensity of the C resonance. This *indirect* NOE is negative. We shall meet an example of this in the next section.

In this connection there is a further aspect which has not yet been mentioned: all spin systems require time for the NOE to build up and decay. The rates of these processes depend on the relaxation times. This fact must be taken into consideration in any experiment aimed at utilizing the increase in intensity from the NOE (Chapters 8 and 9). The indirect NOE in multi-spin systems cannot build up until after the *direct* NOE, which in our example is that between A and B, has developed. This time difference between the appearance of the direct and the indirect NOE can sometimes be a useful aid in solving structural problems.

NOE experiments can be used to estimate internuclear distances by measuring the fractional increase coefficient η. However, such quantitative applications of NOE are very difficult. Occasionally too, NOE experiments are used to determine the dipolar contribution T_{1DD} to the spin-lattice relaxation.

10.2.4 From the One-Dimensional to the Two-Dimensional Experiments, NOESY and ROESY

Figure 10-3 A shows the pulse sequence for the one-dimensional NOE experiment. As can be seen, the experiment begins with the selective saturation of a particular resonance by irradiation at the appropriate frequency v_2. This preparation phase which precedes the 90° pulse and the data acquisition phase (FID) usually lasts several seconds. During this time a new steady state with altered energy level populations becomes established, resulting in the observed NOE.

The pulse sequence for the two-dimensional NOESY experiment is shown in Figure 10-3 B (see also Section 9.4.6). The pulse sequence $90_{x'}^{\circ} - t_1 - 90_{x'}^{\circ}$, which precedes the mixing period Δ, establishes a non-equilibrium spin distribution (see Figure 9-28 B, vector diagram d). During the mixing period Δ that follows, the system undergoes relaxation, by both the cross-relaxation and spin–lattice relaxation mechanisms. However, the changes in signal intensities that constitute the NOE are due solely to the cross-relaxation mechanism. In the NOESY procedure t_1 is increased in a stepwise sequence, and each different value of t_1 results in a different starting condition at the beginning of the mixing period Δ. Thus, in contrast to the one-dimensional experiment, the NOESY experiment involves a transient condition rather than a steady-state condition [2]. An example of a NOESY spectrum will be discussed in Section 10.4 (see Figure 10-6) As already explained in Sections 9.4.6 and 10.2.1, the relative signs of the amplitudes of the cross peaks in the NOESY experiment depend on the correlation time τ_C, and under some conditions the cross peaks may disappear. This difficulty is avoided in the ROESY experiment. It

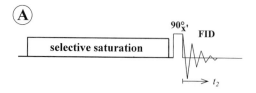

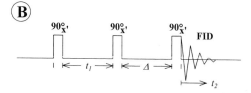

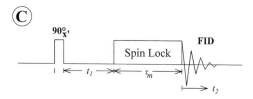

Figure 10-3.
Pulse sequences for
A: the one-dimensional NOE
experiment
B: the two-dimensional NOESY
experiment
C: the two-dimensional ROESY
experiment.
(Versions not employing field
gradients; cf. also Section 9.4.6,
Fig. 9-28). t_1 is the variable and is
altered by successive increments;
Δ is the fixed mixing time.
τ_m is the fixed duration of the spin
lock.

differs from NOESY in an important respect, namely that the
signal amplification is achieved by cross-relaxation between
transverse magnetization components during the spin lock period (see Section 9.4.6). In this experiment the signs of the amplitudes of the cross peaks are always negative and are now unaffected by τ_C, while those of the diagonal peaks are always
positive. However, one still needs to consider the effect of the
magnet field strength, and especially that of the spin lock duration τ_m. Apart from the question of the signs as already discussed, the appearance of the ROESY spectrum is similar to
that of the NOESY spectrum.

10.3 Experimental Aspects

All one-dimensional NOE experiments require one to measure changes in signal intensities, but in most applications it is
sufficient just to prove qualitatively that there is an effect.
Pulse techniques offer a particularly elegant way of obtaining
such evidence. One first records the normal spectrum then the
spectrum with NOE, and subtracts one from the other. The difference spectrum contains only those signals for which there is a
difference in intensity between the two spectra with and without
NOE. A negative signal always appears at the position of the

irradiating frequency, since in the NOE spectrum this signal is saturated and therefore absent, whereas in the spectrum recorded without saturation the signal persists, and it becomes inverted when the difference spectrum is generated.

Since in practice one uses samples which are as dilute as possible, many FIDs must be added together to obtain a difference spectrum with a reasonable signal-to-noise ratio. In such cases the data are always recorded in an alternating fashion, e.g. by taking eight FIDs without NOE followed by eight with NOE and repeating this to finally generate the difference spectrum. In practice, in order to record a spectrum without NOE, instead of switching off the decoupler one merely shifts its frequency v_2 to another point within the spectral region, such as the TMS signal or that of the residual $CHCl_3$ in the deuterochloroform solvent. By this procedure any changes which occur during the measurement affect the normal and NOE spectra equally, so that good quality difference spectra can be obtained even in long accumulations. The difference spectrum reproduced in Figure 10-5 B was recorded by this method.

As already explained, the NOE depends on dipole-dipole relaxation. Other intra- and intermolecular relaxation processes reduce the fractional increase, sometimes even to zero. Consequently a number of rules must be observed in NOE experiments, relating especially to the preparation of the samples:

- The sample must not contain any additives or impurities that are paramagnetic; for example, oxygen must be removed by careful degassing.
- The solvent should, if possible, contain no protons – for best results deuterated solvents should be used.
- The sample should be dilute and of low viscosity.
- If it is desired to look for an NOE between a CH_3 group and a single proton, one should always saturate the CH_3 resonance and measure the signal intensity for the single proton rather than the other way round? Why? The relaxation of the protons of a methyl group is determined mainly by the interactions between the methyl protons themselves, and consequently the NOE signal enhancement for these is smaller – usually only a few percent – or it may even disappear completely.

It should be noted that an NOE can usually only be detected if the distance between the dipolar-coupled nuclei is less than 5 Å, and ideally it should be less than 3 Å. However, it must be appreciated that the enhancement of a signal in a one-dimensional NOE experiment depends not only on the internuclear distance but on whether or not equilibrium has been fully established before recording the FID.

10.4 Applications

In Section 10.2 we established that the NOE signal enhancement factor depends on the distance between the dipolar-coupled nuclei and on the correlation time τ_c. From this fact the following potential areas of application emerge:

- elucidation of molecular constitution and conformation
- aiding assignments
- investigating molecular motions.

In the following discussion we will leave aside the last of these three, as such molecular mobility studies call for a detailed quantitative analysis of the NOE results, a task fraught with many difficulties. Most experiments of this type concern investigations of macromolecules, by observing ^{13}C resonances while saturating the ^1H resonances.

Here we will deal only with the first two areas, taking examples from ^1H NMR spectroscopy in which intramolecular NOE is used.

We have already met three such cases in the introduction to this chapter. There the positive NOEs that were measured gave an indication of the spatial proximity of protons and methyl groups. In compound **1** we even found a signal enhancement ($+45\%$) which is only a little below the maximum possible value of 50%. Evidently the dipole-dipole coupling in this semi-clathrate is particularly large. In dimethylformamide (**2**) and 3-methylcrotonic acid (**3**) the NOE experiments led to a correct assignment of the methyl signals. Since the signal of the isolated proton is only enhanced when a particular one of the two methyl signals is saturated, this must correspond to the *cis* methyl group in each case.

Whereas in these NOE experiments, which were performed a long time ago using the CW method, quantitative results on the signal enhancement factors were also given, most of the more recent work on structural elucidation by NOE has been limited to qualitatively demonstrating the existence of an NOE. This fact is connected with the availability now of the *difference spectroscopy* technique (Section 10.3), which is easily implemented when using pulse methods. The rest of the examples serve to further illustrate the wide range of possible applications of NOE and NOESY measurements.

Example 1:

○ The psychiatric drug chloroprothixene (**4**) metabolizes in humans and animals. Chemical and mass spectrometric studies [3] have shown that in the principal metabolite a hydrogen atom in one of the two benzene rings has been replaced by a hydroxyl group. The analysis of the 300 MHz ^{1}H NMR spectrum (Fig. 10-4) indicates that this OH group must be attached at either the 5- or the 8-position. An NOE experiment showed that the correct position is on C-5. Saturating the resonances of H-1' ($\delta = 5.84$), which is one of the protons in the side-chain, produced a positive NOE for the doublet of doublets at $\delta \approx 7.02$, which had previously been assigned to either H-5 of H-8 on the basis of its chemical shift and splitting pattern (splitting by one *ortho* and one *meta* coupling).

The experiment also showed that the side-chain is in the *Z*-configuration. If the molecule were in the *E*-configuration one would have expected an increase in the intensity of the H-1 signal. Such an effect is in fact found on synthesizing 6-hydroxychloroprothixene (**5**), which was expected, on the basis of the synthetic route used, to have the *E*-configuration. In this compound saturating the H-1' resonance did in fact give an enhancement of the signal of the proton on C-1.

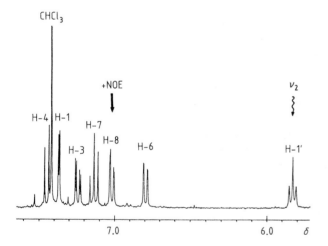

4

5

CHCl$_3$

+NOE

ν_2

H-4 | H-1

H-7

H-3

H-8

H-6

H-1'

7.0   6.0 δ

Figure 10-4.
Portion of the 300 MHz ^{1}H NMR spectrum of chloroprothixene (**4**) in CDCl$_3$ with assignments. When the H-1' resonances ($\delta = 5.84$) are saturated the intensity of the doublet of doublets at $\delta = 7.02$ is increased due to the NOE.

Example 2:

○ Figure 10-5 A shows the 250 MHz ^{1}H NMR spectrum of methicillin (**6**). All the signals can be assigned fairly easily from their chemical shifts and coupling patterns, with the exceptions of the methyl signals A and B. Saturating the methyl signal at $\delta \approx 1.7$ resulted in a positive NOE for the H-3 signal ($\delta \approx 4.25$). This can be clearly seen in the NOE difference spectrum (Figure 10-5 B). (In difference spectra it is often not possible to completely eliminate residual signals from the strong resonances of the solvent and the methyl groups. Also a large negative signal always occurs at the irradiation frequency (Section 10.3).)

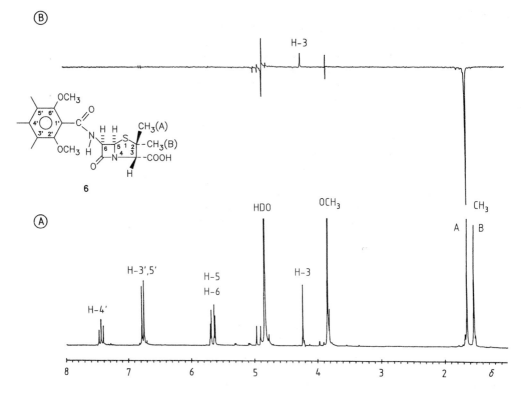

Figure 10-5.

A: 250 MHz ^{1}H NMR spectrum of methicillin (**6**) in 0.2 M sodium acetate buffer (D_2O; pD 7.0), with assignments. On saturating the methyl signal ($\delta \approx 1.7$) the NOE difference spectrum shows an increase in the intensity of the H-3 signal ($\delta \approx 4.25$). The negative signal in the NOE difference spectrum corresponds to the irradiating frequency. Strong signals, such as the residual solvent signals (HDO) or those of the methyl groups, are often found to be not exactly cancelled to zero in the difference spectrum.

Example 3:

○ An interesting application to a structural problem was described by Hunter et al. [4]. When styrene is polymerized in the presence of 4-methoxyphenol one obtains, in addition to the polymer, a 1:1 adduct (**7**). This compound is obtained by the formal addition of a styrene molecule to 4-methoxyphenol. However, the question of whether the addition occurs at C-2 or C-3 could not be answered from either the ^{1}H or the ^{13}C NMR spectrum. An NOE experiment provided a decision in favor of the structure shown. Irradiating the OCH$_3$ resonance gave an increase in the intensities of the signals of the ring protons H^A and H^B. From this we must conclude that both positions *ortho* to the OCH$_3$ group are unsubstituted. In contrast the signals of the third ring proton (H^C) showed a negative NOE! This is a case of an indirect NOE in a multi-spin system, as described in Section 10.2.

In a further NOE experiment it was shown that saturating the OH resonance increased the intensity of the H^C signal, providing additional evidence for the position of substitution.

300

Example 4:

○ As a final example we will analyze the two-dimensional NOESY spectrum of the disaccharide **8** (Figure 10-6) with a particular aim in mind, namely to look for correlation peaks between the protons of rings I and II, and thus possibly learn something about the relative positions of the rings. The 500 MHz ^{1}H NMR spectrum, with the assignments that we already know from the TOCSY spectrum of Figure 9-27, is shown at the upper and left-hand edges of Figure 10-6. We are especially interested in looking for correlation peaks for the protons nearest to the glycosidic linkage, namely H-1' and H-2' in ring I, and H-3, H-4, H-5, H-6a and H-6b in ring II.

Looking along a horizontal line through the H-1' signal at $\delta = 4.64$ (on the vertical scale) we find correlation peaks corresponding to H-2', H-3' and H-5'. We certainly expect to find correlations such as these within the same ring, but they are of no interest in relation to our problem. However, there are also correlation peaks for H-4 and H-3. From these we can conclude that H-1' is near to H-3 and H-4.

Starting now from the H-2' signal at $\delta = 3.06$, we find, in addition to the correlation peaks for H-1', H-3' and H-4' which are of no interest in the present context, further peaks corresponding to protons in ring II, namely H-5, H-6a and H-6b. No correlation between H-2' and H-4 is found.

We need not discuss these results in great detail here. However, we can already conclude that the preferred conformation is a twisted one, in which the relative orientation of the planes of the two rings (as defined by the atoms O, C-2, C-3 and C-5) is such that H-2' and H-4 are far apart. We are not dealing here with a rigid molecule, of course, so this conclusion refers not to a fixed structure but to an energetically preferred conformation.

Whereas the earliest molecules investigated by NOE were relatively small, it is now also possible to obtain good results from large molecules of all kinds. Nowadays one cannot imagine determining the structures of natural products without the help of NOE and NOESY experiments. For example, the stereochemical structures of many peptides and proteins in solution have been determined in this way, sometimes involving the analysis of several hundred correlation peaks. In cases where it has been possible to compare the results with those from X-ray crystallographic analysis, the agreement has been excellent (see Section 13.3).

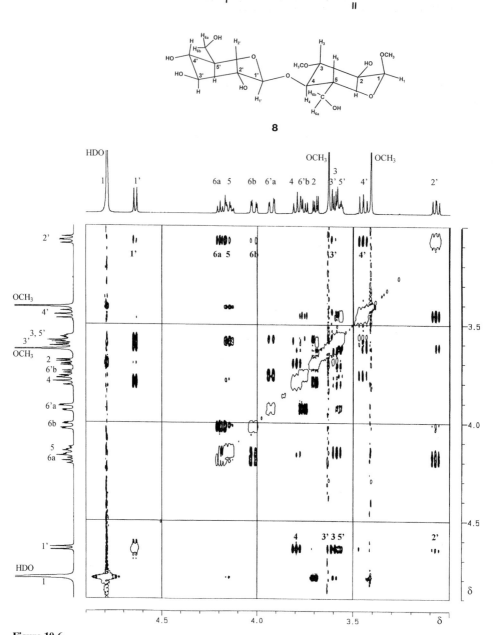

Figure 10-6.

NOESY spectrum of α-methyl-3-O-methylcellobioside (**3**) in D$_2$O. The 500 MHz ^{1}H NMR spectrum is shown at the top and left-hand edges, with assignments. The H-1 resonance is hidden under the HDO resonance of the solvent. Only the cross-peaks on the horizontal lines starting from H-1′ and H-2′, the two protons adjacent to the glycosidic bond in ring I, are labeled. The correlation peaks between protons of rings I and II give clear and reliable information about the relative orientation of the two rings.

(Experimental conditions:

approx. 20 mg in 0.5 ml D$_2$O; 5 mm sample tube; 128 measurements with t_1 altered in 400 μs increments; mixing time $\varDelta$=2 s; each measurement with 16 FIDs and 1 K data points; total time approx. 2 h.)

10.5 Bibliography for Chapter 10

[1]   F. A. L. Anet and A. J. R. Bourn, *J. Amer. Chem. Soc. 87* (1965) 5250.

[2]   D. Neuhaus and M. Williamson: *The Nuclear Overhauser Effect in Structural and Conformational Analysis.* New York: Wiley-VCH, 2000, p. 258 ff.

[3]   U. Breyer-Pfaff, E. Wiest, A. Prox, H. Wachsmuth, M. Protiva, K. Sindelar, H. Friebolin, D. Krauß and P. Kunzelmann, *Drug Metab. Disp. 13* (1985) 479.

[4]   B. K. Hunter, K. E. Russell and A. K. Zaghloul, *Can. J. Chem. 61* (1983) 124.

Additional and More Advanced Reading

D. Neuhaus and M. Williamson: *The Nuclear Overhauser Effect in Structural and Conformational Analysis.* New York: Wiley-VCH 2000.

J. H. Noggle and R. E. Schirmer: *The Nuclear Overhauser Effect, Chemical Applications.* New York: Academic Press, 1971.

J. K. M. Sanders and J. D. Mersh: Nuclear Magnetic Double Resonance; The Use of Difference Spectroscopy. In: *Prog. Nucl. Magn. Reson. Spectrosc. 15* (1982) 353.

11 Dynamic NMR Spectroscopy (DNMR)

11.1 Introduction [1–3]

When nuclei exchange positions between two positions A and B with different shielding values, how does this affect the NMR spectrum?

An example from the earliest days of dynamic NMR spectroscopy will serve to illustrate this problem, and to give some idea of why DNMR spectroscopy is one of the most important areas of application of NMR spectroscopy.

In the ^1H NMR spectrum of dimethylformamide (**1**) at $+22.5°C$ two methyl signals are observed at $\delta = 2.79$ and 2.94 (Fig. 11-1). Above $+100°C$ these two signals broaden, then at $120°C$ they coalesce into one broad band. On raising the temperature further this again becomes a narrow peak, whose position is exactly midway between the two original peaks.

From this temperature-dependent behavior of the ^1H resonances it can be concluded that at room temperature the two methyl groups are differently shielded, whereas at higher temperatures they become equivalent. The reason for this is well known and has long been understood. The CN bond has a high proportion of double bond character, which results in the rotation being hindered, so that the methyl groups A and B are in different magnetic environments. The representation in terms of mesomeric canonical forms makes this clear (Scheme I, A and B).

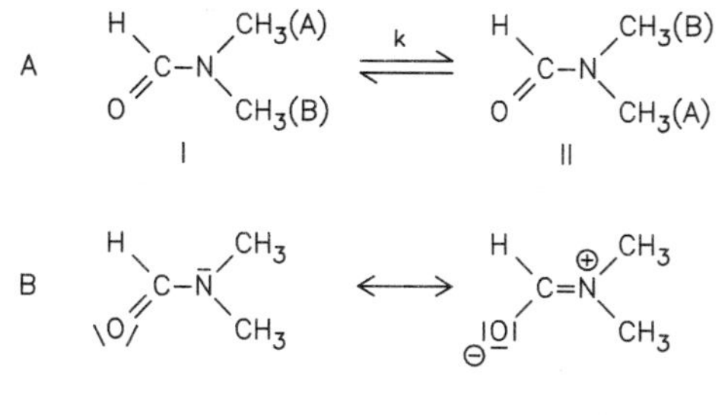

Scheme I

When the temperature is raised the barrier to rotation is overcome, and if the rate at which the two methyl groups

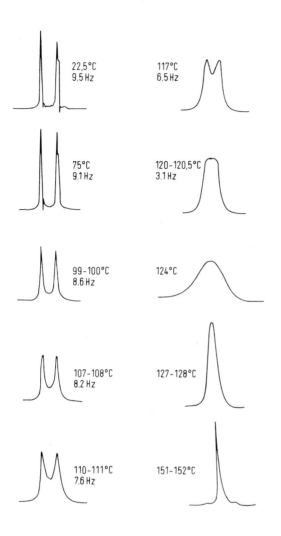

Figure 11-1.
56.4 MHz ^{1}H NMR signals of the methyl protons in dimethylformamide (**1**), recorded at different temperatures. For slow site exchange two peaks are obtained, whereas for fast exchange there is only one. In the intermediate range the signals are broadened. The coalescence temperature T_c is 120° C.

exchange places becomes sufficiently rapid they can no longer be distinguished by NMR spectroscopy. Only one signal is then observed in the spectrum, at

$$\bar{v} = \frac{v_A + v_B}{2}$$

The two extreme cases – slow exchange and rapid exchange – can be understood from what we already know, but why do the signals become broad in the intermediate temperature range? Theory provides an explanation of this, and the relevant results will be considered in the next section.

The dynamic process exemplified by dimethylformamide (**1**) is the simplest case of a *first-order reversible reaction*, which is described by two rate constants:

$$I \underset{k_{II}}{\overset{k_I}{\rightleftharpoons}} II \quad \text{with} \quad k = k_I = k_{II}$$

The rate constants k_I and k_{II} for the forward and reverse reactions are here equal, since the two rotamers I and II in the equilibrium are equal in energy.

If one of the methyl groups of dimethylformamide is replaced by another substitutent, e.g. by a benzyl group (Scheme II), the two rotamers are no longer equal in energy, and the rate constants for the forward and reverse reactions are unequal ($k_I \neq k_{II}$).

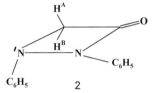

Scheme II

For slow rotation ($k \approx 0$) one therefore finds two methyl signals, whose intensities depend on the proportions of I and II in the equilibrium mixture. Raising the temperature causes rapid positional exchange, and the two signals coalesce. However, the coalesced signal is no longer in the middle, but is at the center of gravity (centroid):

$$v = x_I v_A + x_{II} v_B$$

where x_I and x_{II} are the mole fractions of I and II. Since $x_I + x_{II} = 1$, we have:

$$v = x_I v_A + (1 - x_I) v_B$$

Up to now we have considered site exchange between methyl groups whose ^{1}H resonances are not split into multiplets by spin-spin coupling. Figure 11-2 shows two spectra of 1,2-diphenyldiazetidinone (**2**), a molecule in which a pair of mutually coupled protons H^A and H^B exchange positions. In the spectrum recorded at $-55°$C the geminal proton pair H^A and H^B give a four-line AB-type spectrum. However, at a temperature of $+35°$C these four lines have already coalesced, i.e. the time-averaged shieldings of the two nuclei have become equal. We shall return to the mechanism of this process in Section 11.3.3.

Rapid dynamic processes do not always result in single peaks, as they have in the systems we have considered up to now, but they always lead to a simplification of the spectrum. A complicated example of this is shown in Figure 11-3.

The 90 MHz ^{1}H NMR spectrum of benzofuroxan (**3**) at $-54°$C is of the ABCD type, showing that all the four protons are chemically different (Fig. 11-3 A). At $+53°$C (Fig. 11-3 B) the

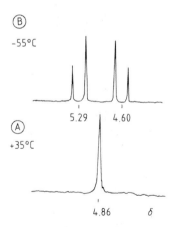

Figure 11-2.
60 MHz ^{1}H NMR signals of the ring protons H^A and H^B in 1,2-diphenyldiazetidinone (**2**).
A: At $+35°$C in acetone-d_6.
B: At $-55°$C in CDCl$_3$;
$J_{AB} = 14$ Hz.

307

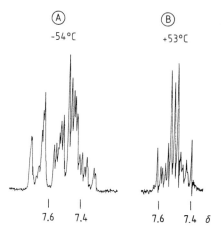

Figure 11-3.
90 MHz ¹H NMR spectrum of
benzofuroxan (**3**).
A: At −54° C in CDCl₃.
B: At +53° C in acetone-d₆. At the
higher temperature one finds a
symmetrical AA′BB′ spectrum
([AB]₂), whereas at low temperatures the spectrum is asymmetrical
(ABCD).

complex ABCD spectrum is replaced by a symmetrical spectrum of the AA′BB′ (or [AB]₂) type. The only way in which
this can occur is if the proton pairs each become equivalent
through a dynamic exchange process. This process is illustrated
in Scheme III [4].

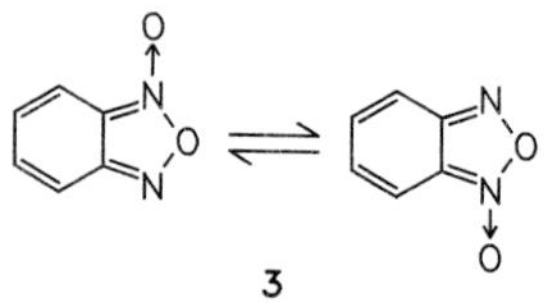

Dynamic NMR spectroscopy can also be used in the same
way to study *intermolecular exchange reactions*. The effects on
the NMR spectra are the same whether an intramolecular or
an intermolecular process is involved.

Some classic examples of intermolecular processes are the
proton exchanges which occur in aqueous solutions of acids,
alcohols, thioalcohols and amines. In a methanol–water mixture, for example, only one common signal is observed for the
methanol OH proton and the water protons:

$$CH_3OH + HOH \rightleftharpoons CH_3OH + HOH$$

Analogous behavior is observed for all other compounds
which have easily exchangeable protons.

Dynamic processes are most commonly investigated using
¹H NMR spectroscopy, but resonances of other nuclides such
as ¹⁹F, ³¹P, and especially ¹³C, can also be used. Such measurements are often carried out as an extension of ¹H NMR studies,
as the greater range of chemical shifts for these nuclides allows
peak coalescences to be observed at higher temperatures than
for protons.

A very effective method for obtaining qualitative evidence of
a proton exchange process is the two-dimensional EXSY
(**EX**change **S**pectroscop**Y**) experiment, the basic principles of
which were covered in Section 9.4.6 along with the NOESY
experiment. The advantages of the EXSY method will be seen
in Section 11.3.7, where it is used to study the behavior of an
ethanol–water mixture in acetone.

Scheme III

11.2 Quantitative Calculations

Figures 11-1 to 11-3 provided qualitative evidence, from the changes in the spectra with temperature, that dynamic processes were occurring in the samples studied. However, the results can also be analyzed quantitatively. From the temperature dependence of the resonance lines one can obtain rate constants and activation parameters. DNMR spectroscopy was responsible for an important step forward, in that it makes it possible to study reactions that are too slow for optical spectroscopic methods but too fast for investigation by classical chemical methods, i.e. reactions with rate constants k in the range 10^{-1} to 10^3 s^{-1}. The methods used to determine rate constants and activation parameters are described in the sections which follow.

11.2.1 Complete Line-shape Analysis

It was shown in Chapter 4 that NMR spectra can be calculated theoretically. Such calculations tell us the number of peaks present and their widths and intensities (amplitudes), in other words the entire spectral curve or *line-shape*. Here the line-shape is understood to mean the functional relationship between signal intensity and frequency. When dynamic processes are involved the line-shape depends also on the exchange rate k, and thus on the time τ_1 spent by the nuclei in a particular environment. For a first-order reaction we have:

$$k = \frac{1}{\tau_1} \qquad (11\text{-}1)$$

A *complete line-shape analysis* (CLA) allows us to determine the values of k and τ_1.

In particularly simple cases – for example when there is no coupling between the exchanging nuclei – the line-shapes can be calculated by classical methods using the modified Bloch equations [3, 5]. In complicated cases a solution is only possible by quantum-mechanical methods [6]. The following account will briefly outline the analytical procedure in an example of a first-order reaction; in this example, as in the rotation about the CN bond in dimethylformamide (1), there is an exchange between two sites with different magnetic shielding values.

Analytical procedure

- First one must record a number of spectra at different temperatures. For a complete line-shape analysis these might consist, for example, of three spectra at temperatures for which the dynamic process is very slow (one of which should give little or no exchange), then five to ten spectra in the region of the coalescence temperature, i. e. corresponding to the middle range of exchange rates, and finally three further spectra at temperatures giving rapid exchange.

- The second step is to analyze the spectrum in the absence of site exchange, so as to obtain the chemical shifts, coupling constants and line-widths (half-height widths $\Delta v_{1/2}$). If the spectrum consists of only two lines, as in dimethylformamide (**1**) (considering here only the methyl protons, since the formyl proton signal is irrelevant), or if it is a four-line AB spectrum, as for H^A and H^B in 1,2-diphenyldiazetidinone (**2**), these parameters can easily be determined. However, for benzofuroxan (**3**) the ABCD-type spectrum can only be analyzed by using computer programs (LAOCOON, WIN-DAISY).

- The third step is to calculate spectra for exchange rates in the coalescence region where broadened signals occur. To do this one uses the previously determined chemical shifts, coupling constants and line-widths, and varies the rate constant k. When the calculated spectrum visually matches one of the experimentally measured spectra, this gives the rate constant k for the temperature at which that spectrum was recorded. From n such spectra we thus obtain n different k-values with their corresponding temperatures. Calculations of this sort must usually be carried out on a computer with a suitable program.

- In the fourth and last stage of the analysis one calculates spectra for fast exchange rates. Often, however, this step is unnecessary. Of the three examples discussed above, only the analysis of the spectrum of benzofuroxan (**3**) would require this.

The method described above, based on visually comparing the calculated and observed spectra, is very tedious. Binsch et al. therefore developed the DNMR5 program, which not only allows one to calculate the spectrum for a given set of values of chemical shifts, coupling constants and rate constant(s), but also provides for automatic iterative matching of the calculated and experimental spectra. Details of this method can be found in the appropriate literature [6, 7].

Although complex dynamic spectra such as that of benzo-furoxan are more difficult to analyze than the simpler ones, they have the advantage that the results are more precise. It is necessary to match, by direct visual comparison or by automatic iteration, the positions, widths and intensities of a large number of lines. This requires quite exact fitting of the parameters, and in these cases the rate constants can be determined very precisely, the values being unaffected by other errors of measurement.

Finally it should be noted that one can also analyze dynamic processes which involve the exchange of nuclei between several different sites, the line-shape then being a function of several rate constants [3].

11.2.2 The Coalescence Temperature T_C and the Corresponding Rate Constant k_C

In interpreting an exchange process it is often not necessary, or even sensible, to invest the time and effort needed for a complete line-shape analysis. Often it is sufficient to know only the order of magnitude of the rate constant at room temperature; for example, one might wish to know whether two isomers in dynamic equilibrium can be separated at room temperature, or whether this must be carried out at a reduced temperature. Cases such as this can be treated in a simpler way than by full line-shape analysis!

To see this let us again consider the spectra of **1** which are shown in Figure 11-1. At 22.5°C we observe two methyl peaks, but at 150°C only one. An important spectrum for our purpose is that recorded at about 120°C, in which the two signals just coalesce. For this *coalescence temperature* T_C the rate constant k_C is given by:

$$k_C = \frac{\pi \, \Delta v}{\sqrt{2}} = 2.22 \, \Delta v \qquad (11\text{-}2)$$

Here Δv is the separation in Hz between the two signals in the absence of exchange.

Equation (11-2) is only valid provided that:
- the dynamic process occurring is first-order kinetically,
- the two singlets have equal intensities, and
- the exchanging nuclei are not coupled to each other.

However, even if these conditions are not fulfilled exactly, Equation (11-2) is often a good approximation for estimating k_C.

Δv is determined experimentally from spectra recorded at temperatures which are as far below the coalescence temperature as possible. If this is not possible, for example, because the low temperatures needed cannot be reached, an estimated value of Δv is used.

Example:
○ For dimethylformamide (**1**) we have:
 $\Delta v = 9.5$ Hz (at 60 MHz resonance frequency); $T_C = 393$ K (120°C); this gives:
$$k_{393} = 21 \text{ s}^{-1}$$

An important fact must be mentioned here, namely that k_C is determined by the magnitude of Δv, which is in turn proportional to the resonance frequency (i.e. to the magnetic flux density). Consequently the value of k_c at a higher resonance frequency is greater than that at a lower frequency, which means that at the higher frequency the coalescence occurs at a higher temperature. Thus T_C is not a constant, but is a quantity which depends on the observing frequency. The following rules of thumb apply for this situation:

- The higher the observing frequency of the spectrometer, the higher is the coalescence temperature.
- Doubling the observing frequency shifts T_C upwards by about 10 K.

For an exchange process between two nuclei A and B with a mutual coupling J_{AB}, an equation analogous to (11-2) applies. The rate constant k_C at the coalescence temperature is then given by:

$$k_C = 2.22 \sqrt{\Delta v^2 + 6J_{AB}^2} \qquad (11\text{-}3)$$

Example:
○ For **2** (Fig. 11-2) we have $\Delta v = 41$ Hz (at 60 MHz), and $T_C = 272$ K; therefore:

$$k_{272} = 119 \text{ s}^{-1}$$

11.2.3 Activation Parameters

11.2.3.1 The Arrhenius Activation Energy E_A

By applying the Arrhenius equation (11-4) it is possible to determine graphically the activation energy E_A for the dynamic process being studied. Plotting $\ln k$ against $1/T$ yields a straight line whose gradient is E_A/R.

$$k = k_0 \, e^{-E_A/RT} \qquad (11\text{-}4)$$

$$\ln k = \ln k_0 - \frac{E_A}{RT}$$

T = temperature in K
k_0 = frequency factor
R = universal gas constant
    = 1.9872 cal K^{-1} mol^{-1}
    = 8.3144 J K^{-1} mol^{-1}

To determine E_A by this method involves a lot of work, since:

- as many spectra as possible must be measured over a wide range of temperatures, and
- the rate constants k for each of these spectra must be calculated from a complete line-shape analysis (Section 11.2.1).

11.2.3.2 The Free Enthalpy of Activation $\Delta G^{\ddagger}$

From the Eyring equation (11-5) one can determine the free enthalpy of activation $\Delta G^{\ddagger}$:

$$k = x \frac{k_B T}{h} e^{-\Delta G^{\ddagger}/RT} \tag{11-5}$$

$$\Delta G_C^{\ddagger} = 4.58 \, T_C \left(10.32 + \log \frac{T_C}{k_C} \right) \text{cal mol}^{-1} \tag{11-5a}$$

$$= 19.14 \, T_C \left(10.32 + \log \frac{T_C}{k_C} \right) \text{J mol}^{-1}$$

k_B = Boltzmann constant = 3.2995×10^{-24} cal K^{-1}
$\qquad\qquad\qquad\qquad = 1.3805 \times 10^{-23}$ J K^{-1}

x = transmission coefficient
$\qquad$ (usually assumed to be exactly 1)

h = Planck constant = 1.5836×10^{-34} cal s
$\qquad\qquad\qquad = 6.6256 \times 10^{-34}$ J s

To calculate the free enthalpy of activation $\Delta G_C^{\ddagger}$ from Equation (11-5a), we need only *one* k-value and *one* temperature. For this we can use the relatively easily determined pair k_C and T_C.

Example:
○ For dimethylformamide (**1**), with $T_C = 393$ K and $k_{393} = 21$ s^{-1}, we have

$$\Delta G_{393}^{\ddagger} = 20.9 \pm 0.2 \text{ kcal mol}^{-1} \, (87.5 \pm 0.8 \text{ kJ mol}^{-1})$$

○ For 1,2-diphenyldiazetidinone (**2**), with $T_C = 272$ K and $k_{272} = 119$ s^{-1}, we have

$$\Delta G_{272}^{\ddagger} = 13.3 \pm 0.2 \text{ kcal mol}^{-1} \, (54.4 \pm 0.8 \text{ kJ mol}^{-1})$$

Equation (11-6) gives the relationship between the free enthalpy of activation $\Delta G^{\ddagger}$, the enthalpy of activation $\Delta H^{\ddagger}$ and the entropy of activation $\Delta S^{\ddagger}$:

$$\Delta G^{\ddagger} = \Delta H^{\ddagger} - T\Delta S^{\ddagger} \tag{11-6}$$

If we insert this expression into the Eyring equation (11-5) and take logarithms (base 10), assuming $x = 1$, we obtain:

$$\log \frac{k}{T} = 10.32 - \frac{\Delta H^{\ddagger}}{19.14\ T} + \frac{\Delta S^{\ddagger}}{19.14} \qquad (11\text{-}7)$$

Thus, plotting $\log(k/T)$ against $1/T$ yields $\Delta H^{\ddagger}$ and $\Delta S^{\ddagger}$.

For monomolecular reactions the relationship between the enthalpy of activation and the Arrhenius activation energy is:

$$\Delta H^{\ddagger} = E_A - RT \qquad (11\text{-}8)$$

Up to now we have assumed that the two isomers I and II which are in dynamic equilibrium have equal energies. If this is not the case, the equilibrium mixture contains different amounts of I and II. The equilibrium constant K is then calculated from Equation (11-9), the relative concentrations of I and II being in most cases easily determined by integration of the corresponding NMR signals.

$$K = \frac{[\text{I}]}{[\text{II}]} \qquad (11\text{-}9)$$

One can then obtain the difference ΔG_0 between the free enthalpies of the two isomers, by using Equation (11-10):

$$\Delta G_0 = -\ RT\ln K \qquad (11\text{-}10)$$

11.2.3.3 Estimating the Limits of Error

The rate constants and the activation parameters $\Delta G^{\ddagger}$, $\Delta H^{\ddagger}$, $\Delta S^{\ddagger}$ and E_A, like any other experimentally determined quantities, are subject to errors. Possible sources of error are inaccuracies in measuring the following quantities:
- the frequency separation Δv,
- the coupling constants J,
- the line-widths $\Delta v_{1/2}$,
- the absolute temperatures T, and
- the coalescence temperature T_C.

One must take into account the fact that Δv and $\Delta v_{1/2}$, and to a lesser extent the coupling constants J, depend on the temperature and the solvent. However, the main sources of error are in measuring the temperatures T and the coalescence temperature T_C. Absolute values of T can seldom be measured with an accuracy better than $\pm\ 2$ K. On the other hand, the accuracy with which T_C can be determined depends on the spectrum; the more complex the spectrum, the greater the inaccuracy in T_C. Here too one must assume an error range of at least $\pm\ 2$ K.

A thorough discussion of the question of errors would not be appropriate here; the problem is treated in detail in the literature [2, 3, 6]. However, to give an idea of what can be expected, listed below are realistic estimates of the order of magnitude of the errors in the parameters used in this chapter.

k and k_C: $\pm 25\%$ (or greater)

$\Delta G^{\ddagger}$: ± 0.2 kcal mol^{-1}; ± 0.8 kJ mol^{-1}

$\Delta H^{\ddagger}$, E_A: ± 1 kcal mol^{-1}; ± 4.2 kJ mol^{-1}

$\Delta S^{\ddagger}$: ± 2 to 5 cal mol^{-1} K^{-1}

However, one should not use these estimates as a substitute for a proper error analysis in any individual case.

11.2.4 Rate Constants in Reactions with Intermediate Stages

In intramolecular dynamic processes the interconversion often takes place via intermediate stages with other conformations. This will be illustrated here by a classic example, that of the ring inversion in cyclohexane. However, instead of C_6H_{12}, which provided the first example of ring inversion, discovered in 1960 by Jensen et al. [8], we will consider the incompletely deuterated cyclohexane $C_6D_{11}H$ (**4**) [9, 10], as its NMR spectra are easier to analyze. At room temperature a single peak is observed, whereas at temperatures below $-60°C$ there are two. The spectra (shown in Ref. [11]) are thus analogous to those in Figure 11-1, when recorded with simultaneous decoupling of deuterium. We find $T_C = 212$ K and $\Delta v = 28.9$ Hz (at 60 MHz). From these values we obtain a rate constant $k_C = k_{212} = 64$ s^{-1}.

The interpretation of these experimental results seems straightforward. At room temperature the molecules undergo rapid interconversion between one chair conformation and the other. During this process the single remaining proton in the cyclohexane-d_{11} molecule switches between the axial and equatorial positions, giving an averaged signal. When the sample is cooled the ring inversion process is frozen, and the protons in the two positions are now distinguishable by NMR spectroscopy.

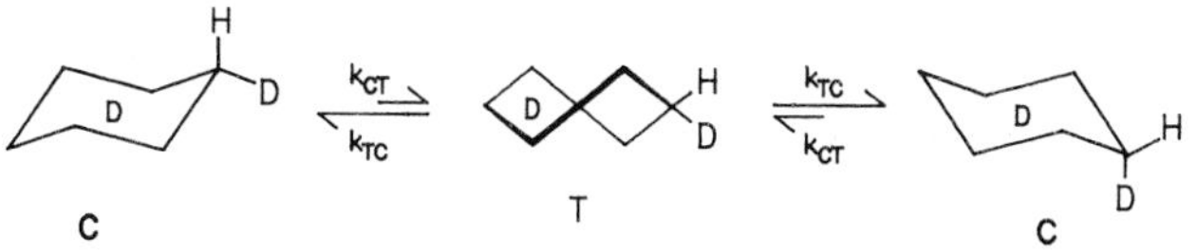

Scheme IV

Examination of molecular models and theoretical calculations indicate that the ring inversion takes place via an intermediate twisted boat conformation (Scheme IV).

In this intermediate conformation, which corresponds to a local minimum on the potential energy curve, the cyclohexane molecule has equal probabilities of reverting to its original chair conformation or switching to the inverted chair structure. From this fact we can conclude that the rate constant k_{CT} for the chair-to-twist interconversion is twice the observed rate constant k_C for the inversion:

$$k_{CT} = 2 \, k_C = 128 \text{ s}^{-1}$$

Using Equation (11-5) we then obtain $\Delta G^{\ddagger}_{212} = 10.2 \pm 0.2$ kcal mol^{-1} (42.6 $\pm$ 0.8 kJ mol^{-1}). This value corresponds to the difference between the free enthalpy of the $C_6D_{11}H$ molecule in the chair conformation and that of the transition state through which the molecule passes in changing to the twisted conformation.

11.2.5 Intermolecular Exchange Processes

As has already been mentioned in Section 11.1, typical examples of intermolecular exchange processes are the proton exchange which occurs in acids, alcohols and amines. In every case only a single time-averaged signal is found for the exchanging protons. The quantitative interpretation of such spectra to determine rate constants is usually difficult, since reactions of this kind are always of second or higher order. This means that the rate constants depend not only on the temperature but on the concentrations of the reactants. Moreover, a quantitative treatment of the spectral data is in many cases prevented by a lack of knowledge of the mechanism of the exchange reactions.

Despite these difficulties, there are many reports in the literature of systems which have been analyzed quantitatively [12].

11.3 Applications

There are two reasons why dynamic NMR spectroscopy has continued to develop so rapidly since even the earliest days of NMR:

- the NMR spectrum provides an excellent means, often the only one, for distinguishing between the starting materials and end-products of a reaction, and
- NMR spectra allow one to observe changes occurring in reactions with rate constants k in the range from about 10^{-1} to 10^3 s^{-1}, corresponding to $\Delta G^{\ddagger}$-values between 5 and 25 kcal mol^{-1} (20–100 kJ mol^{-1}). The values indicated here for the limits are determined by the range of temperatures over which measurements can be made, and, for low temperature measurements, by the solubilities of the compounds.

Most spectrometers allow the sample temperature to be varied from $+200°$C to $-100°$C, although in exceptional cases it can be reduced to $-150°$C or even lower. However, despite significant improvements in instruments during the last four decades, measurements at temperatures below $-80°$C have still not become a matter of routine.

The examples described below represent a cross-section of the wide range of applications of DNMR spectroscopy. Other applications, such as investigations of metal complexes, molecular rearrangements, proton transfer processes and many other types of reactions can be found in the comprehensive reviews by Jackman [1] and Oki [2].

11.3.1 Rotation about CC Single Bonds [13, 14]

In considering rotation about CC single bonds we can distinguish between three situations, namely those where:
- both carbon atoms have sp^3 hybridization
- one carbon atom has sp^2 and the other sp^3 hybridization
- both carbon atoms have sp^2 hybridization.

11.3.1.1 C(sp³)—C(sp³) Bonds

Typical examples of molecules with rotation about a CC single bond between two sp³-hybridized carbon atoms are those of ethane and its derivatives. In these compounds the rotation is so fast at room temperature – even when bulky substituents are present – that only time-averaged signals over the three energetically most favored rotamers are observed (see Scheme V). Although the rotation in ethane itself cannot be frozen out even at low temperatues, this is possible for many substituted ethanes. Nevertheless, except when very severe steric hindrance is present, the coalescence temperatures are well below −50° C.

Halogenated ethanes and *t*-butyl-substituted ethanes have been particularly thoroughly studied. Depending on the nature and number of substituents, $\Delta G^{\ddagger}$-values between 5 and 15 kcal mol^{-1} (20–60 kJ mol^{-1}) are found.

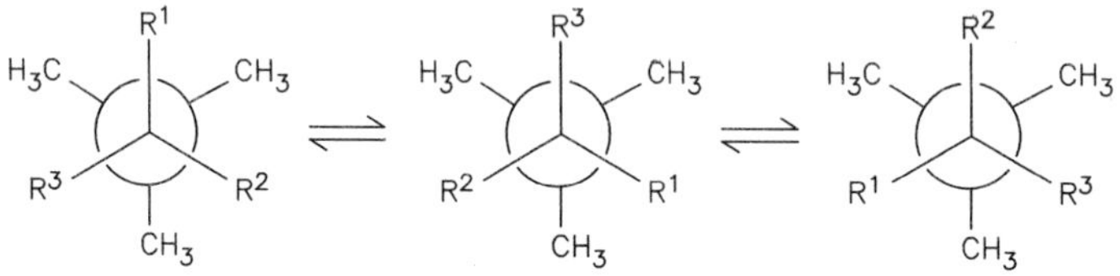

Scheme V

11.3.1.2 C(sp²)—C(sp³) Bonds

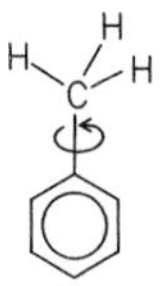

Here we will consider as an example the rotation of the methyl group in toluene (**5**). This rotation cannot be frozen out even at temperatures as low as −150° C. It only becomes possible to measure a C(sp²)—C(sp³) rotational barrier when hindering groups are present, as in the naphthalene derivative **6**. For this compound one finds $T_C = 228$ K (at 60 MHz) and $\Delta G^{\ddagger}_{228} = 12.7 \pm 0.2$ kcal mol^{-1} (53.1 ± 0.8 kJ mol^{-1}) [13].

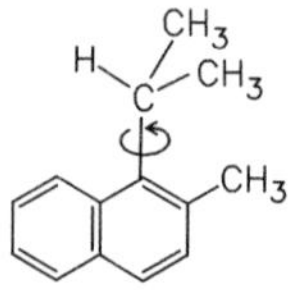

In these cases the sp²-hybridized carbon atom belongs to an aryl ring. However, the rotation is also partly hindered when the sp²-hybridized carbon atom belongs to a CC double bond or to a carbonyl group.

Experiments of this sort often enable conclusions to be reached about the preferred conformation.

5

6

318

11.3.1.3 C(sp²)−C(sp²) Bonds

It was shown a long time ago that in biphenyl derivatives bulky substituents hinder the rotation about the CC bond between the two phenyl rings (atropic isomerism, or occurrence of conformational enantiomers). An example of this type is the biphenyl derivative **7**, which has already been mentioned in Section 2.4.2 (Fig. 2-16). Here the rotational barrier is determined not only by the steric effects discussed above, but also by the electronic effects of the substituents. $\Delta G^{\ddagger}$-values varying from 14 to more than 25 kcal mol^{-1} (60–100 kJ mol^{-1}) have been measured [2].

Other examples in this class are the hindered rotations in chiral derivatives of butadiene [2], and in aromatic aldehydes and ketones. In the case of benzaldehyde (**8**), for example, a $\Delta G^{\ddagger}$-value of 7.9 ± 0.2 kcal mol^{-1} (33 ± 0.8 kJ mol^{-1}) has been determined [14].

HO(CH$_3$)$_2$C   H

H   C(CH$_3$)$_2$OH

7

8

11.3.2 Rotation about a Partial Double Bond
[15, 16]

The rotation about the CN bond in amides (**9**), already described in Section 11.1, is the best known example of hindered rotation about a partial double bond, but there are many others. Thioamides (**10**), amidines (**11**), enamine derivatives (**12**) and aminoboranes (**13**) all show similar behavior.

Amides: There have been many studies of the effects of substituents on the rotation barriers in this class of compounds. In the dimethyl amides (**9**), which are most suitable for such studies from the NMR standpoint, the highest free enthalpy of activation found is that for dimethylformamide (**1**), with a value of 20.9 ± 0.2 kcal mol^{-1} (87.5 ± 0.8 kJ mol^{-1}, T_C = 393 K). Introducing any substituent R in place of the formyl hydrogen atom reduces the barrier. For R = C$_6$H$_5$ the free enthalpy of activation $\Delta G^{\ddagger}_{298}$ is found to be 15.0 ± 0.2 kcal mol^{-1} (62.7 ± 0.8 kJ mol^{-1}). For R = t-C$_4$H$_9$ the value of $\Delta G^{\ddagger}_{298}$ is reduced to 12.2 ± 0.2 kcal mol^{-1} (51.0 ± 0.8 kJ mol^{-1}).

Thioamides (**10**): In thioamides the $\Delta G^{\ddagger}$-values are 2–4 kcal mol^{-1} (8–17 kJ mol^{-1}) higher than in the amides with the same substituents. Attempts have been made to explain these higher barriers as being due to the contribution of a dipolar canonical form (Scheme VI). It appears that the overlap between the 2p and 3p molecular orbitals is less effective than that between the two 2p orbitals in the amides.

O   CH$_3$

R   N   CH$_3$

9

319

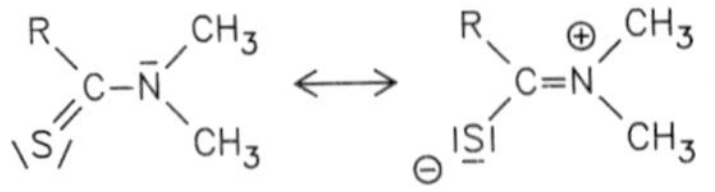

10

Scheme VI

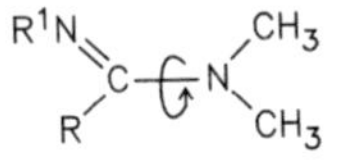

Amidines **11**: In amidines the barrier to rotation about the C−N(CH₃)₂ bond is much smaller than in the amides. $\Delta G^{\ddagger}$-values of 12–14 kcal mol⁻¹ (50–60 kJ mol⁻¹) are found. It appears that this difference is caused by a reduction in the amount of mesomeric stabilization in the ground state.

Enamines: In enamines the mesomerism between the nitrogen and the double bond is too small to give hindered rotation effects detectable by NMR spectroscopy. However, the situation changes when the double bond has a carbonyl group next to it, as in the molecule **12**, giving an unsaturated amide structure. On the basis of the proposed canonical structure with a C=N bond (Scheme VII) one expects to find that the rotation about the CN bond is hindered as in amides. This was in fact found to be the case, giving for **12** a free enthalpy of activation $\Delta G^{\ddagger}_{267} = 13.5 \pm 0.2$ kcal mol⁻¹ (56.5 ± 0.8 kJ mol⁻¹) [17].

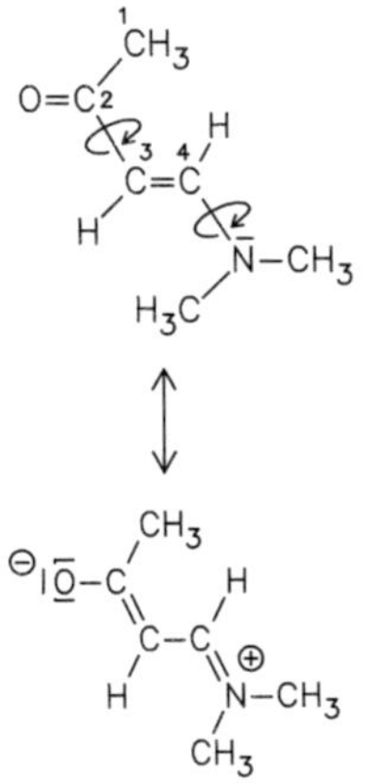

For this class of compounds the rotation about the CN bond is of less interest than that about the bond between C-2 and C-3, as the mesomeric canonical forms indicate that this too should be hindered. This was confirmed experimentally for **12** [18], and the following values were found when using 1,1-dichloroethylene as the solvent:

$$\Delta G^{\ddagger}_{232}(\text{s-cis} \rightarrow \text{s-trans}) = 11.5 \pm 0.2 \text{ kcal mol}^{-1}$$
$$(48.1 \pm 0.8 \text{ kJ mol}^{-1})$$
$$\Delta G^{\ddagger}_{232}(\text{s-trans} \rightarrow \text{s-cis}) = 11.1 \pm 0.2 \text{ kcal mol}^{-1}$$
$$(46.4 \pm 0.8 \text{ kJ mol}^{-1})$$

The equilibrium between the *s-cis* and *s-trans* conformers is markedly dependent on the temperature and the solvent. In compound **12** the preferred form is the *s-cis* conformer. Substituents are found to have a considerable influence on the height of the rotational barrier.

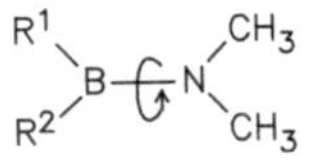

Aminoboranes (**13**): In aminoboranes the NB bond has a high degreee of double bond character. For the rotational barrier in these compounds one finds $\Delta G^{\ddagger}$-values of 15–23 kcal mol⁻¹ (60–100 kJ mol⁻¹), which are comparable with, or even higher than, those of amides [19, 20].

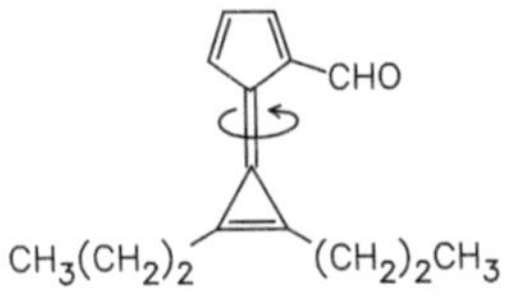

CC Double Bonds: Usually it is not possible by NMR spectroscopy to detect rotation about a CC double bond, as the barrier is too high. Exceptions to this are found when appropriate types of substituents are introduced, as in **14** [21]. The $\Delta G^{\ddagger}$-value for the rotation about the exocyclic double bond in this compound was found to be only 18–19 kcal mol⁻¹

11

12

Scheme VII

13

14

(75–80 kJ mol^{-1}). T_C varies considerably with the solvent, showing values between 339 and 387 K. (Further examples can be found in Ref. [13]).

11.3.3 Inversion at Nitrogen and Phosphorus Atoms

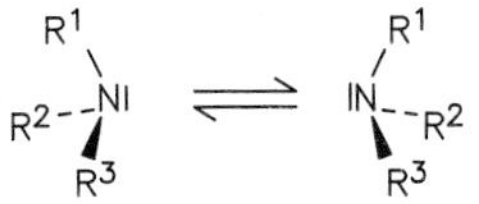

Nitrogen: Nitrogen compounds with three substituents have a pyramidal structure. If the substituents are all different, two enantiomers exist. Usually, however, these cannot be separated, as the molecule undergoes rapid inversion (Scheme VIII).

However, the barrier to inversion can be increased by choosing appropriate substituents, and the inversion can then be studied by NMR spectroscopy. Although the spectra of the enantiomers are identical in achiral solvents, the inversion can be observed by introducing a prochiral group such as benzyl or isopropyl into the molecule (see Section 2.4). Thus, for compound **15** (as a solution in *n*-hexane) the temperature dependence of the methylene proton resonances have a coalescence temperature T_C of 257 K and a value of 12.7 ± 0.2 kcal mol^{-1} (53.1 ± 0.8 kJ mol^{-1}) for $\Delta G^{\ddagger}_{257}$ [22].

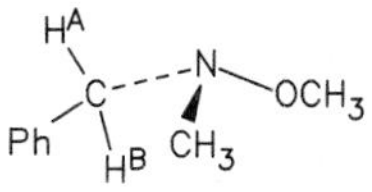

When the nitrogen atom is in a three-, four- or six-membered ring, the inversion is slower. We have already come across an example of this in the introduction to this chapter, namely the inversion in 1,2-diphenyldiazetidinone (**2**, Fig. 11-2), for which T_C = 272 K and $\Delta G^{\ddagger}_{272}$ = 13.3 ± 0.2 kcal mol^{-1} (55.7 ± 0.8 kJ mol^{-1}). At low temperatures the inversion in this compound must occur only slowly, at least for one of the two nitrogen atoms, and this is assumed to be N-1 [23]. The unusually high nitrogen inversion barrier in this case can perhaps be attributed to a transition state which is energetically unfavorable, owing to a parallel orientation of the two doubly occupied p-orbitals.

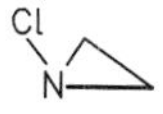

In *N*-chloroaziridine (**16**) the inversion is so strongly hindered that it is not possible to reach the coalescence temperature, as decomposition occurs first (T_C > 180° C).

Phosphorus: The inversion of trivalent phosphorus is very much slower than that of nitrogen, to such an extent that it cannot be observed by NMR spectroscopy. Nevertheless, if a chlorine atom is bonded to phosphorus, as in the chiral compound **17**, the spectrum at high temperatures indicates that there is an inversion process resulting in interconversion between the two enantiomers. However, the process involved here is not a rearrangement of the substituents on the phosphorus atom, but a chlorine exchange of the Walden inversion type [24].

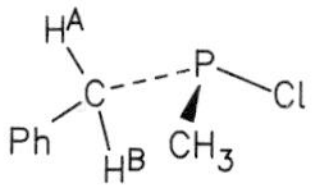

Scheme VIII

15

16

17

11.3.4 Ring Inversion [25]

DNMR spectroscopy opened up a new field of conformational analysis, namely the study of conformational equilibria in ring compounds, including carbocyclic and heterocyclic ring systems of both saturated and unsaturated types. The starting point for these studies was the work of Jensen and his collaborators [8] on the temperature dependence of the ^{1}H resonances in cyclohexane. These investigations and the spectra of cyclohexane-d_{11} (**4**) have already been discussed in Section 11.2.4.

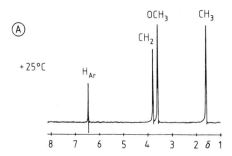

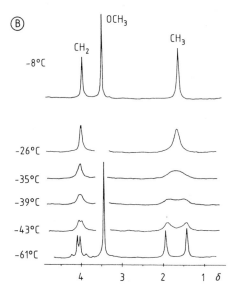

Figure 11-4.
100 MHz ^{1}H NMR spectra of the 2,4-benzodithiepin derivative **18**.
A: Spectrum at 25°C in CS$_2$.
B: Spectra at −8, −26, −35, −39, −43 and −61°C in CS$_2$/pyridine.

Another example of a ring inversion is shown in Figure 11-4. The ^{1}H NMR spectrum of the 2,4-benzodithiepin derivative **18** at +25°C (Fig. 11-4) consists of four signals, which can readily be assigned on the basis of their chemical shifts and relative intensities (see assignments shown above the peaks). This spectrum, consisting only of single peaks, indicates either a high degree of symmetry in the molecular structure, or a rapid inversion of the unsaturated seven-membered ring, which would cause an apparent equivalence of the geminal methyl group pair and of the methylene protons on C-1 and C-5. Examination of molecular models shows that there exists no rigid conformation of the dithiepin ring in which there is equivalence of the four methylene protons and of the two methyl groups at the 3-position, unless this occurs fortuitously. The spectrum can therefore only be accounted for by a rapid equilibration between different conformations of the seven-membered ring.

The series of low temperature spectra shown in Figure 11-4 B tells us something about the nature of this dynamic process. When the sample is cooled to $-61°$C the protons of the two methyl groups at the 3-position give two peaks at δ-values of about 1.4 and 1.9, while the CH_2 protons give an AB-type spectrum of four peaks at $\delta \approx 3.9$ to 4.3. Freezing the intramolecular dynamic process thus enables us to distinguish between the geminal methyl groups and also between the geminally paired protons. But what are the conformers involved in this interconversion?

In principle the most stable conformation could be the chair (C), the boat (B), or the twisted boat (TB) form (Scheme IX). Of these three the twisted boat conformation can be eliminated straight away on the basis of the low temperature spectra, as it has C_2 symmetry, which would make the two methyl groups on C-3 equivalent.

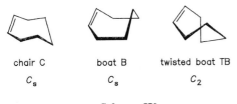

| chair C | boat B | twisted boat TB |
| C_s | C_s | C_2 |

Scheme IX

Furthermore, the predicted shift difference Δv between these two methyl groups for the boat conformation, on the basis of the ring current model (Section 2.1.2.2), is well over 100 Hz (at a resonance frequency of 100 MHz), and is thus inconsistent with the observed shift of 52 Hz.

All the experimental and theoretical results indicate that the energetically preferred conformation is a chair, and we are therefore observing an interconversion between two chair conformations of equal energy, which occurs through a ring inversion (Scheme X).

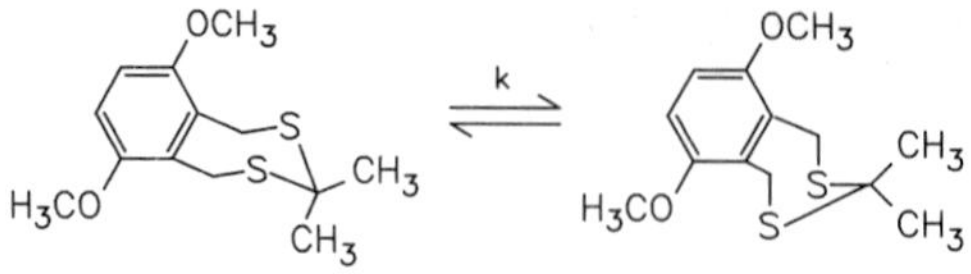

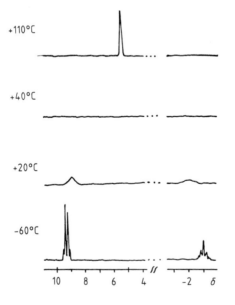

Scheme X

From the observed values $T_C = 236 \pm 2$ K and $\Delta\nu = 52$ Hz (at 100 MHz) we obtain:

$$k_{236} = 116 \pm 30 \text{ s}^{-1} \quad \text{and} \quad \Delta G^{\ddagger}_{236} = 11.5 \pm 0.2 \text{ kcal mol}^{-1}$$
$$(48.1 \pm 0.8 \text{ kJ mol}^{-1}).$$

As has already been shown for the example of cyclohexane-d_{11} (Section 11.2.4), in interpreting kinetic NMR data one needs to take account of intermediate conformations. The existence of these can seldom be proved experimentally, and for this reason they are often mistakenly left out of the analysis. A better and more exact procedure is to first calculate the reaction paths and the energetically favored conformations, then to interpret the experimental results in the light of the insight thereby obtained into the mechanism of the dynamic process [25, 26].

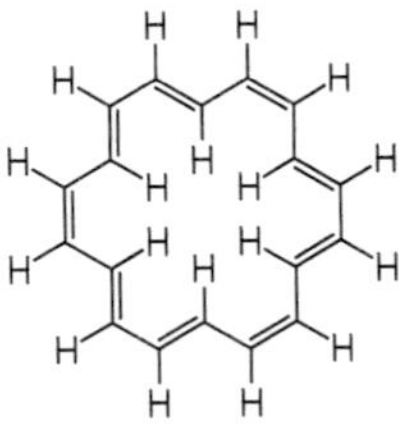

Figure 11-5.
60 MHz ^{1}H NMR spectra (redrawn) of [18]-annulene (**19**), at +110, +40, +20 and −60° C in tetrahydrofuran-d_8 (for original spectra see Ref. [27]).

The spectrum of [18]-annulene (**19**) has already been discussed in connection with the ring current effect (Section 2.1.2.2). The six inner protons are strongly shielded ($\delta = -1.8$), whereas the twelve outer ones are comparatively weakly shielded ($\delta = 8.9$). However, in Section 2.1.2.2 no mention was made of the line-widths of the signals. Figure 11-5 shows that in the spectrum recorded at $20°C$ the signals are very broad. This indicates that an exchange process is occurring. In fact these signals become even broader when the sample temperature is raised slightly, and at $40°C$ they can no longer be seen! However, by the time we reach $110°C$ we have a sharp single peak at $\delta = 5.29$.

If, on the other hand, we reduce the temperature to about $-60°C$, the two signals at $\delta = -3$ and $+9.3$ become narrowed to such an extent that even the couplings to the non-equivalent neighboring protons are resolved [2]. It appears that at low temperatures the molecule is planar, and has aromatic character. At high temperatures, on the other hand, the annulene ring undergoes rapid changes of conformation, and the inner and outer protons exchange positions, so that all 18 protons show the same time-averaged shielding value. At the same time a valence tautomerism occurs.

11.3.5 Valence Tautomerism

The best known example of a valence tautomerism being demonstrated by NMR spectroscopy is the rearrangement process in bullvalene (**20**). The ^1H NMR spectrum of **20** at $+120°C$ consists of a single sharp peak at $\delta = 4.5$. On cooling the sample to room temperature the line-width increases to several hundred Hz. When the temperature is further reduced this broad band splits, and at $-59°C$ one obtains two multiplets, one with a relative intensity of 4 at $\delta \approx 2$, and the other with a relative intensity of 6 at $\delta \approx 5.6$ (the spectra can be found in Ref. [28]). This temperature-dependent behavior of the ^1H spectrum is accounted for by a Cope rearrangement (Scheme XI), which occurs so rapidly at high temperatures that all the protons experience an averaged shielding, and the spectrum consists of only one peak. At low temperatures, on the other hand, the interconversion is slow, and separate signals are observed for the chemically non-equivalent protons.

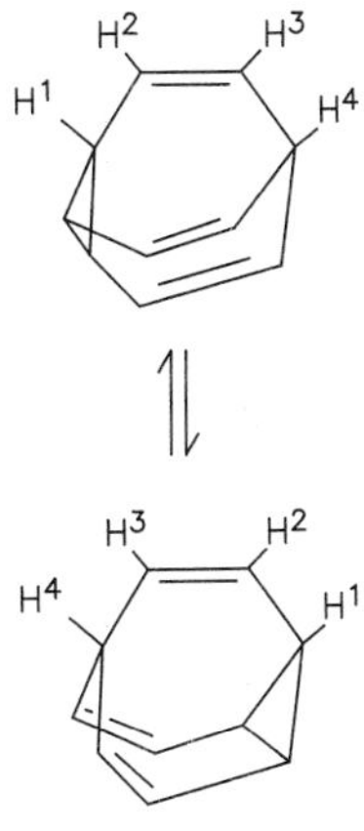

This valence tautomerism can also be observed in the ^{13}C spectrum [29]. At $141°C$ (for a ^{13}C resonance frequency of 25 MHz) a single peak is found at $\delta = 86.4$, whereas at $-60°C$ there are four: two for the olefinic carbon nuclei at $\delta = 128.3$ and 128.5, one for the carbon nuclei of the three-membered ring at $\delta = 21.0$, and one for C-4 at $\delta = 31.0$. In the coalescence region (about 40 to $50°C$) the line-width is of the order of 4000 Hz!

20

Scheme XI

Other examples of valence tautomerism can be found in the literature [2]; one of these, that of [18]-annulene (**19**), has already been mentioned.

11.3.6 Keto-Enol Tautomerism

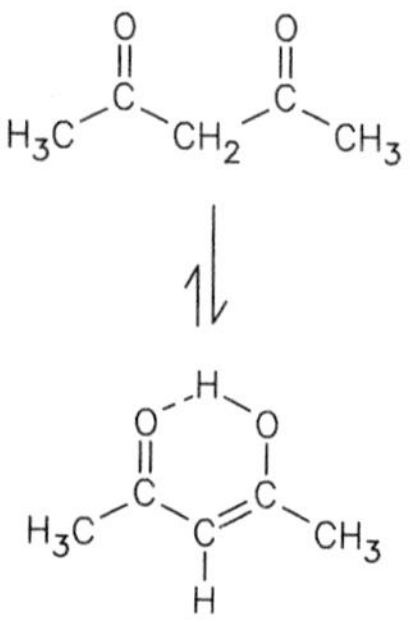

β-Diketones form stable enols. The classic example for which the keto-enol tautomerism has been intensively studied is that of acetylacetone (**21**).

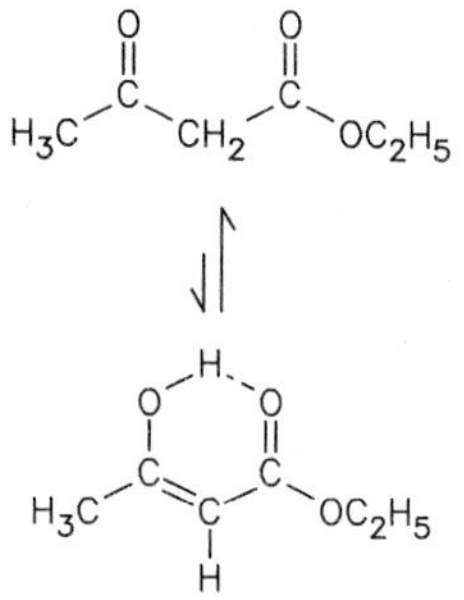

The equilibrium can be studied in an elegant way by ^{1}H NMR spectroscopy, as the interconversion between the tautomers is slow on the ^{1}H NMR time scale, so that separate signals are obtained for molecules in the enol and keto forms (Fig. 11-6). One can unambiguously identify the signal of the olefinic protons in the enol at $\delta = 5.5$, and that of the methylene protons in the ketone at $\delta = 3.5$, and a comparison of their intensities shows that the equilibrium mixture contains more molecules in the enol than in the keto form. From this the assignment of the two methyl signals in the region of $\delta \approx 2$ also follows on directly: the smaller of these signals belongs to the keto form and the larger to the enol. The ratio of the tautomers is best determined from the integral curve using the methyl signals; the enol : keto ratio is found to be about 80 : 20 (see expanded region of spectrum). The signal of the strongly acidic hydrogen-bonded OH proton is at $\delta = 15.5$. Increasing the temperature or adding a base causes the tautomers to interconvert more rapidly, and the spectrum then consists only of time-averaged signals (see also Section 11.3.7).

In the acetylacetic ester **22**, in contrast to acetylacetone, the ketone predominates in the equilibrium mixture (approximately 90 % ketone to 10 % enol). Here, evidently, the enol form is not so energetically favored as in acetylacetone.

Scheme XII

21

Scheme XII

22

Scheme XIII

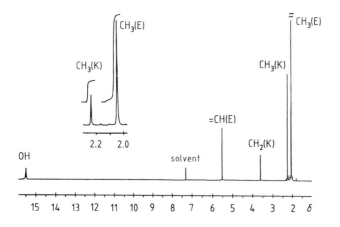

Figure 11-6.
250 MHz ^{1}H NMR spectrum of acetylacetone (**21**) in CDCl$_3$ at about 22° C. The methyl region is shown expanded and integrated. (E = enol, K = ketone).

326

11.3.7 Intermolecular Proton Exchange

The kinetics of proton exchange have long been of interest to chemists, since acids and bases play an important role in many reactions. It was therefore natural that ^{1}H NMR spectroscopy, which allows one to directly observe the protons involved, should be applied to this problem.

Under normal circumstances protons bonded to oxygen or nitrogen undergo such rapid exchange compared with the NMR time scale that these processes cannot be studied by NMR spectroscopy. However, in pure alcohols or solutions of alcohols in dimethylsulfoxide or acetone the situation is different, as can be seen, for example, in the 500 MHz NMR spectrum of ethanol shown at the top and left-hand edges of Figure 11-7. At room temperature the mean bonding lifetime τ_1 of the protons on oxygen is long enough to allow the coupling between the OH and CH$_2$ protons to be seen; the OH signal ($\delta = 3.3$) is split into a triplet. (The relationship between the observed coupling and the lifetime has already been discussed in Section 3.6.3.) Nevertheless, the two-dimensional EXSY spectrum shown in Figure 11-7 provides evidence that a slow exchange process is still occurring (see also Section 9.4.6). The diagonal peaks are of no interest, but we find two off-diagonal correlation peaks at δ-values of (2.8,3.3) and (3.3,2.8). These show that exchange occurs between the OH protons of the alcohol and the protons of the small amount of water present in the acetone solvent, but at a rate too slow to cause a noticeable broadening in the one-dimensional spectrum. The EXSY results do not enable us to determine the rate of the exchange process.

The proton exchange rate can be increased by progressively adding more water. In the one-dimensional spectrum one observes first a broadening of the alcohol OH and water signals, and also of the CH$_2$ signal. Ultimately, for very fast exchange, one obtains only a single averaged peak for all the OH protons. Their coupling to the CH$_2$ protons is then no longer observed, and the CH$_2$ resonance itself becomes a simple quartet caused by the coupling to the three protons of the methyl group. Raising the temperature has a similar effect in accelerating the proton exchange.

Acidifying the sample increases the rate constant for the exchange by many orders of magnitude:

$$CH_3CH_2OH + \overset{\oplus}{H}OH_2 \rightleftharpoons CH_3CH_2\overset{\oplus}{O}\diagup^{\textstyle H}_{\textstyle H} + H_2O$$

For this reason the spectra of carboxylic acids and phenols in aqueous solution always contain only one signal for all the exchangeable protons.

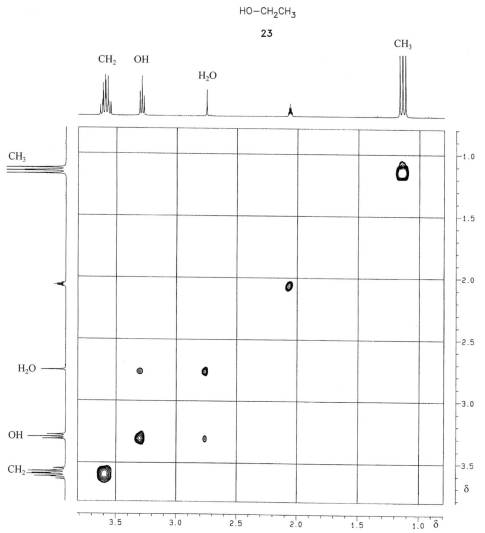

HO−CH$_2$CH$_3$

23

Figure 11-7.
EXSY spectrum of ethanol (**23**) in acetone-d$_6$. The 500 MHz ^{1}H NMR spectrum is shown at the top and left-hand edges, with assignments. The CH$_2$ and OH protons are coupled, showing that the intermolecular proton exchange is slow on the time-scale of the one-dimensional NMR spectroscopic measurement. However, the two cross-peaks lying off the diagonal show that the ethanol OH protons do in fact undergo exchange with the small amount of H$_2$O present in the solvent.
(Experimental conditions:
approx. 10 mg in 0.5 ml acetone-d$_6$; 5 mm sample tube; 128 measurements with t_1 altered in 334 μs increments; mixing time Δ=1 s; each measurement with 8 FIDs and 0.5 K data points; total time approx. 40 min.)

The proton exchange process in the dibenzylmethylammonium ion (**24**) [30] deserves special mention here. In a concentrated HCl solution of the ion at 25°C a coupling is observed between the proton on the nitrogen and the methylene protons, showing that the exchange of the NH proton is slow. However,

328

if the HCl concentration is reduced this coupling disappears. The NH protons must now be exchanging positions rapidly. What makes this process particularly interesting is that the proton exchange involves a Walden rearrangement at the nitrogen atom (Scheme XIV). Here again, as in Section 11.3.3, the exchange process can be studied by observing the methylene proton signals, since at fast rates of inversion the AB spectrum changes to a single peak. (For other proton exchange reactions see Ref. [12].)

$$\underset{\textbf{24}}{Ph\text{--}H^BH^AC\text{--}\overset{H_3C}{\underset{Ph\text{--}H^BH^AC}{\overset{\oplus}{\underset{\diagup}{N}}}}\text{--}H \underset{+\,H^+}{\overset{-\,H^+}{\rightleftharpoons}} \left[\overset{\diagdown}{\underset{\diagup}{N}}I \rightleftharpoons I\overset{\diagup}{\underset{\diagdown}{N}} \right] \underset{-\,H^+}{\overset{+\,H^+}{\rightleftharpoons}} H\text{--}\overset{\oplus}{N}\overset{CH_3}{\underset{CH^AH^B\text{--}Ph}{\text{--}CH^AH^B\text{--}Ph}}}$$

Scheme XIV

11.3.8 Reactions and Equilibration Processes

Finally we must mention here one further area of application of dynamic NMR spectroscopy, even though it may appear rather pedestrian: that of following the progress of a reaction with the help of a stop-watch.

All the kinetic processes discussed up to now were studied in their equilibrium state. However, conventional chemical reactions proceed (slowly) towards an equilibrium, or to complete conversion of the reactants. In such a reaction one does not observe changes in the line shapes in the NMR spectrum, but only changes in the combination of signals in the reaction mixture and their intensities. It is also possible to follow the reaction quantitatively by integrating appropriate signals. Here we will consider an example from biochemistry, namely the enzymic hydrolysis of the synthetic substrate p-nitrophenyl-α-glucoside (**25**) by α-glucosidase extracted from baker's yeast (Scheme XV).

Figure 11-8 A to C shows the spectra of the starting compound **25**, and of the incubation solution after approximately five minutes, then after about an hour. The hydrolysis leads first to the formation of α-glucose (**26**, Fig. 11-8 B), which then undergoes mutarotation to give an equilibrium mixture of α-glucose (**26**) and β-glucose (**27**), with the spectrum shown in Figure 11-8 C. All three compounds – the starting compound and α- and β-glucose – can easily be distinguished in the ^{1}H NMR spectra by means of the characteristic signals of the anomeric proton H-1.

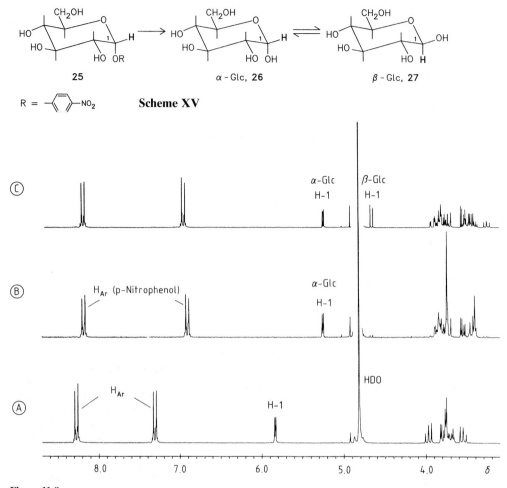

Scheme XV

R = —⟨ ⟩—NO₂

25 → α - Glc, 26 ⇌ β - Glc, 27

Figure 11-8.
250 MHz ^{1}H NMR spectrum of an incubated solution (in D$_2$O) of 10 μmol 4-nitrophenyl-α-D-glucopyranoside (**25**) and 6 units (IU) [31] of α-glucosidase from baker's yeast in 0.5 ml 50 mM KCl solution, pD 6.7; 22° C; 16 K data points. A: Pure substrate **25**, 16 FIDs. B: Incubated solution after about 5 min, 24 FIDs; the substrate **25** has already undergone quantitative decomposition to α-glucose (**26**) and p-nitrophenol. C: Incubated solution in the mutarotational equilibrium, 24 FIDs.

The first reaction step, the enzymic hydrolysis, is an irreversible first-order reaction, and the second is a first-order equilibrium reaction. Both these reactions can be optimized by choosing suitable conditions, and they can be monitored quantitatively by observing the changes in the signal intensities.

The experiment revealed that in the enzymic scission the configuration at the carbon atom C-1 of the glucose is preserved. (The purpose of the experiment was to obtain an answer on this point, not to measure the rate constants.) For solving stereochemical problems of this kind, NMR spectroscopy is virtually the only available method.

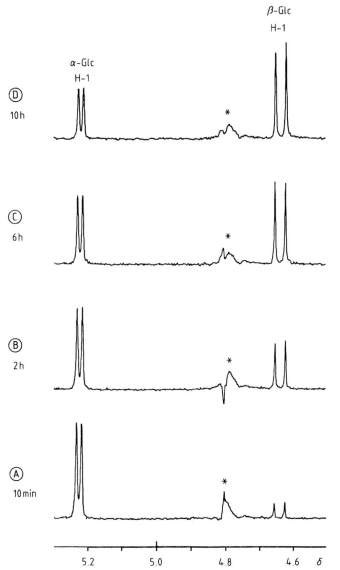

β-Glc
H-1

α-Glc
H-1

Ⓓ
10 h

Ⓒ
6 h

Ⓑ
2 h

Ⓐ
10 min

5.2   5.0   4.8   4.6   δ

Figure 11-9.
Portions of 250 MHz ¹H NMR
spectra of α- and β-glucose (**26** and
27). The initial solution contained
10 μmol α-glucose in 0.5 ml 50 mM
KCl-solution; pD 6.7; 20°C; 8 K
data points; 16 FIDs in each case
with suppression of the solvent
signal [32];
*denotes the residual HDO signal
of the solvent. Spectra A to D
were recorded after 10 min, 2 h,
6 h and 10 h respectively. Integration of the signals in spectrum D
gives proportions of 40 % α- to
60 % β-glucose.

Figure 11-9 shows a series of expanded spectra from which
quantitative results can be calculated. A separate experiment
was carried out to investigate the mutarotation of glucose, starting from α-glucose (**26**). The four expanded 250 MHz ¹H NMR
spectra (Fig. 11-9, A to D) show the region which contains the
H-1 signals. The HDO signal was almost completely suppressed
[32]; under these conditions there is no difficulty in performing
the integration and thereby determining the rate constants.

11.4 Bibliography for Chapter 11

[1] L. M. Jackman and F. A. Cotton (Eds.): *Dynamic Nuclear Magnetic Resonance Spectroscopy.* New York: Academic Press, 1975.

[2] M. Oki (Ed.): *Applications of Dynamic NMR Spectroscopy to Organic Chemistry.* Deerfield Beach: VCH Publishers, 1985.

[3] J. Sandström: *Dynamic NMR Spectroscopy.* New York: Academic Press, 1982.

[4] G. Englert, *Z. Naturforsch. Teil B 16,* (1961) 413.

[5] H. S. Gutowsky: "Time-Dependent Magnetic Perturbations" in Ref. [1], Ch. 1.

[6] G. Binsch: "Band-Shape Analysis" in Ref. [1], Ch. 3.

[7] D. S. Stephenson and G. Binsch, *J. Magn. Reson. 30* (1978) 145 and *32* (1978) 145.

[8] F. R. Jensen, D. S. Noyce, C. H. Sederholm and A. J. Berlin, *J. Amer. Chem. Soc. 82* (1960) 1256 and *84* (1962) 386.

[9] F. A. L. Anet, M. Ahmed and L. D. Hall, *Proc. Chem. Soc.* (1964) 145.

[10] F. A. Bovey, F. P. Hood, E. W. Anderson and R. L. Kornegay, *Proc. Chem. Soc.* (1964) 146.

[11] F. A. Bovey: *Nuclear Magnetic Resonance Spectroscopy.* New York: Academic Press, 1969, p. 191.

[12] E. Grunewald and E. K. Ralph: "Proton Transfer Processes" in Ref. [1], Ch. 15.

[13] A. Mannschreck and L. Ernst, *Chem. Ber. 104* (1971) 228.

[14] S. Sternhell: "Rotation about Single Bonds in Organic Molecules" in Ref. [1], Ch. 6.

[15] Ref. [2], Chs. 2 and 3.

[16] L. M. Jackman: "Rotation about Partial Double Bonds in Organic Molecules" in Ref. [1], Ch. 7.

[17] J. Dabrowski and L. Kozerski, *Chem. Commun.* (1968) 586.

[18] J. Dabrowski and L. Kozerski, *J. Chem. Soc. B* (1971) 345.

[19] D. Imbery, A. Jaeschke and H. Friebolin, *Org. Magn. Reson. 2* (1970) 271.

[20] H. Friebolin, R. Rensch and H. Wendel, *Org. Magn. Reson. 8* (1976) 287.

[21] A. S. Kende, P. T. Izzo and W. Fulmor, *Tetrahedron Lett.* (1966) 3697.

[22] D. L. Griffith and J. D. Roberts, *J. Amer. Chem. Soc. 87* (1965) 4089.

[23] E. Fahr, W. Rohlfing, R. Thiedemann, A. Mannschreck, G. Rissmann and W. Seitz, *Tetrahedron Lett.* (1970) 3605.

[24] D. Imbery and H. Friebolin, *Z. Naturforsch. Teil B 23* (1968) 759.

[25] F. A. L. Anet and R. Anet: "Conformational Processes in Rings" in Ref. [1], Ch. 14.

[26] Ref. [3], Ch. 10.

[27] F. Sondheimer, I. C. Calder, J. A. Elix, Y. Gaoni, P. J. Garratt, K. Grohmann, G. Di Maio, J. Mayer, M. V. Sargent and R. Wolovsky: "The Annulenes and Related Compounds" in *Aromaticity*. Spec. Publ. No. 21, Chemical Society, London 1967, p. 75.

[28] G. Schröder, J. F. M. Oth and R. Merenyi, *Angew. Chem. Int. Ed. Engl. 4* (1965) 752.

[29] J. F. M. Oth, K. Müllen, J.-M. Gilles and G. Schröder, *Helv. Chim. Acta 57* (1974) 1415.

[30] M. Saunders and F. Yamada, *J. Amer. Chem. Soc. 85* (1963) 1882.

[31] IU = International Unit; 1 IU is defined as the quantity of an enzyme which releases 1 μmol of 4-nitrophenol per minute at 30° C from the substrate **25**.

[32] For suppressing the signal of water and HDO we used the microprogram "Multiple Solvent Suppression" given in the *Aspect 2000 NMR Software Manual 1*, Bruker, Karlsruhe-Rheinstetten.

Additional and More Advanced Reading

Refs. [1] to [3].

G. Binsch: The Study of Intramolecular Rate Processes by Dynamic Nuclear Magnetic Resonance. In: *Topics in Stereochemistry,* E. L. Eliel and N. L. Allinger (Eds.) *3* (1968) 97.

H. Kessler: Detection of Hindered Rotation and Inversion by NMR Spectroscopy. *Angew. Chem. Int. Ed. Engl. 9* (1970) 219.

G. Binsch and H. Kessler: The Kinetic and Mechanistic Evaluation of NMR Spectra. *Angew. Chem. Int. Ed. Engl. 19* (1980) 411.

E. Kolehmainen: Novel Applications of Dynamic NMR in Organic Chemistry. In: *Annual Reports on NMR Spectroscopy,* G. A. Webb (Ed.), Vol. 51, Kidlington, Oxford: Elsevier Science, 2003, p. 1.

12 Shift Reagents

12.1 Lanthanide Shift Reagents (LSRs)

12.1.1 Fundamentals

Reference has already been made several times to the undesirable effects of paramagnetic impurities, which shorten the relaxation times and thereby cause line broadening. On the other hand, in Section 1.6.3.2 we saw, in connection with integration in ^{13}C NMR spectroscopy, that the interaction with paramagnetic chelate complexes can sometimes serve a useful purpose by suppressing the nuclear Overhauser effect. A fact not mentioned so far is that paramagnetic substances also cause a shift of the signals. Since not all the resonances in a molecule are affected to the same extent, such shifts can be used as an aid to spectral analysis.

The paramagnetic ions of nickel and cobalt were the first to be used, but these experiments suffered from severe line broadening, which outweighed the benefits from the shift effects. Only when Hinckley [1] discovered, in 1969, that paramagnetic lanthanide ions gave shifts without significant line broadening, did this effect develop into a technique holding great promise for the NMR spectroscopist. An example from ^1H NMR spectroscopy is illustrated in Figure 12-1.

The upper part of the figure (A) shows the 90 MHz ^1H NMR spectrum of hexan-1-ol (**1**) in CCl_4. Spectrum B was recorded after adding the tris(dipivaloylmethanato)-Eu(III) chelate complex **2** to the solution. Here we will take the assignments indicated in B as correct without discussing the details. Two facts about the spectrum are worth remarking on:

- The addition of the shift reagent reduces the shieldings of all the protons, i.e. all the signals are shifted to lower field; at the same time all the CH_2 peaks become separated, in contrast to the previous situation where isolated signals were observed only for OH, OCH_2 and CH_3.
- The shifts increase with the proximity of the protons to the oxygen atom of the alcohol. The signal of the OH proton, which is bonded directly to the oxygen, is shifted by more than 20 ppm.

$$\overset{1}{HO}-\overset{1}{CH_2}\overset{2}{CH_2}\overset{3}{CH_2}\overset{4}{CH_2}\overset{5}{CH_2}\overset{6}{CH_3}$$

1

$(CH_3)_3C$ ... $C(CH_3)_3$

$Eu_{1/3}$

$Eu(DPM)_3$

2

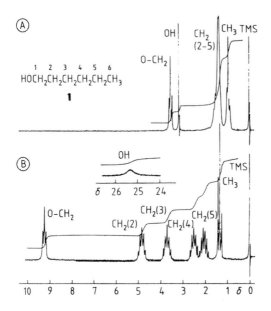

Figure 12-1.
A: 90 MHz ^{1}H NMR spectrum of hexan-1-ol (**1**) in CCl$_4$.
B: Spectrum of **1** after adding the shift reagent Eu(DPM)$_3$ (**2**).
(30 mg of **1** to 50 mg of Eu(DPM)$_3$).

How can these experimental results be accounted for by theory? Since the effect is related to the presence of paramagnetic ions, it is reasonable to suppose that it involves an interaction between the nuclear spins and the spin of the unpaired electron. We must distinguish between two types of interaction, both of which cause shifts of the signals, namely the *contact interaction* and the *pseudocontact interaction.* Both depend on the formation of a complex between the substrate S and the paramagnetic metal ion L; in solution a dynamic equilibrium is established between the free components and the complex:

$$L + S \rightleftharpoons LS \qquad \text{Example:} \quad R{-}\overset{\displaystyle H}{\overset{\diagup}{O}}\ldots\ldots Eu(DPM)_3$$

In the complex some of the spin density of the unpaired electron becomes transferred to the substrate molecule owing to the contact interaction. As the electron spin densities at the positions of the nuclei under observation are widely different, the resulting shift effect is not the same throughout the molecule. In saturated compounds the ^{13}C nuclei whose resonance positions are most affected are those at the α- and β-positions relative to the complexing center (e.g. O, N or S), whereas in conjugated systems, as a result of the greater degree of delocalization of the spin density, the resonances of more distant ^{13}C nuclei are also affected. Also one expects to find significant differences between the shifts for ^{1}H and ^{13}C resonances. The *contact term* is usually negligible in ^{1}H NMR spectroscopy, but this is not the case for ^{13}C.

336

Since our discussion of the effects and their applications will be mainly restricted to ^{1}H NMR spectroscopy, we must now turn our attention to the *pseudocontact term*, which is the most important contribution in the case of proton resonances. The pseudocontact interaction is the name used to describe a dipolar interaction between the magnetic dipole field of the unpaired electron and that of the nucleus being observed. This interaction is transmitted through space. The resulting shift Δ_{Dip} in the resonance frequency depends on the geometry of the complex, and is given by Equation (12-1).

$$\Delta_{Dip} = K \frac{3\cos^2 \vartheta - 1}{r^3} \tag{12-1}$$

In this equation r is the distance between the paramagnetic ligand L and the nucleus R being observed (^{1}H or ^{13}C, see sketch), and ϑ is the angle between the line LR and the axis L−O of the complex. K is a constant whose value depends on the magnetic dipole moment of the paramagnetic metal ion. In order for Equation (12-1) to be valid the paramagnetic center in the complex must be symmetrical about the L−O axis.

It follows from Equation (12-1) that:
- the shift effect decreases in inverse proportion to r^3;
- the value of Δ_{Dip} is independent of the nuclide being observed (^{1}H or ^{13}C in this case);
- Δ_{Dip} can be positive or negative, depending on whether $(3\cos^2 \vartheta - 1)$ is greater than or less than zero.

The experimentally measured chemical shift δ_{exp} is a weighted average between the chemical shift δ_S in the free substrate and the chemical shift δ_{LS} in the lanthanide ion – substrate complex LS, the weighting factors being the respective mole fractions x_S and x_{LS}:

$$\delta_{exp} = x_S\delta_S + x_{LS}\delta_{LS}, \quad \text{with} \quad x_S + x_{LS} = 1 \tag{12-2}$$

12.1.2 Applications and Quantitative Interpretation

From the above description of the technique, two areas of application can be identified, namely:
- simplifying complicated spectra, and
- determining accurate geometrical data for the LS complex, and hence for the molecule of interest.

Putting it simply, we thus have two types of applications, qualitative and quantitative. To determine geometrical data by applying equation (12-1) it is essential to carry out the experiments very carefully. Moreover, the analysis of the experimental data is a lengthy process, and consequently such quantitative investigations are not often undertaken.

The overwhelming majority of LSR experiments are limited to qualitative observation of the shifts. We have already met such an application in Figure 12-1. The advantages of the method are obvious:

- overlapping signals can be separated
- assigning the signals is made easier
- it becomes possible to integrate signals which would otherwise overlap
- decoupling experiments can be carried out.

It is often sufficient to record a single spectrum, without bothering to weigh out an exact quantity of the LSR; for example, only one spectrum was needed in the case of hexan-1-ol (**1**, Fig. 12-1). In other cases it is necessary to record several spectra with different LSR:substrate ratios, so as to be able to identify the individual signals by following the progress of their shifts.

Figure 12-2 shows a series of such spectra for the pure L-enantiomer of 1-phenylethylamine (**3**). The LSR used here was the chiral europium complex **4**. The chirality of the molecules does not concern us here, but in Section 12.2 we shall take a more detailed look at this example.

Figure 12-2 A shows the spectrum of **3**, with assignments, and with no LSR added. In spectra B to H the concentration of added LSR was steadily increased. It can be seen that all the signals are shifted to the left, i.e. the proton shielding is reduced, the effect being greatest for the NH_2 signal. In spectrum D the NH_2 signal has already disappeared, as it lies outside the range shown here, at $\delta = 12.08$. The quartet for the CH proton can be seen up to a higher LSR concentration, but in spectrum H, for equimolar amounts of the substrate **3** and the LSR **4**, this too has shifted outside the range shown.

Even the signals of the *ortho*, *meta* and *para* ring protons become clearly separated through the effect of the LSR, the shift effect being greatest for the *ortho* protons. At the higher LSR concentrations all the signals are broadened. This is seen especially clearly for the CH and methyl group protons. Also signals due to the reagent **4** now start to appear (spectrum H).

It is useful to plot, in a single diagram, the chemical shifts for all the signals against the concentration of the added shift reagent. As the shifts are proportional to the amount of LSR added, the experimental points lie on straight lines. From such a diagram one can determine whether or not any of the signals cross over as the LSR concentration is increased. Also, by

3

Eu(TFC)$_3$

4

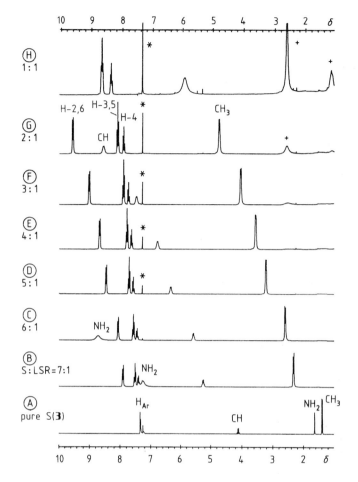

1-phenylethylamine (**3**)

NH₂
|
⁴⟨○⟩¹—CH—CH₃   **3**
  ³ ²

Figure 12-2.
A: 250 MHz ^{1}H NMR spectrum of
1-phenylethylamine (**3**) in CDCl₃.
B to H: Spectra of various mixtures
of **3** with the (chiral) shift reagent
Eu(TFC)₃ (**4**).
In spectrum D the NH₂ signal is at
$\delta = 12.08$, and in spectrum E at
13.96. (+ denotes signal from the
LSR; * denotes residual signal of
solvent.)

extrapolating to zero LSR concentration, one can obtain the
chemical shifts and assignments if these were not already
apparent from the spectrum without the LSR. (In publications
it is usual to quote the shift values for equimolar amounts of
substrate and LSR. These values too can be obtained from a
plot of the kind described above.)

The lanthanide ions that have now become generally adopted
for practical use are the trivalent paramagnetic ions of europium, Eu^{3+}, and praseodymium, Pr^{3+}. Eu^{3+} ions usually
cause a reduction in the shielding (i.e. a shift to greater
δ-values), whereas Pr^{3+} ions cause an increase in shielding
(a shift to smaller δ-values).

So that they can be dissolved in organic solvents, these ions
are used in the form of chelates. The complexing agents most
commonly used to make these chelates are the anions of the
β-diketones 2,2,6,6-tetramethyl-3,5-heptandione (**2**, dipivaloyl-
methane, DPM) and 1,1,1,2,2,3,3-heptafluoro-7,7-dimethyl-
4,6-octandione (**5**, FOD). The signals of the ligand usually

$(CH_3)_3C$ ... $CF_2CF_2CF_3$ with Eu₁/₃ chelate (**5**)

Eu(FOD)₃
5

do not interfere with the spectra, as the ligand protons are very strongly deshielded by the metal ion. In addition to the complexing agents **2** and **5** already mentioned, many others have been described, including perdeuterated and perfluorinated compounds. A few of these shift reagents are commercially available.

The magnitude of the induced shift depends not only on the molar ratio of LSR to substrate, but also on the complex-forming strength of the substrate, i.e. its ability to act as a Lewis base. For LSRs formed from Eu^{3+} and Pr^{3+} ions it is found that the absolute magnitude of the shifts decreases in the order:

$$NH_2 > OH > C=O > COOR > C\equiv N$$

With regard to the question of which LSR should be used with a particular type of substrate, no general rules can be laid down; each problem has to be considered separately. Certain types of compounds present difficulties, for example strong acids and phenols, which decompose chelate complexes. LSRs formed from FOD (**5**) as the chelating agent are stable to a limited extent against weak acids.

Proton resonances of olefins and arenes do not show lanthanide-induced shifts, as these compounds do not form complexes with lanthanide ions. However, a solution to this problem has recently been found. It is known that silver(I) ions (Ag^+) form complexes with compounds which have π-electrons. Because of this, it is found that if one adds Ag^+ ions in the form of AgFOD to a solution containing the substrate and the LSR, shifts are observed even for olefins and arenes. Evidently the bound silver ions serve to transmit the shift effect. A number of applications, mainly using chiral shift reagents, have been described in the literature [2] (see Section 12.2).

The effects observed become very complicated in cases where the substrate contains several functional groups, as in sugars, for example. Despite these difficulties, some elegant results of stereochemical significance have been obtained in the last few years by using chiral LSRs. This special area of application will be discussed in the next section.

12.2 Chiral Lanthanide Shift Reagents

Enantiomers are indistinguishable by NMR spectroscopy. Consequently it is not possible to determine whether the sample is a pure enantiomer or a racemate. However, this difficulty can be overcome by a trick which is well known from chemistry:

that of making diastereomers by using a chiral secondary reagent. It is not essential for a covalent bond to be formed between the two components, as occurs between an optically active carboxylic acid and an alcohol; the formation of a diastereomeric complex is sufficient to make the method work. The secondary reagent can be either a chiral diamagnetic compound (usually the solvent; see Section 12.3) or a chiral paramagnetic lanthanide shift reagent.

Assuming that the substrate S is present as a racemic mixture of S(+) and S(−), the addition of a pure enantiomer of a chiral LSR results in the formation of two diastereomeric complexes. If, for example, the LSR is L(−), we obtain:

$$S(+) + L(-) \rightleftharpoons S(+)L(-) \text{ and}$$
$$S(-) + L(-) \rightleftharpoons S(-)L(-)$$

In the complexes thus formed, the enantiomers S(+) and S(−) can be distinguished by their different resonances.

Figure 12-3 shows an example. Here we used 1-phenylethylamine (**3**) as the chiral substrate, and tris(3-(2,2,2-trifluoro-1-hydroxyethylidene)-*d*-camphorato)europium (Eu(TFC)$_3$, **4**) as the chiral LSR. (The spectra of **3** in the pure L form with different concentrations of this shift reagent have already been given in Figure 12-2. There, however, the objective was merely to show how the shifts depend on the LSR concentration.)

As expected, in the spectrum of the racemate of **3** without any chiral LSR added (Fig. 12-3 A) there is no separation between the signals of the two enantiomers. However, a change is seen when the pure D enantiomer of Eu(TFC)$_3$ (**4**) is added. The proton attached to the asymmetric carbon atom in **3** now gives two quartets with an intensity ratio 1:1 (Fig. 12-3 B). The signals are slightly broadened. Also the methyl resonances have altered. However, instead of the expected two doublets we find only a triplet. Evidently the middle peak is formed by the superposition of a pair of lines, one from each doublet.

Spectrum C in Figure 12-3 results from a sample of **3** containing approximately 80 % of the L enantiomer and 20 % of the D. We again see that the two enantiomers give separate pairs of signals for the methyl protons and the CH proton. By integrating suitable signals one can determine the enantiomer ratio (the so-called optical purity) of a sample; in this case it is best performed by integrating the quartets. However, it is not possible from the studies that have been carried out so far to formulate universal rules as to which enantiomer – L or D – will show the largest shifts.

Chiral LSRs have also been used in an elegant way to study chiral olefins and arenes. In this case, however, an achiral silver salt must also be added, as has already been explained in the previous section. Spectra recorded by this technique for chiral allenes, terpenes and terpene-related hydrocarbons, and also

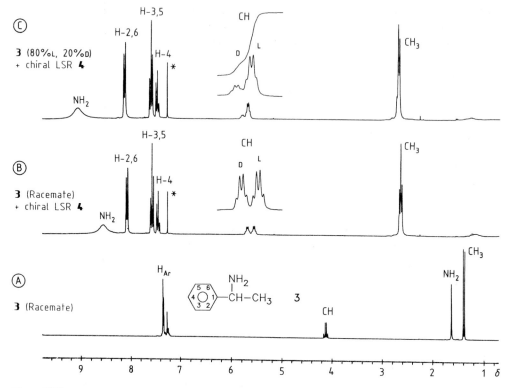

Figure 12-3.
A: 250 MHz ^{1}H NMR spectrum of 1-phenylethylamine (**3**) in CDCl$_3$.
B: Spectrum of the racemate of **3** in the presence of the chiral shift reagent Eu(TFC)$_3$ (**4**) (D form).
C: Spectrum of a mixture of 80 % L enantiomer and 20 % D enantiomer of **3** in the presence of **4**.
In spectra B and C the CH multiplets are shown expanded, and in C they are also integrated.
(∗ denotes the residual signal of the solvent.)

for helicenes, show separate signals for the enantiomers. This has opened up a new area of application of NMR spectroscopy, as the determination of optical purity for these types of compounds, by measuring the specific optical rotation, was previously only possible with great difficulty [2].

12.3 Chiral Solvents

Solvents also often induce shifts in solutes when there is an interaction between the two (Section 6.2.4). However, this does not lead to such solvents being described as shift reagents. Nevertheless, we will now consider the effects produced in chiral compounds by chiral solvents (referred to as CSAs, *chiral solvating agents*), and by chiral reagents in achiral solvents, since experiments with these can yield information of a similar nature to that from chiral LSRs.

If we dissolve the racemate of a chiral substrate (S(+) and S(−)) in an enantiomerically pure solvent with which it associates, L(−) for example, the two *solvation diastereomers* S(+)L(−) and S(−)L(−) are formed. In favorable cases this can result in separate signals being observed for the enantiomers of the substrate.

Solvents capable of being used as CSAs include chiral acids, amines, alcohols, sulfoxides and cyclic compounds. Fluorinated alcohols are often used, as these have acidic character and thus form complexes with basic compounds such as amines. On the other hand, amines are used as CSAs for measurements on organic acids and other acidic compounds. Two particularly important CSAs are 2,2,2-trifluoro-1-phenylethanol (**6**) and 1-phenylethylamine (**3**).

Similar shift effects are often observed even in achiral solvents when a chiral reagent is added to the chiral substrate. An example of this is shown in Figure 12-4. Spectrum A was recorded for the racemate of 1-phenylethylamine (**3**) in CDCl$_3$. To obtain spectra B and C the substrates were dissolved in a mixture of CDCl$_3$ and DMSO-d_6, and a quantity of (+)-2-methoxy-2-(trifluoromethyl)phenylacetic acid (**7**) was added. Spectrum B is again that of the racemate of **3**, while spectrum C is that of an approximate 75 % : 25 % mixture of its L and D enantiomers. As in the case of the chiral LSR (Fig. 12-3), the methyl and CH signals become split, and can be integrated to determine the optical purity.

The induced shifts caused by the chirality of the solvent or of the chiral reagent added depend on the choice of solvent (and any additive used), on the substrate, and on the complexing strength, temperature and concentration ratio. The latter becomes important in cases where the solvent is achiral and the chiral reagent is only present in a low concentration, as in our example.

It is important to note that the shift effect is only observed for the signals of the substrate, not those of the chiral solvent. The reason for this is easy to understand. The substrate molecules S(+) and S(−) are present in the solution in both free and complexed forms, and time-averaging between these environments gives the observed shifts which are different for the two enantiomers. The molecules of the chiral solvent or chiral reagent L(−) are attached first to S(+) then to S(−), and for rapid exchange the effects of S(+) and S(−) on L(−) average to exactly zero.

CF$_3$
|
H−C−OH
|

6

OCH$_3$ O
| //
Ph−C−C
| \
CF$_3$ OH

7

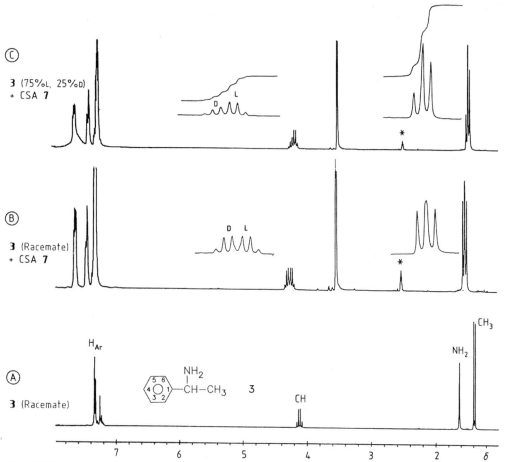

Figure 12-4.
A: 250 MHz ^{1}H NMR spectrum of 1-phenylethylamine (**3**) in CDCl$_3$.
B: Spectrum of the racemate of **3** in the presence of the chiral compound **7** in CDCl$_3$/DMSO-d_6.
C: Spectrum of a mixture of approximately 75 % L-enantiomer and 25 % D-enantiomer of **3** in the presence of **7** in CDCl$_3$/DMSO-d_6.
In spectra B and C the CH$_3$ and CH multiplets are shown expanded, and in C they are also integrated.
(∗ denotes the residual signal of the solvent.)

12.4 Bibliography for Chapter 12

[1] C. C. Hinckley, *J. Amer. Chem. Soc. 91* (1969) 5160.

[2] W. Offermann and A. Mannschreck, *Org. Magn. Reson. 22* (1984) 355.

Additional and More Advanced Reading

R. E. Sievers: *Nuclear Magnetic Resonance Shift Reagents.* New York: Academic Press, 1973.

R. v. Ammon and R. D. Fischer: Shift Reagents in NMR Spectroscopy. *Angew. Chem. Int. Ed. Engl. 11* (1972) 675.

O. Hofer: The Lanthanide Induced Shift Technique: Applications in Conformational Analysis. In: *Topics in Stereochemistry*, E. L. Eliel and N. L. Allinger (Eds.), 9 (1976) 111.

G. R. Sullivan: Chiral Lanthanide Shift Reagents. In: *Topics in Stereochemistry*, E. L. Eliel and N. L. Allinger (Eds.), 10 (1978) 287.

W. H. Pirkle and D. J. Hoover: NMR Chiral Solvating Agents. In: *Topics in Stereochemistry*, E. L. Eliel and N. L. Allinger (Eds.), 13 (1982) 263.

T. C. Morill (Ed.): *Lanthanide Shift Reagents in Stereochemical Analysis.* New York: VCH Publishers, 1986.

13 Macromolecules

13.1 Introduction

The broad field of macromolecules offers great scope for ^{1}H and ^{13}C NMR studies. The problems that can be tackled range from elucidating the compositions, sequences, configurations and chain conformations of synthetic and biological polymers to the kinetics and mechanisms of polymerization reactions. However, in most cases high field spectrometers are essential for such investigations.

The NMR spectroscopist's task is made more difficult by the fact that the signals of large molecules are usually very broad, and it is therefore not possible to resolve small chemical shift differences or splittings caused by couplings. However, raising the sample temperature nearly always makes the resonances narrower and improves the resolution, by simultaneously increasing the mobility of the polymer chains and reducing the solution viscosity.

In the first three parts of Section 13.2, we consider some aspects of structure determination for synthetic polymers using ^{1}H and ^{13}C NMR spectroscopy. However, it should be noted that NMR spectroscopic studies of macromolecules have become much more important in recent years, due to developments in two-dimensional and multidimensional NMR techniques, and especially in solid state NMR spectroscopy. Therefore, Section 13.2.4 discusses the principles of solid state NMR spectroscopy, although without going into details that would take much space and are outside the scope of this book. The final part of the chapter, Section 13.3, describes applications of NMR spectroscopy to investigating the structures of peptides, proteins, nucleic acids, and oligo- and polysaccharides. The importance of such studies is underlined by the fact that in 2002 Kurt Wüthrich of the ETH Zürich was awarded the Nobel Prize for Chemistry in recognition of his groundbreaking achievements in this field.

13.2 Synthetic Polymers

13.2.1 The Tacticity of Polymers

The polymerization of alkenes involves the formation of chains whose stereochemical structures depend on the nature of the monomer, the method of polymerization and the reaction conditions. In the simplest case, the polymerization of ethylene to polyethylene (**1**), a chain of CH_2 groups is formed in which all the protons and all the carbon atoms are equivalent. The ^1H and ^{13}C NMR spectra consist of a single peak in each case.

The ^1H and ^{13}C NMR spectra of polyisobutylene (**2**) are equally simple to analyze, provided that only head-to-tail polymerization occurs, since here all the methyl groups and all the methylene groups are equivalent. The ^1H NMR spectrum consists of two signals with an intensity ratio 3 : 1, and the ^{13}C NMR spectrum of three signals. The resonances of the end-groups, whose shielding values differ from those in the middle of the chain, generally cease to be detectable when the length of the chain reaches about 20 monomer units.

For polymers in which pseudo-asymmetric carbon atoms are formed in the polymerization process, the structures and the spectra are more complicated. Examples of such polymers are polypropylene (**3**), polyvinylchloride (**4**) and polystyrene (**5**).

In these polymers three different types of stereochemical structures are possible. The incorporation of successive monomer units can take place with the same configuration (*isotactic* chains), or with a regularly alternating configuration (*syndiotactic* chains), or with random changes of configuration (*atactic* chains). From the stereochemical formulas shown in Figure 13-1 it can be concluded that these three structural types will give different characteristic NMR spectra.

This can be seen by considering the example of polymethylmethacrylate, PMMA (**6**), whose possible structures correspond to Figure 13-1 with R = CH_3 and R' = $COOCH_3$. Figure 13-2 shows the 220 MHz ^1H NMR spectra of isotactic, syndiotactic and atactic PMMA. In all three cases we can easily arrive at a rough assignment which takes no account of the finer details. The OCH_3 signals are found at $\delta \approx 3.6$, the CH_2 signals between $\delta = 1.5$ and 2.5, and the $C-CH_3$ signals between $\delta = 1$ and 1.4. The intensity distribution is 3 : 2 : 3. The detailed interpretation of the individual spectra can best be understood by considering the projection formulas in Figure 13-1, or better still by examining models.

$$-\!\!\left[\,CH_2\!-\!CH_2\,\right]_n\!\!-$$

1

$$-\!\!\left[\begin{array}{c} CH_3 \\ | \\ C\!-\!CH_2 \\ | \\ CH_3 \end{array}\right]_n\!\!-$$

2

$$-\!\!\left[\begin{array}{c} H \\ | \\ \overset{*}{C}\!-\!CH_2 \\ | \\ R \end{array}\right]_n\!\!-$$

3: R = CH_3
4: R = Cl
5: R = Ph

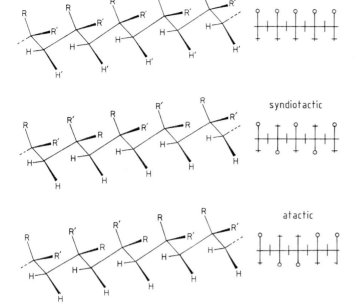

Figure 13-1.
Pictorial and projection formulas showing isotactic, syndiotactic and atactic polymer chains. The configuration at the pseudo-asymmetric carbon atoms is constant in an isotactic polymer chain, regularly alternating in a syndiotactic chain, and random in an atactic chain.

Isotactic PMMA (Fig. 13-2 A): All the OCH_3 groups are equivalent, as also are all the $C-CH_3$ groups, and in each case these give a single peak. The two methylene protons are diastereotopic (Section 2.4.2) and give an AX-type spectrum (Section 4.3).

Syndiotactic PMMA (Fig. 13-2 B): The OCH_3 groups are all equivalent, as are all the $C-CH_3$ groups, and also – in contrast to isotactic PMMA – all the methylene protons. Consequently the spectrum consists of only three peaks.

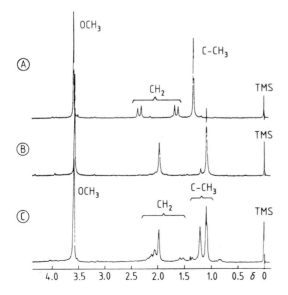

Figure 13-2.
220 MHz ^1H NMR spectra of polymethylmethacrylate (**6**, PMMA) samples of different tacticities in *o*-dichlorobenzene at 100° C.
A: isotactic, B: syndiotactic, C: atactic.

349

There is scarcely any difference between the positions of the OCH$_3$ signals for isotactic and syndiotactic PMMA, but there are large differences for the CH$_2$ and CH$_3$ resonances. The chemical shift of the CH$_2$ protons is strongly influenced by the CRR' groups of the immediately adjacent monomer units, and therefore depends mainly on the diad sequences (pairs of monomer units). For the C−CH$_3$ groups, in which the methyl group is itself directly bonded to a pseudo-asymmetric carbon atom, triad sequences must be taken into consideration. This is made clearer by Figure 13-3.

Atactic PMMA (Fig. 13-2 C): The spectrum of atactic PMMA is more complicated. Although the OCH$_3$ protons still give only one peak, the CH$_2$ signals can no longer be analyzed unambiguously. If we take into account only the diads, we expect to find overlapping signals for CH$_2$ protons in isotactic and syndiotactic diads, with intensities corresponding to their relative probabilities. Since this is not what is found, we need to look at the situation in greater detail; in other words, we must include in our discussion not only the nearest, but also the next nearest neighbors, i.e. we need to look at the tetrad sequences. However, we will not take this part of the analysis any further, because the required information can be obtained more easily from the resonances of the C−CH$_3$ groups. These consist of three peaks of very different intensities in the region δ = 1 to 1.4. The positions of two of these peaks correspond to the CH$_3$ signals of the isotactic and syndiotactic samples, and accordingly we can assign these to CH$_3$ groups in isotactic and syndiotactic triads. Between these lies a third signal, which must belong to methyl groups at changeover positions. The corresponding triads are described as heterotactic (Fig. 13-3).

Integrating the three methyl signals gives the relative frequencies of occurrence of the different triads, and thus the statistical constitution of the polymer. In this case an analysis of the spectrum on the basis of triads is justified, despite the fact that individual signals show signs of a fine structure which indicates pentad effects. This fine structure can be seen more clearly when spectra are recorded at higher resonance frequencies.

In many cases ^{13}C NMR spectroscopy is superior to ^{1}H NMR for investigating polymer tacticity, since:
- spectra recorded with broad-band ^{1}H decoupling contain only single peaks,

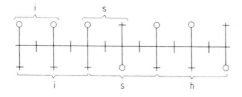

Figure 13-3.
Projection formulas of atactic polymethylmethacrylate (PMMA, **6**) showing diad and triad sequences, i: isotactic diads and triads; s: syndiotactic diads and triads; h: heterotactic triads. (+ = CH$_3$; ○ = COOCH$_3$).

- by recording DEPT spectra one can distinguish between the resonances of CH_3, CH_2, CH and C_q carbon nuclei, and
- long-range steric effects are even more significant in ^{13}C than in ^1H NMR spectra, and consequently tetrad and pentad effects are always seen.

Stereo-block polymers: Polymers in which the chains contain long sections with the same tacticity are known as stereo-block polymers. In this case the NMR spectrum resembles that of a mixture containing polymer chains, each of which has a uniform tacticity.

13.2.2 Polymerization of Dienes

The polymers discussed so far are products of 1,2-linking reactions, for which it is found that head-to-tail polymerization has the highest probability. In dienes, taking butadiene as the simplest example, there is competition between 1,2-polymerization and *cis*- or *trans*-1,4-polymerization (Scheme I). These reactions result in polymers with quite different structures and physical properties.

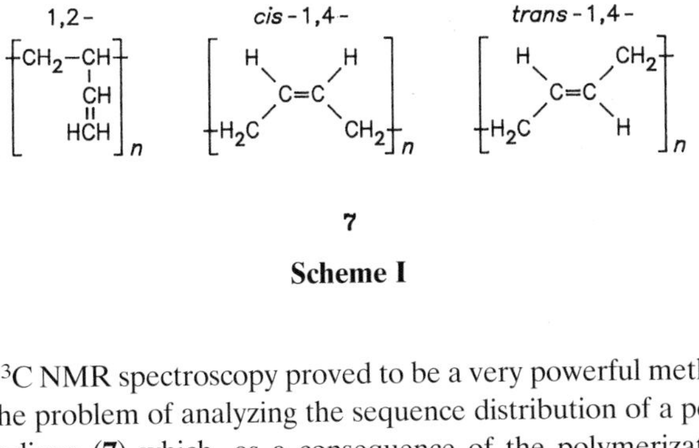

7

Scheme I

^{13}C NMR spectroscopy proved to be a very powerful method in the problem of analyzing the sequence distribution of a poly-butadiene (**7**) which, as a consequence of the polymerization method used, contained both 1,2 and 1,4 linkages.

Figure 13-4 A shows the olefinic ^{13}C resonances of a mixture of pure *cis*- and *trans*-1,4-polybutadienes. Each component gives a single peak.

Spectrum B is that of a predominantly *cis*-1,4-butadiene containing about 20 % of *trans*-1,4 linkages. The polymer whose spectrum is shown in C contains approximately equal numbers of *cis* and *trans* linkages. The signals in B and C can easily be assigned on the basis of the chemical shifts known from spectrum A. Spectra B and C contain two extra peaks marked with a star; these arise from the olefinic carbon nuclei at positions where there is a transition from *cis* to *trans* polymerization [1].

351

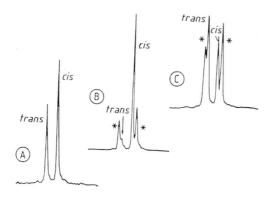

Figure 13-4.
Olefinic carbon region of the 67.88 MHz ^{13}C NMR spectra of 1,4-polybutadienes (**7**) with different proportions of the *cis* and *trans* configurations [1].
A: Mixture consisting of about 60% pure *cis* polymer and 40% pure *trans*.
B: A mainly *cis* polymer with about 20% of *trans* linkages.
C: A polymer with about 50% *cis* and 50% *trans* linkages.
(∗ denotes the signals of olefinic carbon nuclei at transitions between *cis* and *trans* sequences.)

13.2.3 Copolymers

We have already encountered a copolymer, in the wider sense of the term, when we discussed stereo-block polymers. However, copolymers are more commonly understood to be polymers whose chains are made up of at least two chemically different monomer units A and B.

The simplest case is that in which A and B form blocks of uniform constitution: −AAAAA−BBBBB−. The ^{1}H and ^{13}C NMR spectra then consist simply of superimposed spectra of the two (or more) block sequences. If the chain lengths within the blocks are sufficiently great, the signals corresponding to the A−B changeover positions and those of the end-groups become undetectable, as their intensities are two to three orders of magnitude smaller than those of the blocks. However, with the availability of high-field spectrometers offering greater sensitivity, and even more importantly by applying the inverse two-dimensional H,C-correlation methods HMQC and HMBC (see Chapter 9), it has become possible to obtain information about the structures and numbers of changeover points linking the blocks, and about the end-groups [2, 3]. Information of this kind is very important for the synthetic chemist, because the ability to produce polymers with blocks of well-defined structure and length makes it possible to tailor the properties for specific applications.

If instead the A and B units form a random copolymer −AABBABBAAAB−, the spectra become very complicated. This is due to so-called *composite* effects. For example, the central A units in the three triads A**A**A, B**A**A and B**A**B give signals at different positions. If we now also take into consideration the fact that these triads of different chemical constitution can additionally have different configurations – isotactic, syndiotactic or heterotactic – this results in 20 different triads.

A further difficulty is that the triads are present in the chain in different proportions. Here again, the chances of successfully analyzing a spectrum are better for ^{13}C than for ^1H, especially if the ^{13}C NMR data can be combined with information from two-dimensional or multidimensional H,C-correlated spectra. Moreover, the possibilities are not limited to H,C-correlations, since in cases where deuterated or fluorinated derivatives are being studied, one can also record and interpret ^2H,C- or F,C-correlated spectra.

Although sequence analysis in these cases is difficult, one can often determine the chemical composition of the copolymer quite easily by integration. For this purpose ^1H NMR spectroscopy is the most suitable. Figure 13-5 shows the poorly resolved 220 MHz ^1H NMR spectrum of a random copolymer of butadiene and α-methylstyrene. The mole ratio of the monomer units can be obtained directly from the ratio of the areas under the aromatic and olefinic proton signals.

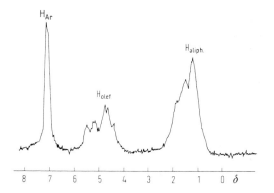

Figure 13-5.
220 MHz ^1H NMR spectrum of a random copolymer of butadiene and α-methylstyrene.

13.2.4 Solid-State NMR Spectroscopy of Polymers

In the previous sections we considered the composition and tacticity of polymer chains. The macroscopic properties of a polymer, in other words its properties as a material, are determined not only by the molecular structure of the macromolecules but also by their arrangement in the solid. Most polymers are amorphous, and the individual molecular chains are disordered, forming tangles, coils, and knots. However, in cases where the stereochemical structure of the chains allows it, polymers can also undergo partial crystallization, forming crystalline regions that are separated by amorphous regions. The properties of the solid polymer, such as whether it is glassy, hard, soft, brittle, or elastic, depend on the chemical composition, the stereochemistry, and the conditions under which it is produced. In a partly crystalline solid polymer, individual polymer chains can extend from one crystalline region, through an amorphous region, and into another crystalline region, and they can even fold back into the first region. Obviously this kind of situation will affect the properties of the polymer, and that the dynamic conditions in the various regions are likely to be very different. For example, conformational changes are relatively easy in amorphous regions, as the chains there are more flexible than in crystalline regions.

With most other methods used for structural characterization of polymers, such as X-ray structural analysis, one usually observes only the crystalline regions. However, as explained above, the amorphous regions have an equally important role in determining the material properties. We have seen in earlier chapters of this book that NMR spectroscopy can yield very detailed information about many different molecular and structural properties of molecules in solution, and about their dynamic behavior. Modern methods of solid-state NMR spectroscopy can also give detailed information of that sort. However, these methods need special equipment, and also a more profound theoretical knowledge. A rigorous and detailed treatment of these aspects would be outside the aims and intended scope of this book, and therefore this chapter is limited to giving a short introduction to the basic features of solid-state NMR spectroscopy for studying polymers.

In general, the NMR spectra of solids consist of very broad bands with no resolved structure. This behavior has three main causes:

1. The nuclear dipoles in a solid interact with each other by direct coupling through space. The strength of the dipolar coupling depends on distance, and is especially large for nuclei that are directly bonded to each other, as in a $^{13}C–H$ fragment, or are separated by only two bonds, as in a CH_2 group.

2. In a solid, whether it is a single crystal, a glass, or an amorphous powder, movement of the atoms is severely restricted or even prevented altogether. As the chemical shift depends on the orientation of the molecule or fragment relative to the external magnetic field (chemical shift anisotropy, CSA), the NMR spectrum of a powder is very broad.

3. If the sample contains nuclei with a spin greater than $1/2$ (thus $I \geq 1$), such as ^{14}N, these have an electric quadrupole moment Q, and the quadrupolar coupling makes a further contribution to the line broadening.

In solutions the dipolar coupling and the chemical shift anisotropy are averaged to zero by the rapid motions of the molecules (see also Section 2.1.1), resulting in narrow resonances. In solids a similar averaging effect can be produced by spinning the sample very rapidly about an axis inclined with an angle of

54.74° with respect to the external magnetic field; this is called the "magic angle". In this MAS (Magic Angle Spinning) technique, the sample is routinely rotated at a frequency of 2 to 18 kHz; the record is more than 50 kHz. The spectra are easiest to interpret when the spinning frequency exceeds the overall width of the anisotropic powder spectrum, that is obtained by measuring a static sample. Although fast spinning of a sample demands a precise mechanical design and a robust construction of the rotor system, solid-state NMR probe heads are common equipment in any NMR facility nowaday, and solid-state NMR is a routine method!

The MAS technique, used under appropriate conditions, yields NMR spectra of solid samples that look similar to those obtained from solutions. Therefore, the method is eminently suitable for studying polymers, and other materials.

The method is not restricted to one-dimensional (1D) NMR spectroscopy – one can also perform 2D experiments such as HMQC or HMBC (see Chapter 9). Studying exchange processes is also possible, and yields much valuable information about the dynamics of polymer chains. Thus, the method extends the scope of polymer investigations to include also the amorphous regions [4].

13.3 Biopolymers

Biopolymers are of enormous importance in nature. Examples that come to mind immediately are proteins, nucleic acids, and polysaccharides. NMR spectroscopy has been used to study all these three classes of materials, often by methods different from those that apply to small molecules. This section explains the principles of using NMR spectroscopy to investigate the structures of biopolymers, with particular reference to the above three classes, although without an in-depth description of the details, which would take up more space than is appropriate for this book. Those details can be found in the literature cited at the end of the chapter [3, 5]. Applications to peptides and proteins are given most attention here, as this is the area where the most intensive research has been carried out and progress has been made.

13.3.1 Peptides and Proteins [5–8]

Peptides and proteins are made up of amino acids, which are linked by peptide bonds (see structural formula, Scheme II). The enormous variety of protein structures results from variations in the sequence of the amino acids. In general nature uses 20 different amino acids, which always have the configuration of the L series. Proteins have many very important roles in living organisms. For example, they can serve as enzymes, which are catalysts that enable reactions to proceed in aqueous solutions at body temperature. However, the many different functions of proteins are not determined solely by the molecular composition (the primary structure), but also depend on the spatial arrangement of the macromolecules, i.e., their secondary and tertiary structures. The usual method of investigating these structural features is X-ray crystallography. However, for that one usually needs crystalline samples, which are difficult to obtain in many cases. Also it must be borne in mind that the three-dimensional structure determined for the solid is not necessarily identical to that existing in solution. Therefore it appears that NMR studies in solution are a good alternative. However, there are serious experimental difficulties with solution spectroscopy, because the spectra of such large molecules contain an enormous number of lines, and the resonances of the individual amino acids overlap considerably. In most cases the required structural information cannot be obtained from an analysis of one-dimensional spectra, and it is necessary to use two-, three-, or multidimensional methods. Moreover, it is absolutely essential to use a spectrometer operating at a proton resonance frequency above 600 MHz, typically 800 MHz or even 900 MHz.

In using NMR spectroscopy to elucidate the structure of a peptide or protein, one hopes to find answers to three questions: which amino acids are present? – in what sequence are they arranged along the chain? – what is the three-dimensional geometry of the chain (the secondary and tertiary structural characteristics).

In the foregoing chapters we have learned about the basic principles of all the homonuclear and heteronuclear one- and two-dimensional NMR techniques that are needed to answer the above questions, namely H,H-COSY, H,C-COSY, TOCSY, NOESY, and ROESY. For the heteronuclear procedures one needs samples that are isotopically enriched in ^{13}C and/or ^{15}N.

First we will consider the basic features of the methods used for determining the constitution and sequence of proteins (polypeptides) with relatively low molecular size (less than 100 amino acid residues in the chain). After that, we will learn about methods for investigating the spatial structure, secondary and tertiary.

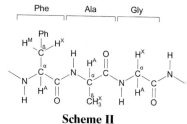

Scheme II

13.3.1.1 Sequence Analysis

The first questions to be addressed are: what amino acids are present in the peptide, and in what sequence? To begin analyzing the ^{1}H NMR spectrum, one can start from the assumption that there is no coupling between protons in neighboring amino acid residues, so that the total spectrum consists of a superimposition of separate spectra produced by the isolated spin systems of individual amino acids. The larger the protein macromolecule, the more complex is the resulting spectrum. However, one is helped by the fact that most of the 20 amino acids used by nature give ^{1}H NMR spectra that are quite characteristic, and are even fairly simple. A few examples will illustrate this. The CHCH$_3$ group of alanine gives an AX$_3$-type spectrum (see Chapter 4). Serine, cysteine, aspartic acid, asparagine, and phenylalanine all give AMX-type spectra, in which the two protons on the β carbon atom are diastereotopic (see structural formula, Scheme II). We have already come across the spectrum of glutamic acid in Figure 9-19 (Section 9.4.2); although the spectrum was not analyzed in detail there, that was unnecessary in the context of the discussion, as the individual groups of signals could be assigned unambiguously. The spectra of the aromatic amino acids histidine, phenylalanine, and tyrosine contain, in addition to the AMX pattern, the resonances of the aromatic protons in their molecules, which are easily recognized. In all the above examples we have assumed that the NH protons are readily exchangeable and therefore do not show couplings to the protons on C-α. That would indeed be the case for the free amino acids, but for amino acids bonded in peptides it is certainly not true, and therefore we have to take into account couplings between the NH and the C-α protons. This makes the spectra somewhat more complicated, but nevertheless still easy to analyze according their spectral type.

Where there is severe overlapping of resonances, it is not generally possible to recognize the signals in the one-dimensional spectrum that belong to a particular spin system, except in the simplest cases. The TOCSY technique can be a very useful aid to analyzing such systems, as also can the HMQC method described in Section 9.4.3, especially in its inverse form. To achieve a complete assignment of the proton resonances, which is essential before proceeding to attempt a full sequence analysis, one must have a suitably equipped spectrometer; moreover, without isotopic enrichment it is only possible for polypeptides or proteins with up to about 100 amino acid residues.

When all the signals in the spectrum have been assigned to the "correct" amino acids, the next step is to determine the sequence of the amino acids. For that, one needs to find connectivities between adjacent amino acids, by detecting correlations transmitted through the amide bonds. There are in principle two methods of achieving that. If only a homonuclear ¹H NMR experiment is possible, one measures the nuclear Overhauser enhancement (the NOESY or ROESY experiments, Sections 9.4.6 and 10.2.4). Alternatively, it may be possible to use the heteronuclear HMBC technique (Section 9.4.4). With the NOE methods, the relevant signals (cross peaks) that one ideally hopes to use are those caused by short-range interactions (see structural formula, Scheme III), such as the NOE between the amide proton and the aliphatic protons attached to the α and β carbon atoms of the chain, which is an NOE between adjacent amino acid residues. However, chain folding (secondary structure) in proteins can bring amino acid groups that are far apart along the chain close to each other spatially, resulting in very small internuclear distances and correspondingly large NOEs. NOESY or ROESY 2D spectra typically contain many hundreds of cross peaks, and therefore it is obvious that, even when the ¹H spectrum has been completely analyzed, the analysis and assignment of NOE data is by no means a trivial problem. However, as a start one can classify the NOEs roughly into strong, medium, and weak, corresponding to inter-proton distances of about 0.25 nm, 0.35 nm, and more than 0.5 nm respectively.

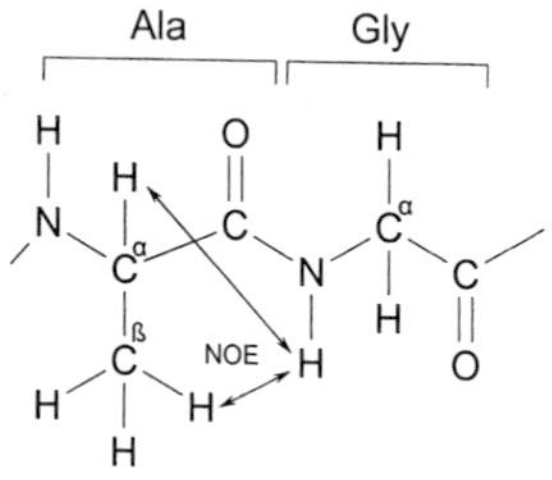

Thus, starting from one amino acid residue that has been determined precisely, one can, on the basis of the observed NOEs, work outward in both directions to determine the entire sequence of a small protein. The procedure described here for sequence analysis is the most sensitive method, as it depends solely on ¹H resonances, not on those of other less sensitive nuclides.

An alternative to the sequence analysis method based on NOE is the two-dimensional HMBC technique (see Section 9.4.4 and structural formula, Scheme IV), which depends on the scalar couplings to heteronuclei, namely ¹³C or ¹⁵N. For this method one needs samples that have high solubility or are enriched in the isotopes ¹³C and/or ¹⁵N. But with this method too one must first achieve a complete analysis of the ¹H NMR spectrum.

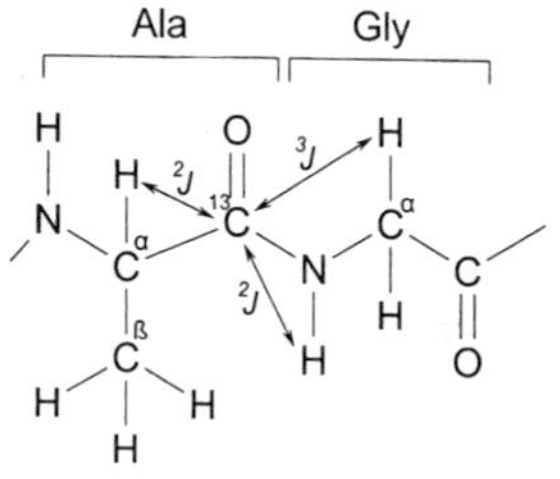

Scheme III

Scheme IV

13.3.1.2 The Three-Dimensional Structure of Proteins

As mentioned earlier, the biological functions of proteins depend not only on their primary structure but also, very importantly, on the stereochemical configuration of the chains. For many years NMR spectroscopy could not offer an effective method for determining the structure of proteins, because of their low solubility and the limited sensitivity of NMR spectrometers. However, with the development of ultra-high-field spectrometers operating at proton frequencies up to 900 MHz, and of multidimensional NMR techniques, gradual progress was made in studying proteins of increasing molecular size. Nevertheless, for studying larger proteins (those with more than about 100 amino acid residues) by the powerful heteronuclear and multidimensional methods, it was found absolutely essential to use proteins isotopically labeled with ^{13}C and/or ^{15}N. The TROSY (Transverse Relaxation Optimized SpectroscopY) technique developed by Wüthrich and colleagues has made it possible to extend protein NMR studies into the 100 kDa range and even beyond, which corresponds to proteins with more than 1000 amino acid residues! [9].

It is now known that peptides with as few as 20–30 amino acids already begin to form secondary structures such as α-helices or β-sheets. However, for larger proteins it is found that as well as ordered structures, there also exist regions that do not have long-range ordered structures of that kind, but in which the individual chain segments differ not only in their structure but also in their mobility. All these features, in addition to the primary structure, affect the NMR spectra and make their analysis a complicated problem.

The most important information about secondary structure is obtained from NOEs. However, in contrast to the situation with sequence analysis, here it is the long-distance inter-residue NOEs that have the main role in the study of secondary structure. These are NOEs between hydrogen nuclei of amino acids that are far apart along the chain (hence "long-distance"), but are spatially close. We know that the NOE factors, as measured by the intensities of the cross peaks in NOESY or ROESY spectra, are very strongly dependent on the distance r between the nuclei (in fact, they are proportional to r^{-6}). However, one cannot always distinguish easily between intra-residue and inter-residue NOEs, because, as already mentioned, protons that belong to amino acids far apart along the chain can be brought close together as a result of the secondary structure. Quantitative interpretation is also made more difficult by overlapping of the cross peaks. Furthermore, there is the fundamental difficulty of measuring the intensities of the cross peaks with sufficient accuracy to calculate distances reliably. However, if one first applies the method to calculate distances between particular hydrogen nuclei within amino acids of the chain, it may be possible to construct models, or to calculate energetically favored structures using suitable programs. For example, one can use the Molecular Dynamics program, which allows one to change bond-lengths, bond angles, and torsional angles. Such a program must also take into account the fact that protein molecules are not rigid. Usually one obtains, as the first result, several structures with the same or similar energies, which must then be refined in an iterative process to obtain the best fit to the experimental data. To obtain a reliable structure for a protein, one should use data on as many NOEs as possible, and thus calculate values for many distances between protons of different amino acids.

In view of the unavoidable uncertainties in the calculated structure, it is appropriate and desirable to use all additional information that can be extracted from the spectra. In Section 3.2.2.1 we learned that in a saturated compound the coupling constant over three bonds, ^3J, depends on the torsional angle. The dependence is given quantitatively by the Karplus equation. If one can measure relevant $^3J(H,H)$, $^3J(H,{}^{13}C)$, or $^3J(H,{}^{15}N)$ couplings, one can apply the Karplus equation to get additional information about the conformation of the peptide chain. As can be seen in the structural formula, Scheme V, with the usual nomenclature for the various torsional angles, to determine φ one needs to know the value of $^3J(HN–CH^\alpha)$. The value of $^3J(H^\alpha H^\beta)$ gives information about the torsional angle χ, and therefore about the conformation of the side-chain. Together with the information about the NOE, it is possible to reach conclusions about the angles ψ and ω. Data obtained in this way can be used for refining the structures calculated by the program mentioned earlier.

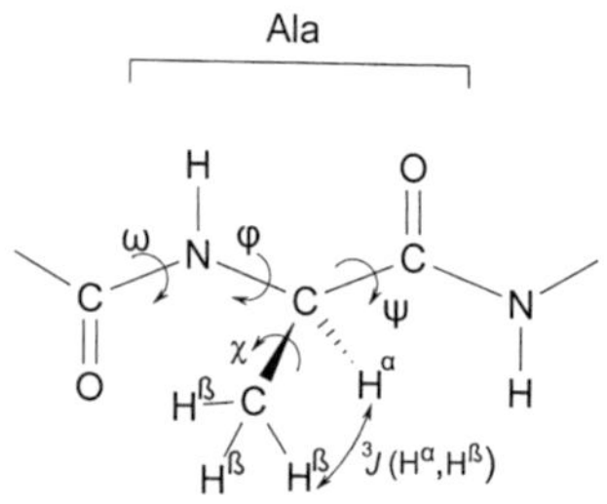

Proteins with conformational mobility present special problems, as the observed values of the NOEs and vicinal coupling constants are averaged values in such cases. Such averaged values cannot be used to calculate reliable structures.

In cases where it has been possible to compare protein structures obtained from X-ray crystallography and from NMR spectroscopy, good agreement has been found [7].

It should be emphasized again here that for a structure determination it is necessary to assign all the signals in the 1D or multidimensional NMR spectrum.

Scheme V

13.3.2 Polynucleotides

Although there has been much research effort in recent years on determining the structures of polynucleotides by NMR spectroscopy, the progress achieved is hardly comparable with that for proteins. The earliest NMR studies were mainly concerned with DNA (deoxyribonucleic acid). Taking this polynucleotide as our example, this section gives a brief account of the main features of NMR investigations in the polynucleotides area.

Nucleic acids have a relatively simple structure. The backbone of DNA consists of a polyester made up of repeated build-

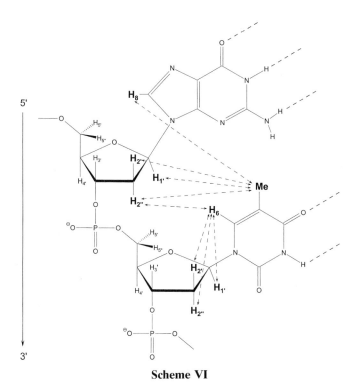

Scheme VI

ing blocks, each of which alternates between a sugar unit (deoxyribose) and a phosphoric acid residue. The phosphoric acid esterifies at the 3' position of one deoxyribose unit and the 5' position of the next one (see structural formula, Scheme VI). Thus, the chain has a free 3' and a free 5' position at its ends, which define its direction. Each sugar unit has a base attached to it at C-1 by an *N*-glycosidic bond. The complete DNA structure is a double helix, which consists of two inter-twined strands of the kind described above, aligned in opposite directions. Structural studies by X-ray crystallography have so far identified three types of double helices, the A, B, and Z types, of which B-DNA is the most important. This B type has a right-handed helix. The bases are directed inward, and form hydrogen bonds between complementary base pairs, which always consist of one purine and one pyrimidine base. These base pairs are adenine(A)–thymine(T), with two hydrogen bonds, and guanine(G)–cytosine(C), with three hydrogen bonds. RNA (ribonucleic acid) has ribose instead of deoxyri-bose, and uracil(U) instead of thymine.

The task for NMR spectroscopy is to reveal information about the sequence of the bases and the structure of the chains (as in the case of proteins), and also about the structure of the DNA double helix (the duplex).

Detailed structural information about DNA, and also about other polynucleotides, is already available from X-ray crystallographic analysis. However, NMR spectroscopy seeks answers to these further questions: is the structure of DNA in solution identical to that in crystals? – if not, can NMR spectroscopy help to elucidate the fine details of the structure in solution? – and what are the most suitable NMR techniques for such studies?

The last question is fairly easy to answer. Isotopic labeling or enrichment of DNA with ^{13}C and/or ^{15}N is only possible at great cost in synthetic work and materials, which generally rules out the heteronuclear methods. That leaves NOE measurements as the method of choice! As we are here concerned exclusively with DNA in the form of the double helix, we can expect to detect only NOEs within individual mononucleotides, where the H–H distances are short. In particular, these will be mainly within individual deoxyribose units, between adjacent nucleotides in the chain (sequential NOEs), and between base pairs that are directly linked through hydrogen bonds (inter-strand short-distance NOEs) [5]. A few of the NOEs that are to be expected on the basis of the internuclear distances are shown in structural formula, Scheme VI. We expect to find relatively few NOEs along the DNA axis. As in the case of proteins, a full analysis of the ^1H NMR spectrum is essential before the cross peaks in the NOESY spectrum can be assigned to the appropriate categories.

This short description has left out many of the more complex details, such as the effects caused by exchangeable protons or by the solvent. A complete and detailed description of the structural analysis would take up much more space than is appropriate for a book of this kind; for that one should refer to the cited literature [5, 8, 10].

13.3.3 Oligosaccharides and Polysaccharides

The most common polysaccharides found in nature are cellulose, amylose, amylopectin, and glycogen. They are all built up from the most energetically favored sugar molecule, glucose, and differ only in the type of glycosidic bonding and the degree of branching. In general, NMR spectroscopy of oligosaccharides and polysaccharides is restricted to ^1H measurements. The above four common polysaccharides give relatively simple ^1H NMR spectra.

In contrast, polysaccharides that are made up of several different monosaccharides usually give spectra that are much more difficult to analyze. Exceptions to that are those heteropolysaccharides that are made up of repeating units consisting of identical oligosaccharide building blocks. These are usually studied most conveniently by breaking down the chain into its individual repeating units, by enzymatic attack if possible, then examining the resulting oligosaccharide.

Some other important natural polymers containing sugar chains are the glycoproteins and glycolipids. Glycoproteins, in which the oligosaccharide chain is bonded to a protein, are important as constituents of cell walls. Here the sugar part of the macromolecule projects from the surface of the cell and has a key role in cell recognition. Glycolipids are constituents of blood-group substances, and here the oligosaccharide part determines the antigen specificity. Pure oligosaccharides are present in the milk of nursing mothers.

The very wide variety of oligosaccharides arises mainly from the different sequences of a small number of monosaccharides, including glucose, mannose, glucosamine, galactose, fucose, N-acetyl-β-D-neuraminic acid, and a few others, and also from the different types of linking (α, β, 1–2, 1–3, 1–4, or 1–6) and from the possibilities for chain branching (see Scheme VII). In glycoproteins the sugar at the end position is usually N-acetyl-β-D-neuraminic acid; we have already met a derivative of it in the previous chapters, where it served as a model compound for describing various NMR methods.

The different biological functions of oligosaccharides are by no means all understood yet. They depend on the structures, and the aim of NMR investigations is to determine those structures.

As in the case of proteins and nucleic acids, the task of the NMR spectroscopist is to find answers to three questions: what sugar structures are present in the oligosaccharide? – how are the monosaccharides linked to each other? – what is the conformation of the chain?

There is no universal recipe for solving these problems. To appreciate the kinds of difficulties that can arise with oligosaccharides, we need only look at the NMR spectra of glucose (Fig. 3-3), of a derivative of cellobiose (Fig. 8-25), and of the N-acetyl-β-D-neuraminic acid derivative already mentioned (**1**, Fig. 9-8). Each monosaccharide contains, in general, seven protons, giving a correspondingly complex pattern, and these partial spectra are superimposed, resulting in an uninterpretable hump, mainly in the region $\delta = 3.4$ to 4.3 ppm. In many cases, derivatization by introducing acetyl groups helps to improve the situation by revealing more detail, even though it may only cause a few peaks to become separated from the hump. Introducing an acetyl group shifts the resonances of directly-bonded ring protons to higher frequencies (i.e., increases their δ-values) by about 1 ppm (see also Section 6.2.5). It should also be mentioned that in practice one can also exchange the OH protons

Neu5Acα2-3Galβ1-3GalNAcα1-OSer
6
|
Neu5Acα2

Scheme VII

with deuterium and use D_2O as the solvent. Another point to note is that oligosaccharides are mobile, especially with regard to rotation about the glycosidic bond, which means that the observed values of coupling constants and NOEs are averaged over several different conformations.

How many sugars are present in the oligosaccharide, and which ones? Fortunately, the signals of the anomeric protons H-1 of different sugars are distributed over a range of δ-values from about 4.5 to 5.5 ppm., and are usually separate from each other and from the signals of the other ring protons. However, they may be obscured by the residual water signal (H_2O + HDO), which falls in this region. By counting the number of H-1 signals and, where these are superimposed, measuring intensity ratios, one can in principle determine the number of monosaccharide units present in the oligosaccharide. In many cases the positions of the H-1 signals enable one to identify the different sugar structures. Also the vicinal coupling constants ^3J(H-1,H-2) give information about the type of glycosidic bonding, in particular whether the monosaccharide units are linked together by α- or β-glycosidic bonds (see Section 3.2.2.1). Even for sugars that cannot be identified by the appearance of an H-1 signal at a particular position, or may not even have an H-1 proton, there are usually other characteristic resonances that reveal their presence - for example, the resonances of H-2 in mannose, or of H-3a and H-3e in N-acetyl-β-D-neuraminic acid.

In order to determine the linking positions, the sequence, and also, where appropriate, the conformation of the chain, one must fully assign the entire ^1H NMR spectrum. That is often a very difficult task because of the overlapping of many resonances, and may even be impossible in some cases. However, one starts with the advantage that each monosaccharide unit is an isolated spin system with either seven protons (as in glucose) or nine (as in the neuraminic acid derivative). The most effective methods for identifying these systems are the one-dimensional and two-dimensional versions of TOCSY, which we met in Sections 8.7 and 9.4.5. The usual starting-points for analyzing the TOCSY spectrum are the resonances or cross peaks of the anomeric proton H-1. As examples of the procedures, Figure 8-28 shows the application of the one-dimensional selective TOCSY experiment to the cellobiose derivative **3**, and Figure 9-27 shows the two-dimensional TOCSY spectrum of the same derivative. It can be seen from these examples that the one-dimensional version (Section 8.7) can sometimes have advantages compared with the two-dimensional version, since in this case one obtains signals from only one sugar. However, in contrast to the simple situation with this disaccharide, for higher oligosaccharides one may need to use special methods, and even three-dimensional techniques.

The linking positions can sometimes be determined from the chemical shifts of the ring protons in the aglycone. However, where that is not possible, the problem may be solved, as in other biopolymers, by recording NOESY or ROESY spectra. Here one is concerned with NOEs between two adjacent monosaccharides, usually between H-1 of the glycone and the proton at the linking position of the aglycone. NOEs between H-1 and other ring protons of the aglycone may also be observed, yielding data about the inter-proton distances, and therefore about the chain conformation. In structural formula, Scheme VIII,

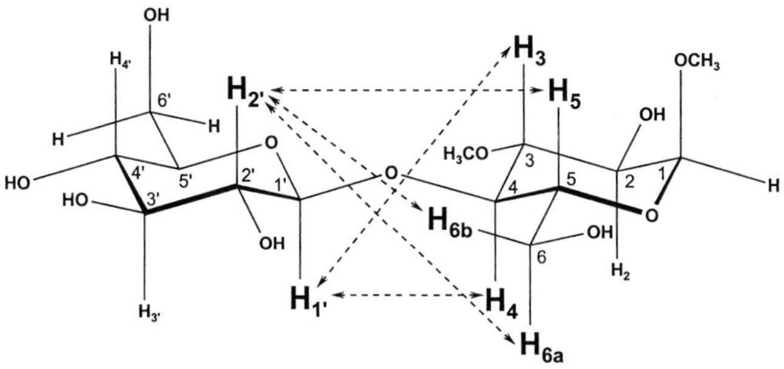

Scheme VIII

showing the same cellobiose derivative as before (see Section 10.4 and Fig. 10-6), these inter-residue NOEs are indicated by arrows. The procedure for complex oligosaccharides is in principle the same as for this disaccharide, although the spectra are, of course, more complicated. However, it should be noted that, unlike the situation for proteins and nucleic acids, one can only obtain information about the sequence of adjacent disaccharide units!

In special cases it is possible to gain more information by the heteronuclear HMBC technique, but one must have a sufficient amount of material for this method. Where this is possible, the heteronuclear couplings between the ^{13}C-1 nucleus of the glycone and the ring protons of the aglycone yield valuable information. If one can measure the three-bond coupling constants $^3J(H,C)$, the Karplus equation can be applied to estimate the angle of the glycosidic bond and thus determine the conformation. A more detailed discussion of the structural analysis of oligosaccharides by NMR spectroscopy can be found in the reference cited [11].

13.4 Bibliography for Chapter 13

[1]   K. F. Elgert: Kunststoffe, Analyse. In: *Ullmanns Encyklopädie der technischen Chemie,* Vol. 15, p. 392, Weinheim: Verlag Chemie, 1978.

[2]   P. L. Rinaldi: Applications of Two-Dimensional NMR to the Characterization of Synthetic Organic Materials. In: [3], p. 841 ff.

[3]   W. R. Croasmun and R. M. K. Carlson (Eds.): *Two-Dimensional NMR-Spectroscopy. Applications for Chemists and Biochemists.* New York: VCH Publishers, 1994, 2nd Edition.

[4]   H. W. Spiess: Multidimensional NMR Methods for Elucidating Structure and Dynamics of Polymers. In: *Annual Reports on NMR Spectroscopy,* G. A. Webb and I. Ando (Eds.), Vol. 34. London: Academic Press, 1997, p. 1.

[5]   K. Wüthrich: *NMR of Proteins and Nucleic Acids,* John Wiley & Sons, New York, 1986.

[6]   H. Kessler and S. Seip: *NMR of Peptides.* In: [3], p. 619.

[7]   H. J. Dyson and P. E. Wright: *Protein Structure Calculation Using NMR Restraints.* In: [3], p. 655.

[8]   D. Neuhaus and M. P. Williamson: *The Nuclear Overhauser Effect in Structural and Conformational Analysis.* New York, Chichester, Weinheim, Brisbane, Singapore, Toronto, Wiley-VCH, 2000, 2nd Edition, p. 550.

[9]   R. Riek: TROSY: Transverse Relaxation-Optimized Spectroscopy. In: O. Zerbe (Ed.): *BioNMR in Drug Research.* Weinheim: Wiley-VCH, 2003.

[10] I. Goljer and P. H. Bolton: *Studies of Nucleic Acid Structures Based on NMR Results.* In: [3], p. 699.

[11] J. Dabrowski: *Two-Dimensional and Related NMR Methods in Structural Analysis of Oligosaccharides and Polysaccharides.* In: [3], p. 741.

Additional and More Advanced Reading

Synthetic Polymers

K. Hatada, T. Kitayama and K. Ute: Application of High-Resolution NMR Spectroscopy to Polymer Chemistry. In: *Annual Reports on NMR Spectroscopy*, G. A. Webb (Ed.), Vol. 26. London: Academic Press, 1993, p. 99.

G. Moad: Applications of Labelling and Multidimensional NMR in the Characterization of Synthetic Polymers. In: *Annual Reports on NMR Spectroscopy*, G. A. Webb (Ed.), Vol. 29. London: Academic Press, 1994, p. 287.

A. E. Tonelli: *NMR Spectroscopy and Polymer Microstructure. The Conformational Connection.* New York: VCH Publishers, 1989.

K. Wüthrich: *NMR of Proteins and Nucleic Acids.* New York: John Wiley & Sons, 1986.

Solid-State NMR

N. J. Clayden: Developments in Solid State NMR. In: *Annual Reports on NMR Spectroscopy*, G. A. Webb (Ed.), Vol. 24. London: Academic Press, 1992, p. 2.

E. O. Stejskal and J. D. Memory. *High Resolution NMR in the Solid State. Fundamentals of CP/MAS.* New York, Oxford: Oxford University Press, 1994.

W. P. Power: High Resolution Magic Angle Spinning – Applications to Solid Phase Synthetic Systems and Other Semi-Solids. In: *Annual Reports on NMR Spectroscopy*, G. A. Webb (Ed.), Vol. 51, Kidlington, Oxford: Elsevier Science, 2003, p. 261.

S. Berger and S. Braun: *200 and More NMR Experiments, A Practical Course.* Chapter 14. Weinheim: Wiley-VCH, 2004.

K. Schmidt-Rohr and H. W. Spiess: Dynamics of Polymers from One- and Two-dimensional Solid-state NMR Spectroscopy. In: *Annual Reports on NMR Spectroscopy*, G. A. Webb (Ed.), Vol. 48. London: Academic Press, An Imprint of Elsevier Science, 2002, p. 1.

Biopolymers

C. H. Arrowsmith and Yu-Sung Wu: NMR of large (> 25 kDa) proteins and protein complexes. In: *Prog. Nucl. Magn. Reson. Spectrosc. 32* (1998) 277.

S. Berger and S. Braun: *200 and More NMR Experiments, A Practical Course.* Chapter 15. Weinheim: Wiley-VCH, 2004.

O. Zerbe (Ed.): *BioNMR in Drug Research.* Weinheim: Wiley-VCH, 2003.

S. S. Wijmenga and B. N. M. van Buuren: The use of NMR methods for conformational studies of nucleic acids. In: *Prog. Nucl. Magn. Reson. Spectrosc. 32* (1998) 287.

14 NMR Spectroscopy in Biochemistry and Medicine

14.1 Introduction

During the last 20 years NMR spectroscopy has helped to provide answers to many biochemical problems, mainly through experiments on *in vitro* model reaction systems. In these studies ^{13}C NMR spectroscopy, because of the larger chemical shifts compared with ^{1}H NMR, has often turned out to have the advantage. However, its great disadvantage – as has already often been mentioned in earlier chapters – is the low detection sensitivity for ^{13}C. In *in vitro* experiments this problem has in many cases been overcome by ^{13}C enrichment.

It is evident from the earlier chapters that the main bene-ficiaries of NMR spectroscopy are chemists and biochemists. However, in the last few years new areas of application have emerged, which have led to rapid developments both in tech-niques and in instruments. These are the NMR studies now being carried out on living organisms, ranging from single cells to whole body investigations on human subjects. This means that the NMR method has now also become of interest to biolo-gists and medical scientists and for medical applications.

What kinds of new knowledge are obtainable from NMR experiments on living organisms? Is it not the case, according to all that we have learned so far, that the main strength of NMR spectroscopy lies in analyzing pure compounds? How is it possible to make NMR measurements on a "sample" as large as a human being? In living cells how can one selectively study the compounds of interest in the presence of the innumer-able other substances that they contain?

Before discussing *in vivo* experiments, we must first see what kinds of information are obtainable from ^{13}C-labeling studies in biochemistry, by considering some examples of conventional NMR spectroscopic investigations carried out to elucidate bio-synthetic pathways.

That will be followed a section on high-resolution *in vivo* NMR spectroscopy. There we will depart from the plan fol-lowed up to now in this book, by discussing ^{31}P resonances in addition to those of ^{1}H and ^{13}C, since many *in vivo* experiments have made use of phosphorus nuclei as probes.

The final part of the chapter will deal with *magnetic resonance tomography* or *magnetic resonance imaging (MRI)* and *magnetic resonance spectroscopy (MRS)* which opens up fascinating prospects for medical research and diagnosis.

14.2 Elucidating Reaction Pathways in Biochemistry

A recurring fundamental question in biochemistry is: how is a particular compound synthesized in an organism? For example, this might be a compound whose structure has already been determined, and one may even already know the starting material or precursor which the bacteria, molds or other microorganisms use for the synthesis.

As a specific example, biochemists already knew that mold cultures of *Cephalosporium acremonium* incorporate acetate ions and valine in the biosynthesis of cephalosporin C. However, it was not yet known where these compounds are incorporated into the molecule, whether or not the synthesis is stereoselective, and whether or not rearrangements occur during the synthesis. These questions can be answered by ^{13}C NMR spectroscopy; in these studies the microorganisms were grown on a nutrient medium containing ^{13}C-labeled compounds. The biochemists had the task of isolating the labeled compounds from the mold cultures in large enough quantities to give satisfactory ^{13}C NMR spectra. The nature of the information that can be obtained from such spectra depends on:
- which precursors have been labeled,
- the molecular positions of the labeling, and
- whether singly or doubly labeled compounds were used.

These questions will be dealt with in the following sections.

14.2.1 Syntheses using Singly ^{13}C-Labeled Precursors

14.2.1.1 Low Levels of ^{13}C Enrichment

Even a few percent of ^{13}C enrichment at specific molecular positions is enough to give a considerable increase in the intensities of the corresponding signals in the ^{13}C NMR spectrum. We have already met this signal enhancement effect in Chapter 6, in connection with its use as an aid to assignment (Section

6.3.6). The same effect is observed in compounds synthesized by microorganisms from ^{13}C-labeled precursors. Cephalosporin C (**1**) provides an example of this. As already mentioned, acetate ions and valine serve as precursors in the biosynthesis.

By carrying out separate experiments in which ^{13}CH$_3$COO$^{\ominus}$, CH$_3$^{13}COO$^{\ominus}$ and valines labeled at either the 1- or the 2-position were used as precursors, the labeled cephalosporin C (**1**) shown in Figure 14-1 was obtained. However, it was still not known which of the carbon atoms 2 and 17 originated from which methyl groups of the valine molecule. To answer this question one of the two valine methyl groups was selectively labeled with ^{13}C. As a result of this the labeled molecules have a second asymmetric center, giving two diastereomers (**2** and **3** in Fig. 14-1); in our example only the R and S configurations at C-3 are of importance. The analysis of the ^{13}C NMR spectra of the cephalosporins synthesized from these labeled valines proved conclusively that the synthesis occurs in the manner shown in Figure 14-1 [1].

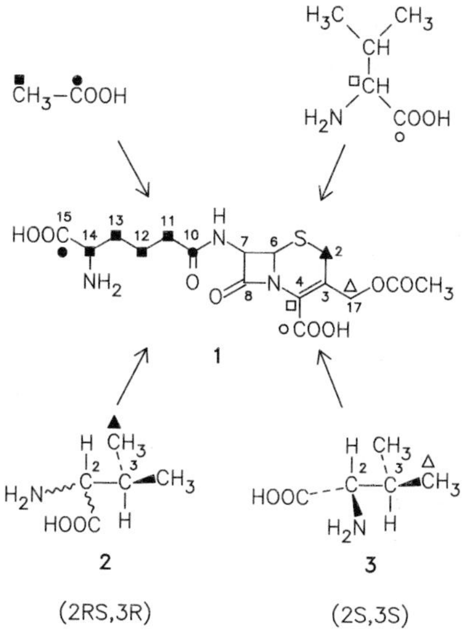

Figure 14-1.
Formula scheme showing the biosynthesis of cephalosporin C (**1**). The filled and open triangles, squares and circles indicate the ^{13}C-labeled positions in the starting materials and in **1**.

Studies of a similar kind have also clarified details of the biosynthetic routes for penicillin, chlorophyll, vitamin B$_{12}$ and many other compounds [1].

It must be emphasized that in these and all the following experiments it is essential to assign every signal in the ^{13}C NMR spectrum, since only then can one determine the exact position in the molecule at which a group is incorporated.

14.2.1.2 High Levels of ^{13}C Enrichment

In the examples described in the previous section the levels of ^{13}C enrichment were only a few percent. Higher levels were unnecessary in those cases, and would even be undesirable, as all that was required was to enhance and thereby pick out certain signals in the spectrum. With suitable preparative methods one can achieve higher levels of enrichment, but this leads to complications caused by $^{13}C,^{13}C$ couplings. However, there is an advantage in this insofar as the $^{13}C,^{13}C$ coupling can give important additional information. To illustrate this we will carry out an imaginary experiment. Let us suppose that a dimerization reaction takes place between two sodium acetate molecules whose methyl groups are labeled with ^{13}C ($^{13}CH_3COONa$). The four possible reaction products are shown in Figure 14-2, with ^{13}C atoms shown as filled circles and ^{12}C atoms by open circles. It is evident that only the first of the four synthetic routes (1) gives two directly bonded ^{13}C nuclei, and thus a mutual coupling. In the three other bonding situations the ^{13}C nuclei are separated by at least one ^{12}C nucleus.

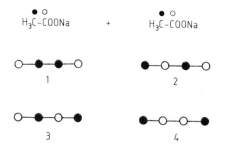

Figure 14-2.
Diagram showing the four possible products from the reaction between two sodium acetate molecules with ^{13}C-labeled methyl groups (●). Only the product 1 contains two adjacent ^{13}C nuclei and thus shows a ^1J (C,C) coupling in its ^{13}C NMR spectrum.

Let us now apply this knowledge to the biosynthesis of cephalosporin C (Fig. 14-1). If one were to use highly enriched $^{13}CH_3COONa$ as a precursor, the ^{13}C NMR spectrum of the resulting compound should show $^{13}C,^{13}C$ couplings between positions 11 and 12, between 12 and 13, and between 13 and 14, as these are the only carbon atoms which originate from the methyl groups of the acetate ions.

From Figure 14-2 we can deduce two rules:
- carbon nuclei at the end positions will never show $^{13}C,^{13}C$ couplings, and will always appear in the spectrum as single peaks, and
- coupled pairs of ^{13}C nuclei must always originate from different acetate groups.

14.2.2 Syntheses using Doubly ^{13}C-Labeled Precursors

The usual starting material for experiments with double ^{13}C-labeling is doubly labeled sodium acetate, $^{13}CH_3^{13}COONa$, as this compound is a precursor in many biochemical reactions. The products obtainable from acetate ions include polycarbonyl compounds, whose importance as intermediates in biosynthetic reactions has long been recognized.

Using highly enriched sodium acetate in biosyntheses by microorganisms often gives compounds with very complicated ^{13}C NMR spectra. In cases where, as in the polycarbonyl compounds, the molecule is synthesized entirely from labeled acetate ions, each ^{13}C nucleus, except those at the terminal positions, can couple to two, three, or even four others. A way of getting around the resulting complexity is to dilute the precursor with normal sodium acetate, $^{12}CH_3^{12}COONa$. The probability of two or more doubly labeled acetate units occurring next to each other is then reduced to an extent which depends on the dilution; during the synthesis unlabeled units become inserted between the labeled ones. It has been found in practice that it is sufficient to dilute in the ratio 1 part doubly labeled to 2.5 parts normal sodium acetate, and no couplings between ^{13}C nuclei derived from two different acetate ions are then apparent (Fig. 14-3).

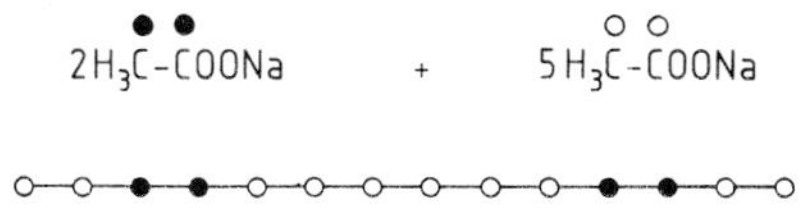

Figure 14-3.
Distribution of acetate moieties in the molecule obtained by a synthesis using a 1:2.5 mixture of doubly-labeled and unlabeled sodium acetate.

On the other hand, ^{13}C,^{13}C couplings are, of course, always seen when the $^{13}C-^{13}C$ unit is incorporated as such. In these cases, single peaks in the spectrum indicate that CC bonds have been broken.

Using this method Holker and Simpson [2] showed that mellein (**4**), a fermentation product from *Aspergillus melleus*, is synthesized from five intact $^{13}C_2$ acetate units (Fig. 14-4).

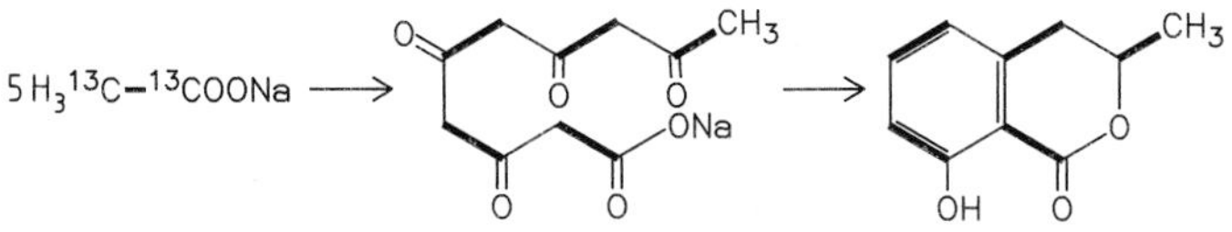

Figure 14-4.
Biosynthesis of mellein (**4**), a fermentation product from *Aspergillus melleus*, from five intact $^{13}C_2$ units [2].

The biosynthesis of bikaverin (**5**), a red pigment isolated from the fermentation product of *Gibberella fujikuroi*, also appears to proceed via a polycarbonyl compound. Figure 14-5 shows two possible conformations of the chain from which the final molecule could be formed.

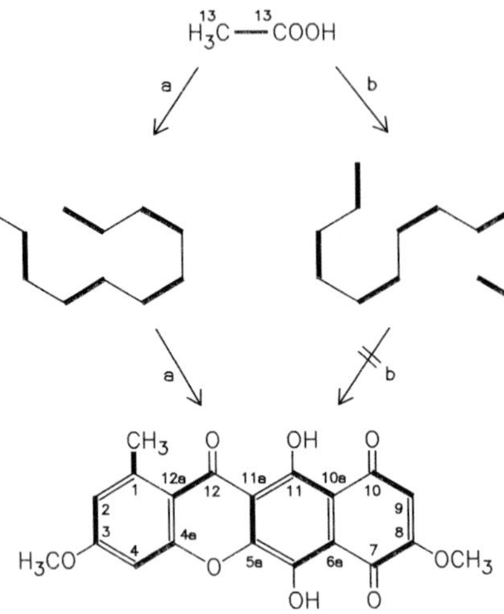

Figure 14-5.
Biosynthesis of bikaverin (5), the fermentation product from *Gibberella fujikuroi*, from nine intact $^{13}C_2$ units. In the ^{13}C NMR spectrum a C,C coupling between carbon nuclei 5a and 11a is observed, and route b can therefore be ruled out [1].

An investigation using doubly labeled sodium acetate showed that nine $^{13}C_2$ units are incorporated into the molecule (for the spectrum see Ref. [1]), thus confirming that a polycarbonyl compound occurs as an intermediate stage. A separate ^{13}C labeling experiment showed that the OCH$_3$ groups are derived from methionine.

The second question of whether bikaverin is synthesized by route 'a' or route 'b' (Fig. 14-5) still remains to be answered. The key to this is the existence or otherwise of a spin-spin coupling between ^{13}C nuclei at positions 5a and 11a. If a coupling between these two positions is found, it indicates that route 'a' is correct; if not, then route 'b' is correct. The spectrum clearly shows a 5a–11a coupling, showing that these two carbon atoms must come from the same acetate ion, and therefore route 'b' can be ruled out.

It can be seen from this example how important it is to assign the signals correctly. In the case of bikaverin this required many deocupling experiments, even involving measurements of long-range C,H couplings. The pair of coupled ^{13}C nuclei were eventually identified by homonuclear ^{13}C decoupling experiments. (For other examples see Refs. [1, 3].)

14.3 High-Resolution *in vivo* NMR Spectroscopy

14.3.1 The Problem and its Solution

The purpose of *in vivo* NMR experiments is to study chemical reactions in living cells from outside the cells in a non-invasive manner. For a long time the aim of carrying out such experiments was frustrated by several formidable problems. The most serious of these are concerned not so much with instruments and techniques as with the intrinsic nature of the systems to be investigated. A living organism, even if it consists of only a single cell, is a heterogeneous sample containing many different substances in extremely low concentrations.

In order to study the molecular processes occurring within cells, it is necessary to achieve a resolution and a sensitivity as high as in normal NMR spectroscopy, where, in contrast to living organisms, one generally deals with homogeneous samples of small volume. In *in vivo* spectroscopy the sample volume may, in the extreme case, be that of an entire human body. Experiments of this kind are only possible by using magnets with an appropriately large bore (i.e. usable field diameter). However, the problems of constructing such magnets have now been overcome.

If large heterogeneous samples are to be studied, some questions come immediately to mind. What is to be observed, and where? Is it possible to assign the NMR signals to definite molecular species? What metabolites are formed in the cell, in the heart, in the brain, or in some other organ? How can a particular region of interest in the organism be selectively studied in a non-invasive way?

At present the best ways of achieving this latter objective employ either surface coils or pulsed field gradients. A surface coils is a single turn of wire which is designed to fit the position to be examined and is applied externally. This acts as the transmitter for the radiofrequency pulse and also as the receiver for the NMR signal. The size of the coil determines the volume examined. The depth of penetration can be adjusted within certain limits by altering the pulse duration, which makes it possible to selectively examine even deep-lying organs. This selectivity is important, as one must avoid including too much of the tissue between the coil and the organ of interest, which would cause excessive background noise in the spectrum. The method of pulsed field gradients achieves volume selectivity in a more elegant and versatile way, and is now widely used in the techniques known as magnetic resonance tomography, MRI, magnetic resonance spectroscopy, MRS, and magnetic resonance microscopy. It will be described in detail in Section 14.4.1.

What kinds of information do the spectra yield? Mainly one obtains chemical shifts and intensities. Coupling constants are seldom measured, since the resolution achieved is not usually good enough to show multiplet splittings.

In the following sections we will begin with an account of some results obtained by ^{31}P NMR spectroscopy, and will then briefly look at what has been achieved by ^{1}H and ^{13}C NMR measurements.

14.3.2 ^{31}P NMR Experiments

Several factors make ^{13}P NMR spectroscopy especially suitable for *in vivo* experiments:
- Organophosphorus compounds play a central role in the energy balance of living organisms.
- The number of organophosphorus compounds that occur in significant amounts is limited.
- ^{31}P resonances are easy to observe. The phosphorus nucleus has a large magnetic moment and a natural abundance of 100 %, and ^{31}P is therefore one of the sensitive nuclides. Furthermore, in many organs the concentrations of organophosphorus compounds are relatively high, which means that one can obtain measurable spectra within a few seconds.

Many of the *in vivo* studies that have been made are concerned with energy metabolism in muscle, with special emphasis on the heart in view of its medical importance. However, experiments of this kind encounter problems caused by the movement of the heart. One therefore has to either accept some line broadening from this cause, or arrange for the start of the signal generating pulse to be triggered at the frequency of the beating of the heart. Here we will look at a simpler example, the ^{13}P NMR spectrum of the forearm of a human subject (Fig. 14-6), which was recorded over a period of about 1.5 min using a surface coil with a diameter of 4 cm.

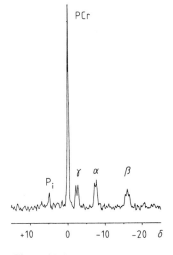

Figure 14-6.
40 MHz ^{31}P NMR spectrum from a human forearm, recorded using a 4 cm diameter surface coil. Signal assignments: P_i: free phosphate ions (inorganic phosphate); PCr: creatine phosphate (**6**); α, β, γ: phosphorus nuclei at the α, β and γ positions in adenosine triphosphate ATP (**7**). P,P couplings give a doublet splitting of the α and γ signals and a triplet splitting of the β signal. The δ-scale is referred to PCr, i.e. δ (PCr) = 0.

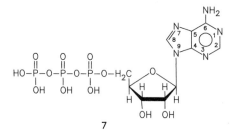

7

6

Before assigning the signals we must consider what compounds can, in principle, be present in the cells. This question can be answered immediately, as the biochemistry of energy metabolism in muscle has been very thoroughly studied. Phosphate ions are transferred from creatine phosphate (PCr, **6**) to adenosine diphosphate (ADP), forming adenosine triphosphate (ATP, **7**) and creatinine (Cr).

$$\text{PCr} + \text{ADP} \rightleftharpoons \text{ATP} + \text{Cr}$$

In the spectrum we do in fact see a single peak due to PCr and three signals corresponding to the three differently bonded phosphate groups at the α-, β- and γ-positions in ATP. These latter signals have splitting patterns corresponding to the P,P couplings: P_α and P_γ are both coupled to P_β and therefore give doublets, while P_β gives a triplet as a result of its coupling to two neighboring phosphorus nuclei. The assignments are shown in Figure 14-6, in which the δ-scale is referred to the PCr signal. The ADP signals are hidden under the P_α and P_γ signals of the ATP. In addition there is a small signal due to free phosphate ions present as inorganic phosphate, P_i. The signals of other organophosphorus compounds present are lost in the noise.

In the ^{31}P NMR spectrum of the brain of a rabbit (Fig. 14-7) we can detect, in addition to the PCr and ATP signals, other signals due to phosphoric acid diesters and glucophosphates.

Basic experiments such as these stimulated numerous other investigations on different types of samples: intact organisms, and isolated or perfused organs such as heart, muscle, brain, liver, kidney, eye, etc.

As an example from the very wide variety of such experiments, the series of spectra in Figure 14-8 shows the changes, compared with normal levels, in the concentrations of energy-carrying phosphorus compounds in the forearm muscle of a human subject following prolonged strenuous work, and their subsequent recovery to normal levels.

A control spectrum was first recorded under normal conditions (Fig. 14-8 A), by accumulating ten FIDs, the same number as in the experiment which followed. The spin system was allowed to relax for about 2 s between successive pulses; the total time for the ten FIDs was 22 s. After the recording of the control spectrum the subject was required to open and close his hand outside the magnet, squeezing a ball repeatedly until the muscle was tired. Immediately after this work phase the four spectra B-E were recorded. Each of these spectra represents an average over the recording time of 22 s. It is clearly seen that immediately after the work phase the PCr signal is considerably reduced, whereas the signal due to the inorganic phosphate P_i has greatly increased. After 66 s relaxation the initial situation is almost restored. This conclusion is based not on comparing the heights of the peaks in the final spectrum and the control

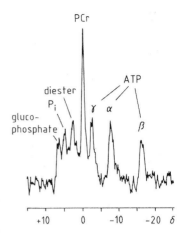

Figure 14-7.

40 MHz ^{31}P NMR spectrum of a rabbit's brain, recorded by resting the animal's head on a 4 cm diameter surface coil. In addition to the signals of free phosphate (P_i), creatine phosphate (PCr) and ATP (α, β, γ), those of phosphoric acid diesters and glucophosphates can be seen. The δ-scale is referred to PCr, i.e. δ (PCr) = 0.

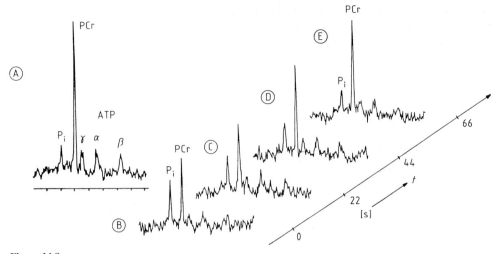

Figure 14-8.
40 MHz ^{31}P NMR spectrum from a human forearm before and after strenuous muscular effort (see text), recorded using a 4 cm diameter suface coil; 10 FIDs; total time per measurement: 22 s; all spectra recorded with same amplification. A: control spectrum before physical effort. B to E: spectra recorded during the relaxation phase. The signal intensities show that immediately after effort the P_i signal has greatly increased and the PCr signal has greatly diminished.

spectrum, but on the integral curves (not shown), since the line-widths in the two spectra are not the same. Whereas there was time to optimize the field homogeneity before recording the control spectrum, this was not possible for the spectra recorded after the work phase. The field homogeneity could not have been adjusted without sacrificing the measurements in the initial stage which is of greatest interest, since it can be seen from these results that the muscle recovers very quickly.

In these experiments it is also found that the separation between the P_i and PCr signals alters. The explanation of this important result emerged from *in vitro* experiments, which showed that the separation is pH-dependent, increasing at higher pH-values! Thus the position of the P_i signal relative to the PCr signal gives a measure of the changes of the pH within the cells. In our example the pH is reduced after the work phase, which is understandable from a biochemical point of view, as muscular exertion results in the formation of lactic acid.

The proportions of the different organophosphorus compounds in the tired muscle differ from those in the normal state. Even more striking effects are found if the muscle is starved of oxygen over a long period (ischemia). This can occur if the blood supply is reduced as a result of a constriction of a blood vessel, or even stopped completely by a blockage. The most familiar example of this is a cardiac arrest. Many *in vivo* NMR experiments on animals have shown that malfunc-

378

tions of individual organs can be immediately recognized. The application of such methods for medical diagnosis is a very promising development for the future.

14.3.3 ¹H and ¹³C NMR Experiments

How can ¹H and ¹³C NMR measurements also be used in this field, so as to extend the scope of such techniques to include the whole range of compounds involved in the chemistry of living organisms?

There remain some difficulties to extending *in vivo* experiments to include these two nuclides, and these are as follows:

- Mixtures are always present, and these often include very complex molecular structures.
- The concentrations of individual components are extremely low.
- The solvent is water in many cases.
- The chemical shift differences are often small, especially for protons.
- The resonances are very broad due to the unavoidable field inhomogeneities in heterogeneous samples.

Nevertheless, some promising approaches are being investigated. Figure 14-9 shows the ¹H NMR spectrum of a human forearm. Despite the broadness of the signals, well separated signals are seen for water (left-hand peak) and the aliphatic protons in the fatty tissue.

A significant advance in this area has been brought about by the nowadays wide availability of high-field spectrometers, firstly owing to their higher sensitivity, and secondly because they give better separation of the ¹H signals obtained from *in vivo* measurements, with typical line-widths of 20–25 Hz. Also it now appears likely that the method of volume selection based on pulsed field gradients will be used in preference to other methods, especially where one wishes to measure ¹H NMR spectra (see Section 14.4.2.2 Magnetic Resonance Spectroscopy, ¹H MRS).

¹³C measurements offer the possibility of obtaining more information on individual molecular structures. In the ¹³C NMR spectrum of a human forearm (Fig. 14-10) the large chemical shifts enable one to clearly distinguish the signals of the carboxyl groups ($\delta \approx 172$), those of double-bonded carbon nuclei ($\delta \approx 130$), the glycerol carbon nuclei ($\delta \approx 60$–75), and those of aliphatic fatty acids ($\delta \approx 10$–35). (In this example it was even possible to assign some of the signals of different carbon nuclei in the aliphatic chains.)

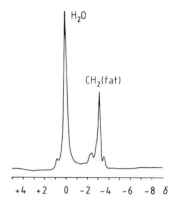

Figure 14-9.
100 MHz ¹H NMR spectrum from a human forearm. The main signals are from the protons in the water and in the fatty tissue; the δ-scale is referred to the water signal. 2 FIDs; total time about 5 s.

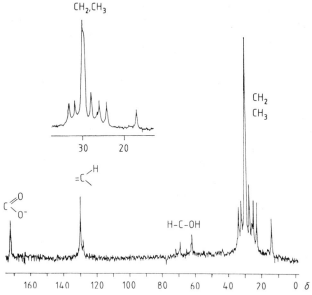

Figure 14-10.
25 MHz ^{13}C NMR spectrum from a human forearm, recorded with ^{1}H BB decoupling, using a 4 cm diameter surface coil. 500 FIDs; total time about 10 min. The signals can be assigned to the carbon nuclei of the fatty acids and of the glycerol moieties. The region containing the CH$_3$ and CH$_2$ signals ($\delta = 10$ to 40) is shown expanded.

For the spectrum shown here, with ^{13}C in natural abundance, it was necessary to accumulate 500 FIDs, requiring a total recording time of about 10 min. The spectrum was recorded using a surface coil with a diameter of 4 cm. For detecting special compounds and reactions one would need to use ^{13}C-labeled materials. In this area too, further development is needed.

14.4 Magnetic Resonance Tomography

The techniques to be described in this section have developed as an offshoot of NMR spectroscopy which has become especially important in medical diagnosis. The methods are known alternatively as magnetic resonance tomography (from Greek *tomos* = section or slice) or magnetic resonance imaging (MRI).

14.4.1 Basic Principles and Experimental Considerations [4–6]

In 1973 Paul Lauterbur started to produce images of large objects by NMR [4]. 30 years later, in 2003, he and Peter Mansfield were awarded the Nobel Prize for Medicine in recognition of their achievements and of the present-day importance of the method in medical research and diagnosis.

For these experiments Lauterbur used proton resonances because of their high detection sensitivity. Since living systems always involve aqueous media, it was natural that the first investigations were concerned with the distribution of water in an organism. One expects to find strong signals from those parts of an animal's body that contain a lot of water, and weak signals from those such as bones which contain little water.

The pioneering studies by P. Lauterbur inspired many physicists and engineers to become engaged in the search to develop instruments and techniques that could be used to measure the water distribution in the human body by observing the ^{1}H resonance, and to present the results as an image. In order to develop the method of magnetic resonance imaging (MRI) for medical applications, many problems concerning the apparatus and the measurement technique had first to be solved, as well as others connected with the processing of the data and the formation of the image. Eventually these were overcome, and whole-body tomographs working at flux densities from 0.2 to 8 tesla now exist, although the magnets used for routine clinical measurements are generally limited to 2 tesla. Instruments with higher-field magnets are mainly used for research. Table 14-1 lists data for various types of magnets that are used.

Instead of the technique of "filtered back projection" that was originally used by Lauterbur for generating images, the usual method now is the *two-dimensional imaging technique with slice selection*. The principle of this can be understood on the basis of what we learned earlier in Chapters 1, 8 and 9 about pulses, their duration and phase angle, the behavior of magnetization vectors, and field gradients.

In Section 8.2.3 we considered the effect of a (pulsed) field gradient on a macroscopic sample such as an NMR sample tube filled with a liquid. In the MRI technique, field gradients of this kind are used to achieve volume selectivity. First we need to understand how this volume selection occurs in one dimension, then we will see how it can be extended to two and to three dimensions. Figure 8-5 showed how the application of a field gradient G_z along the direction of the field B_0, giving additional field contributions g_z, resulted in different resonance frequencies in slices at different heights in the sample, in accordance with the resonance condition $v_n = (\gamma/2\pi)(B_0 + g_z)$. In Figure 14-11 only the sample tube, assumed to be filled with water, is shown. When a gradient pulse is applied, the sample tube experiences the gradient-modified field $B_0 + g_z$ (where g_z depends on the height within the sample). Let us consider the slice 5, for which the resonance frequency is v_5. If we apply a selective r.f. pulse at this frequency, only the resonances of the protons in this one slice are excited. Considering now the situa-

Table 14-1.
Types of magnets used in magnetic resonance imaging and their operating flux densities.

B_0 [T]	Magnet type
0.2–0.3	permanent or conventional electromagnet (resistive)
0.5	superconducting
1.0	"
1.5	"
2.0	"
4.7	"
8.0	"

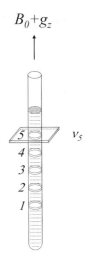

Figure 14-11.
Principle of slice selection by means of a field gradient producing additional field contributions g_z.

381

tion in the MRI technique, such a slice might correspond, for example, to a slice through the head of a patient. How does one produce such a selective pulse? We learned in Section 1.5.1 that a pulse contains a whole band of frequencies, the width of which is proportional to τ_p^{-1}, the reciprocal of the pulse duration. Therefore we must choose a long pulse duration so as to obtain a narrow frequency band. Unfortunately, however, we cannot increase τ_p indefinitely, as it is involved in the expression for the pulse angle $\Theta = \gamma B_1 \tau_p$ [cf. Eq. (1-14)], and for our experiment we need a pulse angle of about 90^0. We might compensate for an increase in τ_p by reducing the radiofrequency field B_1, but that involves a sacrifice of sensitivity, so the scope for that is limited. Therefore the choice of a value for τ_p must be a compromise. In practice a pulse duration of 2–4 ms is commonly used, giving a "soft" pulse with a bandwidth of about 500 Hz. As the strength of the gradient G_z that can be applied is limited by practical considerations, this means that the slice selected by the combination of such a pulse with the field gradient has an appreciable thickness, typically several mm. The field gradient together with the "soft" pulse has now given us the first stage of selection, a *slice selection*, and this first gradient is therefore called the *slice selecting gradient*. If, after switching off this gradient, we were to simply record the FID, the Fourier transformation of this would give us a signal proportional to the quantity of water in the selected slice.

However, in the experiment described here we do not record the FID immediately. Instead we apply a second field gradient pulse G_x (or G_y) in a direction at right angles to the first. To understand the function of this gradient we will now assume that there are two water-filled sample tubes, I and II, both with their axes along the direction of the field B_0, which is the z-direction. As before, we begin the experiment by applying the gradient G_z, and simultaneously we apply a soft $90°$ pulse to select a slice. Figure 14-12 B shows the selected slice viewed from above, i.e. along the $-z$ direction, and the intersections of this plane with the samples I and II. When the second gradient G_x is applied, I and II experience different magnetic fields, and therefore have different resonance frequencies. To be more exact, owing to the finite diameters of the tubes the protons within each sample do not all experience the same magnetic field, and consequently the frequency-domain spectrum obtained by Fourier transformation of the FID consists of two broad lines (inhomogeneous broadening). We shall return later to the problem that this presents. Our spectrum of slice 5 for the two samples corresponds to a projection of the spin densities in the samples onto the x-axis. If we had chosen instead to apply a gradient G_y, we would have obtained a projection onto the y-axis. As the function of the gradient G_x in the above is to generate (or "read out") the frequency-domain spectrum, it is sometimes

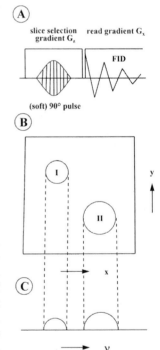

Figure 14-12.
Principle of two-dimensional volume selection.
A: Sequence of gradient pulses, r.f. pulse, and data acquisition.
B: Slice through the hypothetical model (two cylindrical samples) selected by applying the gradient G_z.
C: Projection of the spin density onto the x-axis.

called the "read gradient". Thus we now have a volume selection in two dimensions, the z- and x-directions (see Figure 14-12 C). By rotating the direction of the read gradient around the object, one can obtain projections onto the different directions chosen for the gradient. By this method, using "filtered back projection", Lauterbur was able to calculate an image of an object in a manner analogous to X-ray tomography.

Up to now we have not shown how to achieve selection in the third dimension, the y-direction. For this we need a third field gradient G_y. Figure 14-13 A shows the sequence for the three-dimensional experiment. As before, we consider our simplified model consisting of two cylindrical samples. First the resonances of the protons within a slice of samples I and II are excited selectively by means of a "soft" pulse (slice selection). After switching off the gradient G_z we apply a second gradient G_y. Unlike the sequence described above, we do not record the FID at this stage; instead, after an interval T_{Gy} we switch from the gradient G_y to a gradient G_x. We already know what the gradient G_x does: it enables us to record an FID (with the time dimension t_2) which after Fourier transformation yields a frequency-domain spectrum of the projection of the two samples onto the x-axis. But what is the purpose of the gradient G_y of duration T_{Gy}?

First we will consider just sample I. When the gradient G_y is applied the resonance frequency of the water protons in this sample tube is ν_1. After the time T_{Gy} the spin system will have acquired a certain phase due to the precession of the nuclear spins (i.e., of the macroscopic magnetization vector M_y). This is the principle of the final selection mechanism, and therefore G_y is called the *phase gradient*. If the experiment is then repeated with a different value of T_{Gy}, a different phase state is obtained. A series of measurements are made with different values of T_{Gy}, so that the FIDs recorded after switching off G_y and switching on G_x contain phase information as a function of T_{Gy}. A similar argument applies for the FIDs from sample II. In our two-sample model, of course, we record just a single FID, which contains all the information.

To generate a two-dimensional image one performs N measurements in which T_{Gy} is altered by regular increments. The number of measurements N is usually 256 ($= 2^8$). Performing a Fourier transformation with respect to the time t_2 on the N different FIDs yields N frequency-domain spectra (F_2-dimension), in which the intensities of the signals have a characteristic dependence on T_{Gy}. A second Fourier transformation with respect to T_{Gy} then yields the required image.

To simplify the above discussion we have assumed that all the protons in the selected slice in each sample experience the same magnetic field, and thus have the same resonance frequency. This is not actually the case, because as a result of the gradients

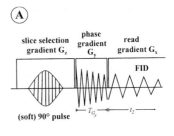

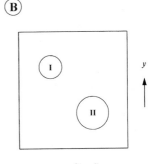

Figure 14-13.
Three-dimensional volume selection using a sequence of three field gradient pulses.
A: Sequence of gradient pulses, r.f. pulse, and data acquisition.
B: Image obtained by triple Fourier transformation of the FIDs.

applied along the x- and y-directions there are phase variations along the y-axis within the slice before the application of the gradient G_x, and variations in frequency during the following gradient pulse G_x. All this information is present in the image obtained from the double Fourier transformation. One aims to achieve high resolution, in other words to make the individual volume elements (voxels) that make up the image as small as possible. The thickness of the slice is also relevant in determining the resolution, and should be as small as possible. It was shown above that in order to achieve a thin slice we need a selective r.f. pulse with a narrow frequency band. The thickness of the slice can also be reduced by increasing the gradient G_z.

The above description has, of course, omitted many of the finer details and complications. Thus, for example. in practice the incrementation of the phase gradient pulse is carried out not by increasing T_{Gy} but by increasing the gradient G_y. Also, in the above description we applied the gradients in the sequence G_z, G_y, G_x, giving a slice selection perpendicular to the field direction, whereas in practice one can choose any direction for selecting a slice through the object; it is only necessary that the three gradients should be at right angles to each other (orthogonal).

It is also possible to perform volume selection by a purely computational method using the computer, without the need for a physical slice selection. One can then generate images of sections through the object in any number of arbitrary directions. For this one uses three field gradients in the x-, y- and z-directions. The measurement begins with the application of a 90° r.f. pulse (*not* a selective pulse!). This is followed by two phase gradient pulses, first a gradient G_x whose duration T_{Gx} is increased by increments, then a gradient G_y, whose duration T_{Gy} is also incremented. Lastly a gradient G_z is applied as the "read" gradient. Each of the recorded FIDs contains phase information as functions of the gradient pulse durations T_{Gx} and T_{Gy}, and this is reflected in the intensities of the signals in the frequency-domain spectrum.

The experiment is very time-consuming, as it is usual to make 256 measurements with different values of T_{Gx}, and for each value of T_{Gx} to make a further 256 measurements with different values of T_{Gy}, making a total of 256^2 individual measurements. Finally, a second Fourier transformation with respect to T_{Gx} and a third with respect to T_{Gy} gives a three-dimensional digital image of the object. From the stored data one can then generate any number of images of arbitrarily chosen sections through the object.

Using the above technique, which requires a triple Fourier transformation, it is possible to examine small objects in very strong magnetic fields and achieve a volume element (voxel) as small as $10 \times 10 \times 10$ μm^3. This high-resolution imaging method is called *NMR microscopy* by analogy to optical microscopy, although the resolution falls far short of that in the optical case [7].

From a medical point of view the ability to determine the water distribution is not particularly exciting. The technique of *magnetic resonance tomography* (or *imaging, MRI*) became more interesting when it was discovered that the relaxation times T_1 and T_2 of the water protons depend in a characteristic way on how the water is bound in the tissue. To make use of this one needs to produce images which show not only the water distribution but also the variations in T_1 and T_2. To understand the principle of the technique, let us remind ourselves of what we learned in Chapter 7 about these two relaxation times. T_1, the spin–lattice or longitudinal relaxation time, is a time constant describing the rate at which the equilibrium magnetization M_0 along the field direction becomes reestablished after the system has been disturbed by a radiofrequency pulse. We learned in Section 7.2.2 that T_1 can be measured by the inversion recovery method which uses the pulse sequence

$$180° - \tau - 90° - \text{FID}.$$

T_2 is the spin–spin or transverse relaxation time, and describes the rate at which the transverse magnetization decays after a radiofrequency pulse. It is measured by the spin–echo experiment (Section 7.2.3), using the pulse sequence

$$90° - \tau - 180° - \tau - \text{echo}.$$

The principles of the inversion recovery experiment, and more especially that of the spin–echo experiment, also play an important role in MRI. If one uses a pulse sequence containing several slice-selecting pulses to generate an image, it is necessary to apply the slice-selecting gradient G_z with each radiofrequency pulse.

An important advance in the medical applications of MRI came when it was found experimentally that the T_1- and T_2-values for water not only differ depending on the tissues in which it is present, but are also affected by whether the tissues are healthy or diseased. This made it possible to generate images with a suitable contrast mechanism so that the exact positions of tumors can be revealed. The images are comparable to those obtained by X-ray tomography, but are of superior quality in many cases.

The contrast between normal and diseased tissues can be increased by introducing contrast agents such as gadolinium-DTPA, which shorten T_1 and T_2. The blood flow through abnormal regions such as tumors is greater than that through the surrounding normal tissues, and consequently they absorb the contrast agent more rapidly. Compared with X-ray tomography the NMR method has the great advantage that the patient is not exposed to harmful radiation, and the experiment can therefore be repeated with no risk. Research so far has not revealed any risk to health from the static magnetic field used in routine MRI. However, it may be necessary to consider limiting the exposure to pulsed field gradients.

As the recording time for a human subject cannot be prolonged indefinitely, it is necessary to decide on a compromise between recording time and resolution. This also explains why only proton resonances, for which the sensitivity is high, can be used as an imaging probe at present. However, there are ways of modifying the procedure so as to shorten the measurement time (see below). In Section 1.5.5 ("Spectrum Accumulation") it was explained that in order to obtain a spectrum with a good signal-to-noise ratio one must record a large number of individual FIDs. The interval T_R (the "time to repeat") between successive cycles must be long enough to allow the system to become fully relaxed before the start of the next pulse cycle. The required time can be as long as many seconds, whereas the active cycle itself, from the excitation pulse to the end of the data acquisition period, is only a small fraction of that, usually less than 0.1 s. By suitably adjusting T_R and the acquisition time it is possible to generate images which show the variation of either T_1 or T_2 through the object (see Figures 14-14, 14-15 and 14-16 in the next section).

In MRI one can use the long waiting time T_R to examine other slices through the object. In this multislice procedure the completion of the first measuring cycle on slice 1 is followed immediately by the selection of the next slice, and so on. In this way it is possible to measure as many as 15 further slices during the time T_R while the protons in slice 1 are undergoing relaxation. Another method of reducing the time needed to obtain images is to extend the normal spin–echo measurement by adding on several more echoes, thereby improving the signal-to-noise ratio for the same measurement time. Alternatively, each echo is recorded with a different phase gradient which is increased by regular increments; this reduces the total measurement time.

14.4.2 Applications

14.4.2.1 Magnetic Resonance Tomography

Some examples will serve to illustrate the potential applications of MR tomography.

Figure 14-14 shows a T_1-weighted image of a knee, obtained by a conventional spin–echo technique. The bright regions arise

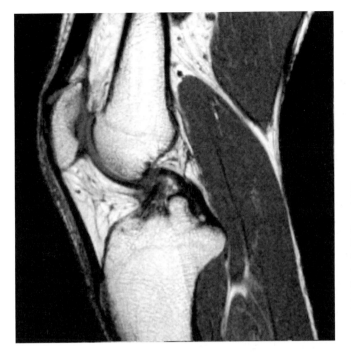

Figure 14-14.
High-resolution musculo-skeletal T_1-weighted image of a knee joint achieved using a conventional spin–echo technique. Magnetic flux density $B_0 = 3$ T; slice thickness 4 mm; measurement time approx. 4 min. (All the images reproduced here were obtained with cryo-magnet tomographs made by Bruker Medical GmbH, D-76275 Ettlingen, Germany.)

mainly from fatty tissues and from fat within the bones, while the dark regions are due to muscles and ligaments. Excellent T_1 contrast, as displayed here, can be obtained fairly easily for the musculo-skeletal system at a variety of field strengths (unlike the situation for other anatomical regions, most particularly the brain). However, slight to moderate alteration of the acquisition timing parameters is mandatory. The image shown here was recorded in about four minutes.

Figure 14-15 shows an image of a coronal slice through the cerebellum and occipital cortex of a human brain. The image is T_1-weighted (i.e., T_1 provides the main contrast mechanism), with excellent differentiation between the gray and white material of the brain. The image was recorded in about 6 minutes at a field of 3 T.

Figure 14-16 shows eight images of a medio-sagittal section through a human skull. These were recorded using an imaging slice 7 mm in thickness. The pulse sequence used for the measurements (a Carr-Purcell-Meiboom-Gill multiple echo sequence as shown in Fig. 14-17) allowed eight echoes to be recorded.

Each of the images reproduced in Figure 14-16 corresponds to one of the eight echoes. The first image is a fairly exact representation of the water distribution in the skull (the bright parts correspond to high water concentrations and the dark parts to low concentrations). It can be clearly seen that the image con-

387

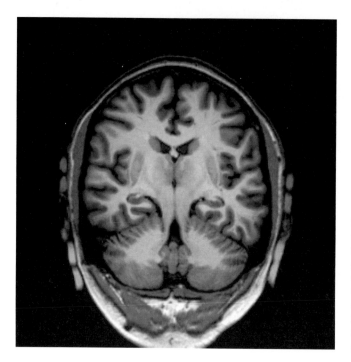

Figure 14-15.
Coronal head image displaying marked T_1 contrast, with excellent grey/white matter differentiation. Obtained using adiabatic inversion with a rapid multi-slice gradient-echo acquisition scheme. Magnetic flux density $B_0 = 3$ T; slice thickness 5 mm; measurement time approx. 6 min.

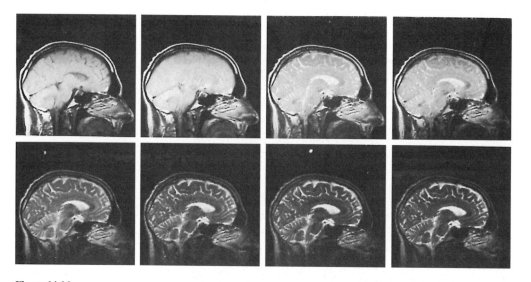

Figure 14-16.
Medio-sagittal section through a human skull. The images were obtained using a multiple echo pulse sequence which allowed eight echoes to be recorded. Each image in the series corresponds to one of the eight echoes. Each of the eight images is of the same 7 mm thick slice, divided for imaging purposes into 256×256 matrix elements. Two FIDs were accumulated for each matrix element. The total time for recording the eight images was about 11 min. Magnetic flux density $B_0 = 1.5$ T.

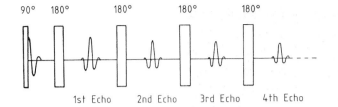

trast changes from one echo to the next, and in some cases more detail becomes apparent. This results from the differences in spin-spin relaxation time T_2 for different types of brain tissue. It can also be seen that the overall brightness of the images diminishes from the first echo to the last one as a consequence of relaxation, since the time from the excitation pulse to the eighth echo is nearly 0.3 s. The total recording time for all the images was 11 min.

In Figure 14-18 the eight echo images of Figure 14.16 are summed together; one of the image elements is identified by a hairline cross. The intensities of all eight echoes for this element have been measured and plotted as a function of time. The resulting curve (shown in the figure) indicates a T_2-value of 101 ms (Section 7.3.2).

Figure 14-19 shows a single coronal slice through the entire abdominal cavity. The liver, spleen and spinal column are clearly visible. Even small diameter blood vessels within the pulmonary vasculature, as well as those alongside the spinal column, are clearly discernible, although the image was not

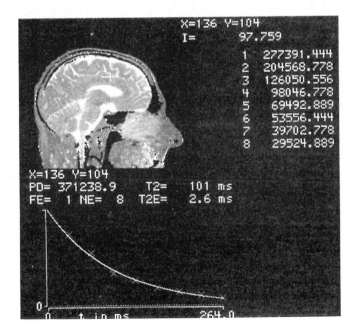

Figure 14-18.
Measurement of the spin-spin relaxation time T_2. In the diagram the intensities (signal amplitudes) of the eight echoes corresponding to the eight images in Figure 14-16 for the point marked with a hairline cross are plotted as a function of time. The computer system allows one to move the hairline cross to any desired position and determine the corresponding T_2-value.

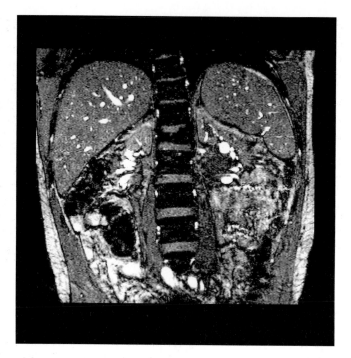

Figure 14-19.
An example of the inflow effect, here displayed not as a true angiogram but rather as a single coronal slice through the entire abdominal cavity. Magnetic flux density $B_0 = 3$ T; slice thickness 2 mm; total scan time approx. 20 s.

recorded under the conditions of an angiogram. The total scan time was intentionally restricted to about 20 seconds, so that data acquisition could be completed within a single breath-hold.

14.4.2.2 Magnetic Resonance Spectroscopy, ^{1}H MRS

For the diagnosis of tumors and for cancer research it is important, from both the clinical and the biochemical standpoints, to be able to recognize the differences between the metabolism of normal cells and that of cancerous cells. The combined use of MR tomography and MR spectroscopy is illustrated below by an example from actual practice.

Figure 14-20 shows three MR images of sections through the head of a patient, in which the images on the left and on the right are identical. These clearly show the presence of a tumor, the position of which is marked by a cross in all three images. The sagittal section image (upper figure) was recorded with T_2 weighting, whereas the images of the coronal and transverse sections (middle and bottom figures respectively) were recorded after introducing a contrast agent (Gd^{3+}) to give T_1 weighting.

390

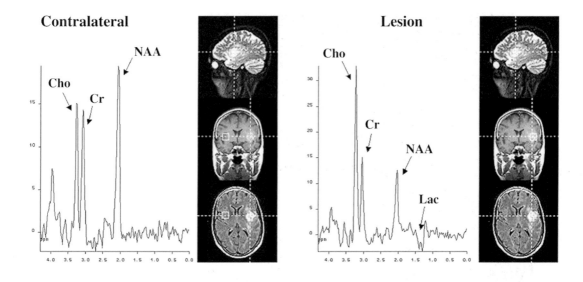

Figure 14–20

Three sectional images through the head of a patient, recorded to locate a brain tumor. From top to bottom: sagittal, coronal, and transverse sections. The white squares in the coronal and transverse sections show the volume selected for recording the ^{1}H NMR spectra, which were obtained by the PRESS (Point-REsolved SpectroScopy) technique [8], with the water signal suppressed by frequency-selective presaturation. Cho = Choline, Cr = Creatine + Phosphocreatine, NAA = N-Acetylaspartate, Lac = Lactate. The δ scale was calibrated using the residual water signal (δ = 4.6 ppm) as reference.

The results were obtained using a Siemens Magnetom Vision Plus instrument at the Düsseldorf University Medical Center (Institute of Diagnostic Radiology).

In the next stage of diagnosis the tumor volume of interest must be selected (see right side of Fig. 14-20), as one wishes to record the ^{1}H NMR spectrum from this region only. NMR signals of the normal tissue surrounding the tumor must be excluded so far as possible. In this example, that was achieved by the "PRESS" (Point-REsolved SpectroScopy) technique, which is a double spin-echo procedure [8]. It is a single-shot method that can be applied with a commercial whole-body MR tomograph instrument (MRI scanner), using the PRESS pulse sequence to select the volume of interest and record its ^{1}H NMR spectrum. The PRESS sequence also includes a frequency-selective presaturation step for suppressing the water signal [9].

For comparison with the latter NMR spectrum, a volume element of the same size in the opposite (contralateral) side of the brain, the healthy side, is examined by exactly the same procedure. In Figure 14-20 the two selected regions, each of which is a 2x2x2 cm^3 cube, are shown by the white squares.

In the spectra one can easily recognize the methyl proton signals of *N*-acetylaspartate (NAA), of creatine (Cr) and phosphocreatine, and of choline (Cho), and in the spectrum of the tumor region (the lesion) one also finds a signal due to lactate (Lac). N-acetylaspartate (NAA) is present at a relatively high concentration in the healthy part of the brain, whereas its concentration in the tumor tissue is greatly reduced.

Phosphocreatine (PCr) acts as an energy-storing medium, which is capable of quickly transferring phosphate to adenosine diphosphate (ADP). During this process creatine (Cr) is formed (see Section 14.3.2). As the methyl proton chemical shifts of PCr and Cr are only slightly different, they cannot be separately resolved by the instruments used in clinical diagnosis, and therefore one measures only the sum of their concentrations (PCr + Cr). This concentration is reduced in a tumor.

Choline is essential for the synthesis of acetylcholine, a neurotransmitter. In a tumor its concentration is greatly increased.

Lactate (Lac) is an end-product of anerobic glycolysis, and in healthy tissue it is normally present only at very low concentrations, so that it cannot be detected under the above conditions. However, increased concentrations can occur in tumors, giving a methyl proton signal at 1.33 ppm, which has a doublet structure due to coupling with the CH proton. This doublet is observed as a negative signal under the conditions of the measurement technique used here.

The changes in the concentrations of NAA, Cho, and Lac as described above are characteristic of the spectrum obtained from a tumor, and can therefore be used in clinical diagnosis [10].

14.5 Bibliography for Chapter 14

[1] A. G. McInnes, J. A. Walter, J. L. C. Wright and L. C. Vining: ^{13}C NMR Biosynthetic Studies. In: *Topics in Carbon-13 NMR Spectroscopy*, G. C. Levy (Ed.), Vol 2, Ch. 3. New York: John Wiley & Sons, 1976.

[2] S. E. Holker and T. J. Simpson, *J. Chem. Soc. Perkin Trans. 1* (1981) 1397.

[3] E. Breitmaier and W. Voelter: *^{13}C NMR Spectroscopy*. 3rd Edition. Weinheim: VCH Verlagsgesellschaft, 1987.

[4] P. C. Lauterbur, *Nature 242* (1973) 190.

[5] P. Mansfield and P. K. Grannell, *J. Phys. C, Solid State Physics 6* (1973) L422.

[6] A. Kumar, D. Welti and R. R. Ernst, *J. Magn. Reson. 18* (1975) 69.

[7] B. Blümich and W. Kuhn (Eds.): *Magnetic Resonance Microscopy. Methods and Applications in Materials Science, Agriculture and Biomedicine*. Weinheim: VCH Verlagsgesellschaft, 1992.

[8] P. A. Bottomly, Ann. NY Acad. Sci., 508 (1987) 333.

[9] A. Haase, J. Frahm, W. Hanicke and D. Matthaei, *Phys. Med. Biol. 30* (1985) 341.

[10] H. Lanfermann, S. Herminghaus, U. Pilatus, P. Raab, S. Wagner and F. E. Zanella, *Klin. Neurobiologie 1* (2002) 1.

Additional and More Advanced Reading

Biochemical Topics:

O. Jardetzky and G. C. K. Roberts: *NMR in Molecular Biology.* New York: Academic Press, 1981.

R. G. Shulman (Ed.): *Biological Applications of Magnetic Resonance.* New York: Academic Press, 1979.

In vivo **NMR:**

D. G. Gadian: *Nuclear Magnetic Resonance and its Applications to Living Systems.* Oxford: Clarendon Press, 1982.

R. A. Iles, A. N. Stevens and J. R. Griffiths: NMR Studies of Metabolites in Living Tissue. In: *Prog. Nucl. Magn. Reson. Spectrosc. 15* (1983) 49.

K. O'Neill and C. P. Richards: Biological ^{31}P NMR Spectroscopy. In: *Annual Reports on NMR Spectroscopy*, Vol. 10A, G. A. Webb (Ed.). London: Academic Press, 1980, p. 133.

MR Tomography:

I. C. P. Smith and L. Stewart: Magnetic resonance spectroscopy in medicine: clinical impact. In: Nucl. Magn. Reson. Spectrosc. 40 (2002) 1.

R. Damadian (Ed.): NMR in Medicine. In: P. Diehl, E. Fluck, R. Kosfeld (Eds.): *NMR Basic Principles and Progress*, Vol.19. Berlin: Springer, 1981.

M. J. McCarthy and M. K. Cheung: Magnetic Resonance Imaging. In: *Annual Reports on NMR Spectroscopy,* Vol. 31, G. A. Webb (Ed.), London: Academic Press, 1995, p. 19.

F. W. Wehrli, D. Shaw and J. B. Kneeland: *Biomedical Magnetic Resonance Imaging. Principles, Methodology and Applications.* Weinheim: VCH Verlagsgesellschaft, 1988.

MR Spectroscopy, ^{1}H MRS:

J. Frahm and W. Hänicke: *Single Voxel Proton NMR: Human Subjects.* In: Encyclopedia of Nuclear Magnetic Resonance. D. M. Grant and R. K. Harris (Eds.), Vol. 7. Chichester: John Wiley & Sons, 1996, p. 4407.

Subject Index

P

Paramagnetic species   37, 163, 167, 298, 335
  – see also Lanthanide shift reagents;
  Oxygen, dissolved
Paramagnetic shielding term   45, 66
Parts per million, ppm   25
Pascal's triangle   31
Pascual-Meier-Simon rule   142
Pauli exclusion principle   103
Pentads   346
Peptides   356–360
Permanent magnet   9
PFG   18, 188 – see Pulsed field gradient   187
pH within the cells   378
Phase coherence   12, 172
Phase correction   16
Phase cycle   216
Phase-sensitive detector/detection   20, 256
Phenols   60
Phosphorus NMR (^{31}P)   38, 81
– *in vivo*   376
Planck constant   2
Platinium NMR (^{195}Pt)   82
Point dipole model   48
Polarization transfer   202–218
Polymers   343–350
Polynucleotides   360–362
Populations (of Energy levels)   6, 12, 174,
  202 ff., 290
Potassium NMR (^{39}K)   81
Precession   4
– classical picture of   4, 6
– frequency   4
Preparation phase   232
Presaturation   217
PRESS (point resolved Spectroscopy)   391
Probe-Head   19
Prochiral center   77, 321
Progressively connected transitions   118
Projection, filtered back   381, 383
Proteins   356–360
Protein structures
– sequence, primary   357
– three-dimensional, secondary   358 ff.
Protonation   51
Proton exchange   327
– alcohols   327
– ammonium ions   329
Pseudo-asymmetric carbon atom (in polymers)   348
Pseudocontact interaction   336
Pulse   9
– angle   10
– duration (length, width)   9
– generator   9

– "hard"   10
– power   36
– repetition rate or time   37
– "soft"   224, 382
Pulsed field gradients (PFG)   18, 187, 217, 375, 381
Pulsed gradient Spin Echo Experiment   200–202
Pulsed NMR-spectrometer   18
Pulsed NMR Method   9

Q

Quadrature detection   21
Quadrupolar broadening   105, 178
Quadrupolar relaxation   163
Quadrupole moment, electric   3, 40, 105
Quantitative analysis   35
Quantum number
– angular momentum   2
– magnetic or directional   4
Quarternary carbon atoms/nuclei   136

R

Radio-frequency (r.f.) field   7
Rate constant   276, 307, 309
Reaction pathways in biochemistry   370–374
Real part of signal   15
Receiver   20
  – see The Pulsed NMR Spectrometer   18
Reduced coupling constant   136
Reference compound   24
Reference spectra   156
Regressively connected transitions   118
Relaxation   13, 161–179, 290, 385
– agents   37
– by chemical shift anisotropy   162
– by paramagnetic species   37, 163, 335
– by quadrupolar nuclei   163
– by scalar coupling   162
– by spin rotation   162, 169
– dipolar   162, 168, 291, 293
– longitudinal (Spin lattice)   13
– mechanisms   162
– transverse (Spin-spin)   13
Relaxation time
– longitudinal or spin-lattice T_1 36, 154,
  162–171, 385
– – measurements   150
– transverse or spin-spin T_2 171–178, 385
– – measurements   173
Repetition rate/time (pulse)   35
Resolution   177, 244, 375
– digital   36
Resonance   7
Resonance condition   7, 23, 381

Index of Compounds

(^{1}H) or (^{13}C) after the page number indicates that the ^{1}H or ^{13}C spectrum is shown.